Birkhäuser

Deformation and Gravity Change

Indicators of Isostasy, Tectonics, Volcanism and Climate Change
Volume III

Edited by
Detlef Wolf
Miguel A. Santoyo
José Fernández

Previously published in *Pure and Applied Geophysics* (PAGEOPH), Volume 169, No. 8, 2012

Birkhäuser

Editors
Detlef Wolf
University of Stuttgart
Institute of Geodesy
Geschwister-Scholl-Str. 24D
70174 Stuttgart
Germany

José Fernández
CSIC - Universidad Complutense de Madrid
Instituto de Geociencias
Plaza de Ciencias 3
28040 Madrid
Spain

Miguel A. Santoyo
Universidad Complutense de Madrid
Departamento de Física de la Tierra,
Astronomía y Astrofísica I
Plaza de Ciencias S/N
28040 Madrid
Spain

ISBN 978-3-0348-0459-2 e-ISBN 978-3-0348-0460-8
DOI 10.1007/978-3-0348-0460-8

Library of Congress Control Number: 2007925494

Cover illustration: Based on Fig. 9 from "Land Water Storage Changes from Ground and Space Geodesy: First Results from the GHYRAF (Gravity and Hydrology in Africa) Experiment" by J. Hinderer, J. Pfeffer, M. Boucher, S. Nahmani, C. de Linage, J.-P. Box, P. Genthon, L. Seguis, G. Favreau, O. Bock, M. Descloitres, and GHYRAF team.

Cover design: deblik, Berlin.

Printed on acid-free paper

Springer Basel AG is part of Springer Science+Business Media

www.birkhauser-science.com

Contents

Pure Appl. Geophys. 169 (2012), 1329–1330

DOI 10.1007/s00024-011-0451-7

Pure and Applied Geophysics

Deformation and Gravity Change: Indicators of Isostasy, Tectonics, Volcanism and Climate Change, Volume III. Introduction

DETLEF WOLF,[1,2] MIGUEL ANGEL SANTOYO,[3] and JOSÉ FERNÁNDEZ[4]

Changes of the deformation and gravity fields are associated with many geodynamic phenomena (e.g., isostatic, tectonic, volcanic, hydrologic, oceanographic or climatic processes), and their measurement using terrestrial or space geodetic techniques have made a basic tool for research and monitoring (WOLF and FERNÁNDEZ, 2007; WOLF *et al.*, 2009). The development of novel and highly precise measuring techniques for monitoring the temporal variations of these phenomena are yielding new insights for their analysis and the study of their effects on the change of the Earth's environment from local to global change. These topics together with the ongoing improvements in the theoretical formulations of the processes and their numerical implementations have reached a level that allows us to model them in an increasingly realistic way.

The third workshop on "Deformation and gravity change: indicators of isostasy, tectonics, volcanism and climate change" was held during February 23–26, 2009 at the Casa de los Volcanes on Lanzarote in Canary Islands, Spain, as were its first and second editions in March 2005 and March 2007 (WOLF and FERNÁNDEZ, 2007; WOLF *et al.*, 2009). This workshop was jointly organized and supported by the International Association of Geodesy (IAG), the Spanish Ministry of Science and Innovation, the Spanish Council for Scientific Research and the Cabildo Insular de Lanzarote. This meeting also served as the first meeting of the members of the International Association of Geodesy (IAG) ICCT Study Group on "Temporal Variations of Deformation and Gravity", open to all colleagues interested in its topic.

The meeting on Lanzarote keeps the tradition established by a series of preceding workshops organized under the auspices of the IAG on "Models of Temporal Variations of the Gravity Field" in Walferdange, Luxembourg (March 17–19, 1997) and Potsdam, Germany (November 23–25, 1998), and on "Dynamic Theories of Deformation and Gravity Fields" in Sopron, Hungary (February 19–23, 2001), and Lanzarote, Spain (February 18–21, 2003; March 1–4, 2005, and March 27–30, 2007) (WOLF and FERNÁNDEZ, 2007; WOLF *et al.*, 2009).

The present volume follows the topical issues in *Pure and Applied Geophysics*, Vol. 164, No. 4 (2007) and Vol. 166, Nos. 8–9 (2009) published as an aftermath of the workshops held on Lanzarote in 2005 and 2007 respectively. This one contains 13 papers addressing different topics: the observation of gravity changes using terrestrial and space techniques; the calibration of gravimeters; the geodetic consequences of past and present ice-mass change; the observation of hydrological signals observed from space; combination of space and surface ocean data to study vertical crustal movements; the correction of radar satellite data using a meteorological model; the deformation modeling; the measurement of interseismic deformation from space; the analysis of GPS measurements; the glacial isostatic adjustment; and the seismic hazard and ground motion of a dam site.

[1] Section 1.5: Earth System Modelling, German Research Centre for Geosciences (GFZ), Telegrafenberg, 14473 Potsdam, Germany.

[2] Institute of Geodesy, University of Stuttgart, Geschwister-Scholl-Str. 24D, 70174 Stuttgart, Germany.

[3] Departamento de Física de la Tierra Astronomía y Astrofísica I (Geofísica y Meteorología), Facultad de Ciencias Físicas, Universidad Complutense de Madrid, Plaza de Ciencias s/n, 28040 Madrid, Spain.

[4] Institute of Geosciences, CSIC-UCM, Facultad de Ciencas Matemáticas, Plaza de Ciencias, 3, 28040 Madrid, Spain. E-mail: jft@mat.ucm.es

We thank the Consejeria de Ciencia y Tecnología of the Cabildo Insular de Lanzarote, the Lanzarote Laboratory for Geodynamics (CSIC-UCM and Cabildo Insular de Lanzarote), the staff of the *Casa de los Volcanes*-in particular, its director Joaquín Naverán as well as Orlando Hernández and Jaime Arranz - and the Timanfaya National Park administration for their support during the workshop. The editorial process of this issue has been done in the frame of the research projects CGL2005-05500-C02-01, 115/SGTB/2007/8.1 and AYA2010-17448, and of the Moncloa Campus of International Excellence (UCM-UPM, CSIC). Useful suggestions by Renata Dmowska during the preparation of this topical issue are greatly appreciated. We also thank all authors for their contributions and acknowledge the assistance of different reviewers.

References

Wolf, D., and J. Fernández (2007) (eds). Pure and Applied Geophysics, *Topical Issue "Deformation and Gravity Change: Indicators of Isostasy, Tectonics, Volcanism and Climate Change."*, *164*(4):633–878.

Wolf, D., P.J. González and J. Fernández (2009) (eds). Pure and Applied Geophysics, *Topical Issue "Deformation and Gravity Change: Indicators of Isostasy, Tectonics, Volcanism and Climate Change. Volume II."* *166*(8/9):1165–1531.

Pure Appl. Geophys. 169 (2012), 1331–1342

DOI 10.1007/s00024-011-0397-9

Pure and Applied Geophysics

Observing Gravity Change in the Fennoscandian Uplift Area with the Hanover Absolute Gravimeter

LUDGER TIMMEN,[1] OLGA GITLEIN,[1] VOLKER KLEMANN,[2] and DETLEF WOLF[2]

Abstract—The Nordic countries Norway, Sweden, Finland and Denmark are a key study region for research of glacial isostasy. In addition, such research offers a unique opportunity for absolute gravimetry to show its capability as a geodetic tool for geophysical research. Within a multi-national cooperation, annual absolute gravity measurements have been performed in Fennoscandia by IfE since 2003. For the Hanover gravimeter FG5-220, overall accuracy of ± 30 nm/s^2 is indicated for a single station determination. First results of linear gravity changes are derived for ten stations in the central and southern part of the uplift area. Comparing with the rates predicted by glacial rebound modelling, the gravity trends of the absolute measurements differ by 3.8 nm/s^2 per year (root-mean-square discrepancy) from the uplift model. The mean difference between observed and predicted rates is 0.8 nm/s^2 per year only. A proportionality factor of -1.63 ± 0.20 nm/s^2 per mm has been obtained, which describes the mean ratio between the observational gravity and height rates.

Key words: Absolute gravimetry, Fennoscandian land uplift, glacial isostatic adjustment (GIA), postglacial rebound (PGR).

1. The Fennoscandian Land Uplift

In the Fennoscandian land uplift area, the Earth's crust has been rising continuously since the last glacial maximum in response to ice deloading. This process is an isostatic adjustment of the Earth's elastic lithosphere and underlying viscous mantle. For a general overview, WOLF (1993) gives a historical review of the changing role of the lithosphere in models of glacial isostasy.

The Fennoscandian rebound area is dominated by the Precambrian basement rocks of the Baltic Shield, which is part of the ancient East European Craton and comprises South Norway, Sweden, Finland, the Kola Peninsula and Russian Karelia. The region is surrounded by a flexural bulge, covering northern Germany and northern Poland, The Netherlands and some other surrounding regions. The bulge area was once rising due the Fennoscandian ice load and, after the melting, sinking with a much smaller absolute value than the uplift rate in the centre of Fennoscandia. Denmark is part of the transition zone from the uplift to the subsidence area. The maximum spatial extension of the uplift area is about 2,000 km in northeast–southwest direction; see Fig. 1 for the approximate shape (after ÅGREN and SVENSSON, 2007). Presently, the central area around the northern part of the Gulf of Bothnia is undergoing uplift at a rate of about 1 cm/year.

Geophysical approaches to study the postglacial rebound are associated with evidence for ancient shorelines and lake level data, knowledge or assumptions about the geometry of the ice sheets (thickness, position), and some Earth model parameters (lithosphere thickness, mantle viscosity). After LAMBECK *et al.* (1998b), the inverse solution for the sea level data includes both ice and Earth model parameters as unknowns. Despite the recent progress in understanding the underlying models, definite models for the isostatic rebound do not yet exist. Lateral rheological variations have to be taken into account to obtain a more realistic glacially induced uplift model (KAUFMANN *et al.*, 2000).

To monitor and investigate the recent land uplift in Fennoscandia, various measurements have been collected since 1892: mareograph records, geodetic levellings, and relative gravity measurements since 1966. With these observations, the capability of terrestrial point measurement techniques to determine

[1] Institut für Erdmessung (IfE), Leibniz Universität Hannover, Hanover, Germany. E-mail: timmen@ife.uni-hannover.de

[2] Helmholtz-Zentrum Potsdam Deutsches GeoForschungs-Zentrum (GFZ), Potsdam, Germany.

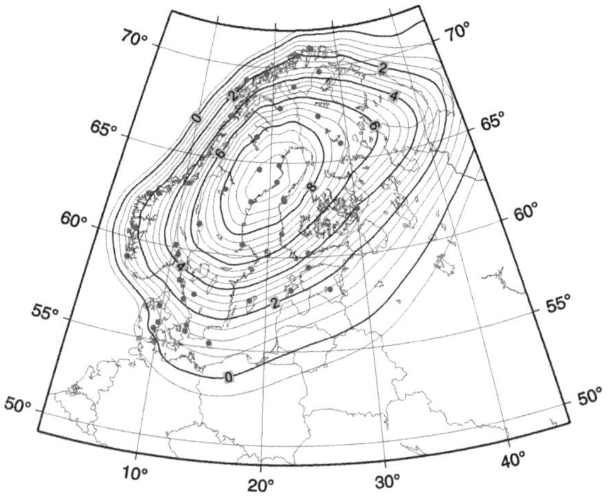

Figure 1
Map of the postglacial uplift of Fennoscandia in mm/year after ÅGREN and SVENSSON (2007) derived from model NKG2005LU, courtesy of ÅGREN. *Dots* indicate the positions of gravity stations of the Nordic absolute gravity project

the land uplift was proven along east–west profiles. They follow approximately the latitudes 65°N (observed 1975–2000), 63°N (1966–2003), 61°N (1976–1983) and 56°N (1977–2003); see EKMAN and MÄKINEN (1996) or MÄKINEN *et al.* (2005). According to EKMAN (1996), the maximum orthometric height change of 1 cm/year in the uplift centre is associated with a maximum gravity change of about -20 nm/s^2 (-2.0 μGal) per year. Based on these numbers, a geoid change of 0.6 mm/year has been derived for the central area.

NAKIBOGLU and LAMBECK (1991) deduced a eustatic sea level rise of 1.15 $\pm$ 0.38 mm/year from tide gauge observations, which has been taken into account for the uplift determination by EKMAN (1996). More recent papers show different results for the eustatic sea level trend; e.g. WÖPPELMANN *et al.* (2007) obtained 1.83 $\pm$ 0.24 mm/year [glacial isostatic adjustment (GIA) corrected] or 1.31 $\pm$ 0.30 [Global Positioning System (GPS) corrected] for the global trend, and MILNE *et al.* (2001) deduced 2.1 $\pm$ 0.3 mm/year (GPS corrected) from tide gauges in the Fennoscandian region. LIDBERG *et al.* (2007) corrected the tide gauge records from EKMAN (1996) by GPS velocities and confirmed the value 1.05 $\pm$ 0.25 mm/year (GIA corrected) derived by LAMBECK *et al.* (1998a).

Since 1993, permanent GPS stations were established in Fennoscandia to implement a further geodetic method with several advantages compared with the classical techniques (continuous data acquisition, homogeneous point distribution, large extension of the measurement area, low cost, three-dimensional survey). In this respect, the Baseline Inferences for Rebound Observations, Sea Level, and Tectonics (BIFROST) project has been based on the GPS technique and geophysical modelling, and has delivered a maximum height change rate (with respect to a geocentric reference ellipsoid) of about 11 mm/year, cf. MILNE *et al.* (2001), JOHANSSON *et al.* (2002), SCHERNECK *et al.* (2003) and LIDBERG *et al.* (2007).

In March 2002 the Gravity Recovery and Climate Experiment (GRACE) gravity satellite was launched to measure the detailed stationary Earth's gravity field and its regional and large-scale variations with time. During the mission duration of GRACE (nearly 8 years already), a temporal geoid change of approximately 5 mm can be expected in the centre of the Fennoscandian land uplift area, corresponding to a gravity change of about 160 nm/s^2 ($\equiv$16 μGal). This is a clear secular gravity change on regional scale, and it is a challenging task to detect this signal from GRACE gravity data most accurately (WAHR and VELICOGNA, 2003). Early results from TAPLEY *et al.* (2004) confirm that this satellite mission is able to resolve geoid variations for a range of spatial scales down to 400 km for particular regions with large signals. They found that the error level in the 2003 solutions was on the order of 2–3 mm for spatial features of about 600 km. Considering the large extension of the land uplift area, Fennoscandia is a suitable application region for GRACE. Vice versa, the temporal gravity field change can also be used for validation of GRACE results. Because the observation of the rebound signal suffers from interference by mass variations due to oceanographic, land hydrology and atmospheric processes, these effects have to be accounted for in GRACE data analysis using appropriate mathematical approaches (e.g. WIEHL *et al.*, 2005). Hence, the combination with other geological and geodetic measurements is inevitable.

2. *Absolute Gravimetry*

Besides the geometrical approaches, terrestrial absolute gravimetry is a further geodetic technique to study land uplift or subsidence. In general, it is applied as a complementary tool to the geometrical methods. The absolute measurements are most sensitive to height changes and provide an obvious way to define and control the vertical height datum. No additional reference points (connection points) at the Earth's surface are needed. Shortcomings of relative gravimetry, such as calibration problems and deficiencies in the datum level definition, can be overcome. Both absolute and relative gravimetry can measure gravity changes between two points, but only the absolute technique by itself solves the ambiguity problem of whether both points are undergoing a decrease or increase with different magnitudes, or one point experiences an increase and the other a decrease. In addition, the accuracy of an absolute gravity net is independent of geographical extension and the covered gravity range. Thus, applications on local, regional and global scales with consistent measurement quality are feasible. Independent verification of displacements measured geometrically by GPS, Very Long Baseline Interferometry (VLBI) and Satellite Laser Ranging (SLR) is possible. A combination of gravimetric and geometric measurements may enable discrimination among subsurface mass movements associated with or without a surface deformation.

The benefit of absolute gravimetry has already been exploited in different scientific projects. The International Absolute Gravity Basestation Network (IAGBN) serves, among other purposes, for the determination of large-scale tectonic plate movements (Boedecker and Fritzer, 1986; Boedecker and Flury, 1995). The recommendations of the Interunion Commission of the Lithosphere on Mean Sea Level and Tides propose regular implementation of absolute gravity measurements at coastal points, 1–10 km away from tide gauges (Carter *et al.*, 1989). The height differences between gravity points and tide gauges have to be controlled by levelling or GPS. In Great Britain, the main tide gauges are controlled by repeated absolute gravity determinations in combination with episodic or continuous GPS measurements (Williams *et al.*, 2001). Overall, absolute gravimetry can be an important research tool for studying geodynamic processes, especially land uplift effects due to postglacial rebound (PGR). Lambert *et al.* (1996) give an overview of the capability of absolute gravity measurements in determining the temporal variations in the Earth's gravity field. In Lambert *et al.* (2001), the gravimetric results for the research of the Laurentide postglacial rebound in Canada are described. Mäkinen *et al.* (2007) compare observed gravity changes in Antarctica with modelled predictions of the glacial isostatic adjustment as well as of the glacier mass balance.

In 2002, the Institut für Erdmessung (IfE) of the Leibniz Universität Hannover received a new FG5 absolute gravity meter from Micro-g Solutions, Inc. (Erie, Colorado), which is a state-of-the-art instrument. The absolute measurements are based on time and distance measurements along the vertical to derive the gravity acceleration at a specific position on the Earth. The term "absolute" is based on the fact that the time and length standards (rubidium clock, helium–neon laser) are incorporated as components of the gravimeter system. The FG5 series is presently the most common gravimeter model, and may be considered as the successor system of the JILA generation (Carter *et al.*, 1994; Niebauer *et al.*, 1995). The influence of floor vibration and tilt on the optical path could largely be removed by the improved interferometer design. The iodine-stabilised laser, serving as the primary length standard, is separated from the instrumental vibrations caused by the free-fall experiments, by routing the laser light through a fibre-optic cable to the interferometer base; see Fig. 2. During a free-fall experiment (drop), the trajectory of a test mass (optical retro-reflector) is traced by laser interferometry over the falling distance of about 20 cm within an evacuated chamber. The "co-falling" drag-free cart provides a molecular shield for the dropped object. The multiple time–distance data pairs collected during the drop (FG5: 700 pairs at equally spaced measuring positions) are adjusted to a fitting function, giving the gravity acceleration g for the reference height above floor level (FG5: $\sim$1.2 m). For the reduction of local noise and other disturbances, 1,500–3,000 computer-controlled

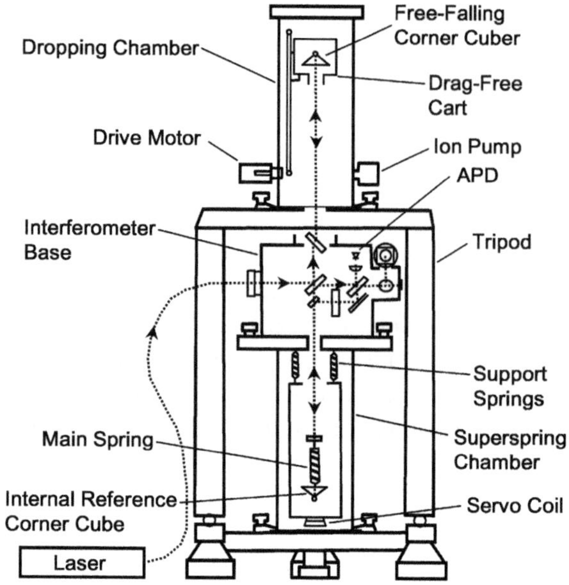

Figure 2
Schematic diagram of the FG5 absolute gravimeter, after Micro-g Solutions Inc. (1999), courtesy of Micro-g Lacoste, Inc.

drops are performed per station determination. Generally, the measurements are subdivided into sets of 50 or 100 drops each, and distributed over 1–3 days. The result of a station determination is the average of all drops, reduced for gravity changes due to Earth's body and ocean tides, polar motion and atmospheric mass movements. For more details, readers may refer to Timmen *et al.* (2008). A good overview about the principles of absolute gravimetry is given, e.g., in Torge (1989) and Torge (1993). Van Camp *et al.* (2005) concentrate especially on long time series of absolute gravity measurements and the inherent uncertainties.

3. *Project Realisation*

To determine recent crustal deformations by absolute gravimetry, secular gravity changes should be measured with precision of about ±5 nm/s^2 per year. This can be achieved by annual measurements over 5–10 years. To exploit absolute gravimetry in combination with GPS for "pointwise" validation of the GRACE results or to support GRACE data evaluation by terrestrial gravimetry, the temporal variations of gravity disturbances (or gravity anomalies) are needed in accordance with the resolution of the GRACE data. Because of the long-wavelength nature of the GRACE results, the terrestrial point results derived from absolute gravimetry have to be reduced for all local effects changing gravity with time. In this connection, a severe problem is subsurface water mass movement (change of groundwater table, temporary water storage in clefts and crevasses). Such impacts are partly considered by the station selection. Moreover, by measuring the absolute gravity value at a station every year over a 5-year period, the impact of groundwater variations is averaged out to a large extent within the computation of the linear gravity rate. In addition, observations of the groundwater table in boreholes and in nearby wells are taken during the absolute gravity surveys and used to assess the disturbing impact. Furthermore, a second favourable averaging effect arises from deriving a spatial mean over a larger area with a few hundred kilometres extension. For that reason, and to allow the determination of an uplift model (mathematical surface model, as in Fig. 1) with possible regional structures, a rather large number of stations have to be observed every year over the whole Fennoscandian area.

A joint project for the annual gravimetric survey of the land uplift in Fennoscandia was established in 2003. The Working Group on Geodynamics of the Nordic Geodetic Commission (NKG) serves as a platform to organise the project. Besides the IfE from Hanover (with the absolute meter FG5-220), the following institutions are participating in the project: Department of Mathematical Sciences and Technology, Norwegian University of Life Sciences (UMB, Ås, FG5-226); Federal Agency for Cartography and Geodesy (BKG, Frankfurt/Germany, FG5-301); Finnish Geodetic Institute (FGI, Masala, FG5-221); Norwegian Mapping Authority (Statens Kartverk/SK, Hønefoss); Onsala Space Observatory (Chalmers University of Technology, Onsala/Sweden); Swedish Mapping, Cadastre and Land Registration Authority (Lantmäteriet/LM, Gävle, FG5-233); Technical University of Denmark, National Space Institute (DTU Space, Copenhagen). FGI procured the new FG5-221 at the beginning of 2003, UMB the FG5-226 in spring 2004, and LM the FG5-233 in autumn 2006. Nearly all absolute gravity sites are co-located with permanent GPS stations, and also tide gauges are in the vicinity of coastal stations. The gravity points, GPS stations and tide gauges are connected locally, using terrestrial surveying techniques, such as levelling, to control the local vertical variations. The employment of more than one absolute gravimeter allows simultaneous (parallel) observations in stations with two sites close to each other and control measurements at identical sites. This strategy increases the network reliability and accuracy because it helps to identify possible offsets of the instruments. To exclude uncertainties introduced by relative gravimetry (e.g. via the measured vertical gravity gradient) into the absolute gravimetric results, the final absolute values are all related to a common height at 1.200 m above the reference mark at floor level, cf. Timmen (2003). Figures 3 and 4 show typical conditions for the absolute gravimetric fieldwork in Fennoscandia.

In Gitlein *et al.* (2008), a summary of the measurements performed from 2003 to 2008 is given. Altogether, 46 different stations were occupied by the group of five participating absolute gravimeters. For e.g. 2008, 33 gravity stations were surveyed at least once, and 11 stations were observed by two or more absolute meters partly simultaneously. Figure 5 shows the stations occupied by the IfE gravimeter FG5-220 in the period from 2003 to 2008. The IfE experts visited 34 different stations in the uplift area with their instrument and performed 84 gravimetric

Figure 3
Station Skellefteå (Sweden) with an absolute gravity pier inside and a temperature-stabilised pillar for continuous GPS outside

Figure 4
Absolute gravimeter FG5-220 of IfE installed at station Östersund

station determinations (2003: 16 stations; 2004: 16; 2005: 15; 2006: 12; 2007: 15; 2008: 10).

4. *Measurement Accuracy and its Control*

The manufacturer of the FG5 system performed an error budget analysis to determine the single instrumental uncertainty contributions through calculations and measurements of known physical effects. In Niebauer *et al.* (1995) a total uncertainty

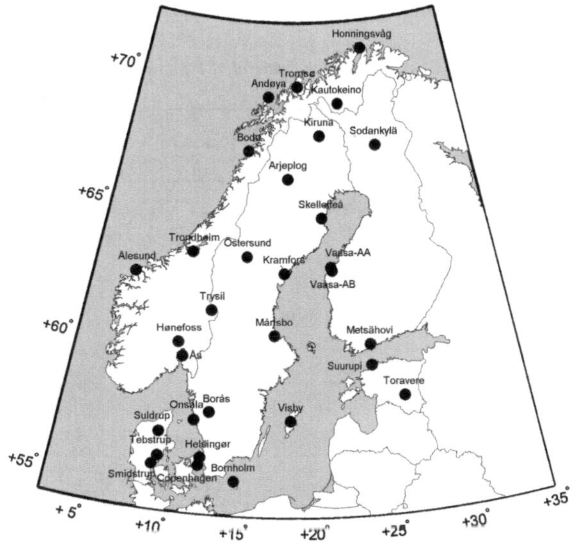

Figure 5
Absolute gravity stations in the Fennoscandian land uplift area occupied by the IfE gravimeter FG5-220 in the period from 2003 to 2008

of 11 nm/s^2 is obtained from the FG5 instrumental error budget. To assess the accuracy of the Hanover absolute gravimeter from the user point of view, the measurement experiences with FG5-220 are used to derive an empirical accuracy estimate. The accuracy and stability have been continuously controlled by comparisons with other absolute gravity meters, and with repeated measurements in several stations after time intervals of some months to a few years. Rigorous control of the absolute accuracy with respect to a "true" gravity value at the moment of an absolute gravity measurement is not possible. The real g-value is not known with superior accuracy, and a "standard" absolute gravimeter which is superior to the state-of-the-art FG5 meters does not exist. Therefore, the empirical accuracy estimate has to be understood as describing the agreement of the instrument's measuring level and its time stability with regard to the international absolute gravity datum definition. Here, the international datum is defined by the physical standards (time and length) and, in addition, as the average result obtained from all operational absolute gravimeters participating in the international comparison campaigns.

Since the 1980s, international comparisons of absolute gravimeters (ICAG) have been performed at the Bureau International des Poids et Mésures (BIPM) in Sèvres, and since 2003, with a 4-year time interval, also at the European Centre of Geodynamics and Seismology (ECGS) in Walferdange, Luxembourg. For the gravimeter FG5-220 of IfE, Table 1 presents the results from the international comparisons in Walferdange in 2003 and 2007 (external comparisons, Francis and van Dam, 2006; Francis *et al.*, 2010), and FG5-220 reference measurements in Bad Homburg (station of BKG, Wilmes and Falk, 2006) from 2003 to 2008. Within 20 nm/s^2, the Hanover FG5 instrument agrees with the internationally realised measuring level. With respect to the FG5-220 observations in Bad Homburg, it has to be mentioned that the differences between the single epochs also contain real gravity changes due to time-varying environmental effects such as seasonal hydrological variations. As shown in Table 1, the six station determinations agree very well, better than expected from empirical estimates, with mean scatter of 11 nm/s^2 only (root-mean-square difference, rms). An instrumental instability cannot be identified. Similar experiences are also gained from the yearly repetition surveys and from the comparisons with the other FG5 absolute gravimeters involved in the Nordic absolute gravity project, to determine the Fennoscandian land uplift, cf. Timmen *et al.* (2006) and Bilker-Koivula *et al.* (2008).

Table 1

FG5-220 absolute gravimeter controlled by external (international) and internal (repetition) comparisons to secure consistent long-term measurement accuracy

	Remarks	Epoch	Δg [nm/s^2] (FG5-220 − mean g)
FG5-220 external comparison			
ICAG2003, ECGS (Francis and van Dam, 2006, Table 16)	13 abs. meters, 14 points, 52 determinations	Nov 2003	−19 SD (single meter) 18
ICAG2007, ECGS (Francis *et al.*, 2010, Table 3)	19 abs. meters, 16 points, 73 determinations	Nov 2007	+24 SD (single meter) 20

	Remarks	Epoch	Δg (FG5-220) [nm/s^2] (single − mean g)
FG5-220 internal comparison			
Bad Homburg (gravimetry lab. of BKG, Wilmes and Falk, 2006)	Reference station for FG5-220 since 2003, point BA	Feb 2003	+9
		Nov 2003	−8
		Apr 2005	+12
		Apr 2006	+7
		Nov 2007	+2
		Sep 2008	−21

The results from parallel measurements in the Fennoscandian land uplift area are summarized in Table 2, which compares observations of FG5-220 with other meters participating in the Nordic absolute gravity project since 2003 (GITLEIN, 2009). Such measurements are normally performed by simultaneous registrations with two gravimeters at adjacent piers during 1 day (Fig. 6), and swapping the places on the next day to start the parallel measurements again.

The overall discrepancy (rms) of the comparisons is 23 nm/s^2, which proves the high accuracy of the absolute gravimetric survey of Fennoscandia. The mean values show that no significant offset of the IfE instrument exists in comparison with the other four absolute gravimeters. Considering that the discrepancies in Table 2 are caused each time by both of the

Table 2

Gravity discrepancies in nm/s^2 from parallel measurements of FG5-220 with other FG5 meters participating in the Nordic absolute gravity project: FG5-101/301 (BKG), FG5-221 (FGI), FG5-226 (UMB)

Bad Homburg		Metsähovi		Onsala		Vaasa	
Feb 2003	12	Aug 03	−10	Oct 04	−5	Aug 03, AA	−29
Nov 2003	37	May 04	19	Oct 04	9	Aug 03, AB	−30
Apr 2005	−26	May 05	18	Oct 05	34	May 04, AB	2
Apr 2006	13	Aug 05	42	Oct 06	29		
Nov 2007	7	Aug 06	−14				
Sep 2008	−19	Jul 07	22				
Mean (nm/s^2)	4		13		17		−19
rms	21		23		23		24

Figure 6
Parallel measurements with FG5-220 (IfE) and FG5-221 (FGI) at station Metsähovi in Finland

participating gravimeters, an instrumental precision of better than 20 nm/s^2 can be assumed for a single gravimeter. The Hanover group estimates a mean absolute accuracy for their FG5-220 of about 30 nm/s^2. This empirical estimate incorporates:

- Instrumental errors, e.g. due to instrumental vibrations or laser instabilities.
- Gravitational "noise" due to incomplete modelling and reduction of gravity variations with time (Earth's body and ocean tides, polar motion, atmospheric mass movements).

5. *Gravity Changes in Fennoscandia*

Repeated observations with the FG5-220 from IfE were performed at 10 stations in Fennoscandia nearly every year from 2003 to 2008. From these results, linear gravity changes were calculated for each station (GITLEIN, 2009). Table 3 summarizes the observational gravity trends and the comparison with the predicted rates of the glacial rebound model provided by KLEMANN (2004). The computations were based on solution algorithms developed by MARTINEC (2000) and HAGEDOORN *et al.* (2007) and use a global ice model with SCAN-II (LAMBECK *et al.*, 1998a) for Fennoscandia. The trends for three stations are presented exemplarily in Fig. 7. All station determinations of the summer/autumn campaign 2003 have not been taken into account because the gravity values seem to be systematically too low by 50–90 nm/s^2 compared with the following years. This effect can only partly be explained by a possible instrumental offset, cf. Sect 4. A real gravity decline is assumed and should be connected to the very dry season in northern Europe. For e.g. northern Germany, both IfE reference stations, Hanover (IfE laboratory, based on glacial sediments) as well as Clausthal (Institut für Geophysik, TU Clausthal, Harz Mountains, bedrock), show a strong decline of the observed gravity values from February to November 2003. In Clausthal the observed gravity acceleration decreased by about 50 nm/s^2 and in Hanover by 100 nm/s^2 (TIMMEN *et al.*, 2008). The monitoring of the groundwater table in Hanover confirmed the large seasonal gravity variation (correlation 90%). A similar effect in Fennoscandia cannot be excluded.

Table 3

Comparison of gravity trends in Fennoscandia derived from IfE absolute gravity measurements AG (n: number of gravity determinations) and from the geophysical model predictions provided by KLEMANN *(2004)*

Station	AG (FG5-220) (nm/s²/year)		n	Geophys. model (nm/s²/year)	Difference (nm/s²/year)
Arjeplog	−8.7	±2.4	4	−11.6	2.9
Copenhagen	1.9	±5.7	6	−1.9	3.8
Kiruna	−11.3	±11.0	4	−9.7	−1.6
Kramfors	−14.4	±2.7	5	−15.8	1.4
Mårtsbo	−15.6	±3.8	5	−11.7	−3.9
Metsähovi	−8.8	±5.2	7	−7.4	−1.4
Onsala	5.0	±5.7	6	−4.2	9.2
Östersund	−14.8	±10.8	5	−12.5	−2.3
Skellefteå	−18.8	±3.8	5	−16.6	−2.2
Vaasa AB	−12.2	±7.8	5	−14.4	2.2
Mean diff.					0.8
rms diff.					3.8

The standard deviations ($1-\sigma$ estimates) of the observational trends have been calculated from the single gravity determinations (equally weighted). Both results agree well with respect to the AG accuracy estimate

From Table 3, a decrease in gravity due to land uplift is evident at almost all stations. The largest gravity changes were found around the uplift centre as depicted in Fig. 1. In Copenhagen, close to the zero uplift line, the obtained gravity rate is nearly zero. Overall, the regional rebound signal is clearly visible, but still seems to be disturbed by environmental mass variations, e.g. at station Vaasa AB. From the experiences over the last 5 years, the hydrological changes are considered as a main contributor, which is also indicated by the water level observations of the reservoirs and wells close to some of the absolute stations. The largest discrepancy from the predicted results has been found for station Onsala. The measurements do not indicate land uplift. Up to now, this is not understood. In the literature, HAAS *et al.* (1997) found larger discrepancies for the coastal station Onsala when determining atmospheric loading parameters from geodetic VLBI data. They suspect un-modelled effects due to wind-driven ocean loading as a possible reason. The absolute gravimetric time series has to be extended.

Overall, the observational trends from FG-220 are in good agreement with the predicted results. The obtained standard deviations seem to be realistic estimates for the accuracy of the deduced secular gravity changes. The disturbances caused by unaccounted-for hydrological effects are cancelled out in the trends to some extent due to the annual gravity measurements. Thus, absolute gravimetry has shown its capability to observe the Fennoscandian land uplift within the rather short time span of 4–5 years.

6. Ratio between Observational Gravity and Height Rates

JACHENS (1978) gives an introduction to and an overview of the relationship of observed temporal gravity variations and elevation changes for a fixed point on the Earth's surface. In tectonically active areas such as Fennoscandia with a still ongoing postglacial rebound (PGR), the ratio between gravity rate and height rate $\dot{g}/\dot{h}$ depends on two contributing factors: (1) vertical displacement of the observation point along the free-air gravity gradient, and (2) temporal variation of the density distribution of materials in the subsurface. The combination of geometrical and gravimetrical observations may help to separate the effects of both contributors and can serve as an observational constraint for geophysical research on the mechanism of crust formation and on the rheology of Earth's mantle and crust. From a simple theoretical contemplation, a ratio of -1.7 nm/s² per mm is obtained when assuming a free-air gradient (1) of -3.1 nm/s² per mm for the vertical surface displacement and an ongoing mass increase with density 3,300 kg/m³ in the upper mantle (Bouguer plate approximation) which yields an effect (2) of 1.4 nm/s² per mm for the variation in the density field.

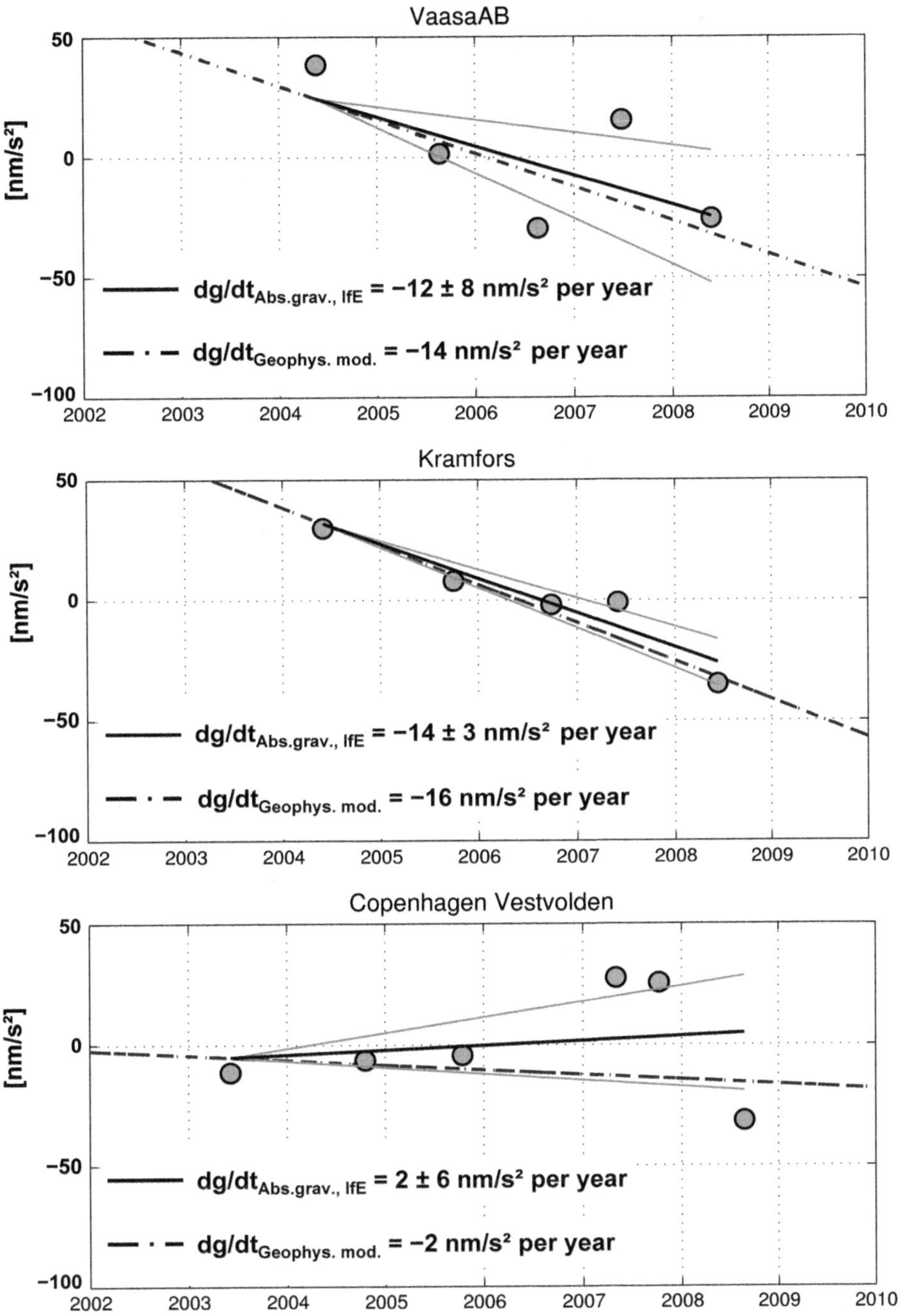

Figure 7

Linear gravity changes for three stations in Fennoscandia derived from absolute gravity measurements of IfE and compared with the trends from the model predictions provided by KLEMANN (2004). *Grey lines* beside the IfE trends indicate the standard deviation of the corresponding trend line

As an approximation, WAHR *et al.* (1995) provide a ratio (or proportionality factor) of -1.5 nm/s^2 per mm for a postglacial rebound signal. The authors assumed a free-air gradient (1) of -3 nm/s^2 per mm for the surface shift and derived from theoretical considerations a relation (2) of 0.65 mm per nm/s^2

($\equiv 1.5$ nm/s^2 per mm) for a viscoelastic response of the solid Earth to a surface load (Maxwell solid). In Mäkinen *et al.* (2005), the results from relative gravimetry surveys of the Nordic land uplift lines (east–west profiles) have been combined with various estimates of the uplift (levelling, continuous GPS, GIA models), which gives ratios between -1.6 and -2.0 nm/s^2 per mm.

Table 4 contains gravimetric trends measured by IfE together with BIFROST GPS height rates as published by Lidberg *et al.* (2007). Both kinds of temporal variations have been merged to derive the proportionality factor between gravity and height rate for the single stations. Only the eight sites with a clear uplift signal have been considered. Assuming a geographically constant ratio for Fennoscandia, an average value of -1.63 ± 0.20 nm/s^2 per mm is obtained as a weighted mean. This result is in a good agreement with the estimates given in Mäkinen *et al.* (2005), and with the approximation provided by Wahr *et al.* (1995).

7. Summary and Conclusions

With respect to the FG5-220 absolute gravity surveys since 2003, the achievements in the Fennoscandian land uplift area may be described as follows:

Table 4

Proportionality factor $\dot{g}/\dot{h}$ between gravity and height rate

Station	d*g*/d*t* (abs.grav.) (nm/s^2/year)	d*h*/d*t* (BIFROST) (mm/year)	$\dot{g}/\dot{h}$ (nm/s^2/mm)
Arjeplog	-8.7 ± 2.4	7.7 ± 0.2	-1.14 ± 0.31
Kiruna	-11.3 ± 11.0	6.4 ± 0.3	-1.78 ± 1.72
Kramfors	-14.4 ± 2.7	10.2 ± 0.5	-1.41 ± 0.27
Mårtsbo	-15.6 ± 3.9	6.7 ± 0.2	-2.32 ± 0.59
Metsähovi	-8.8 ± 5.2	4.3 ± 0.2	-2.07 ± 1.21
Östersund	-14.8 ± 10.8	8.3 ± 0.2	-1.79 ± 1.30
Skellefteå	-18.8 ± 3.8	9.6 ± 0.2	-1.95 ± 0.40
Vaasa AB	-12.2 ± 7.8	8.6 ± 0.2	-1.41 ± 0.91
Mean (unweighted)			-1.73 ± 0.14
Mean (weighted)			-1.63 ± 0.20

The BIFROST GPS results are from Lidberg *et al.* (2007). For the weighted mean and its accuracy, the standard deviations in the last column are used as weights

- The whole uplift network includes more than 40 absolute gravity stations, mostly co-located with permanent GPS points. From 2003 to 2008, the IfE absolute gravimetry team performed 84 gravity determinations at 34 different stations.
- The results from the IfE absolute gravimeter FG5-220 have been compared with results from international comparisons, and from other instruments participating in the Fennoscandia uplift project, as well as with results from repetition measurements at the German reference station in Bad Homburg. An overall accuracy of approximately ± 30 nm/s^2 is indicated for a single station determination with FG5-220. This empirical estimate includes not only the instrumental errors but also residual errors from uncertain reduction models (tides, polar motion and atmospheric mass movements). Subsurface water variations may cause gravity changes of a few times 10 nm/s^2 or even more than 100 nm/s^2 at some special stations over a year, which is not considered in the given accuracy estimate.
- A check of the instruments by parallel and reference measurements is essential. Especially for projects with the highest accuracy demands over large areas, e.g. the Nordic absolute gravity project, more than one absolute gravimeter should be employed to increase the reliability of the results and to detect instrumental offsets. This procedure improves the absolute accuracy of the whole network.
- Based on comparisons with rates predicted by geophysical modelling, the FG5-220 absolute gravity measurements for 2003/2004–2008 delivered reasonable and reliable gravity trends and accuracy estimates. The predicted rates and the observational trends agree within 3.8 nm/s^2 per year (rms difference).
- A mean proportionality factor $\dot{g}/\dot{h} = 1.63 \pm 0.20$ nm/s^2 per mm has been deduced from gravimetric and GPS observations. The result agrees well with the assumption of a Bouguer plate approximation with a mass increase ($\sim$3,300 kg/m^3) in the upper mantle.
- Absolute gravimetry has shown its capability to observe the secular gravity variations in the Fennoscandian PGR area within a time span of 4–5 years.

Acknowledgments

We appreciate the cooperation and the great efforts of the Nordic Geodetic Commission (NKG) and its Working Group on Geodynamics. We gratefully acknowledge the essential support of a number of colleagues from the Nordic countries and from Germany. The following institutions are participating in the joint project: Department of Mathematical Sciences and Technology, Norwegian University of Life Sciences (Ås); Federal Agency for Cartography and Geodesy (Frankfurt); Finnish Geodetic Institute (Masala); Norwegian Mapping Authority (Hønefoss); Onsala Space Observatory, Chalmers University of Technology; Swedish Mapping, Cadastre and Land Registration Authority (Gävle); Technical University of Denmark, National Space Institute (Copenhagen). The research has been supported generously by the German Research Foundation (DFG) through the research grants MU 1141/3-1, 3-2 and 3-3 ("Geotechnologien").

References

Ågren, J., and Svensson, R., Postglacial land uplift model and system definition for the new Swedish height system RH 2000, Reports in geodesy and geographical information systems (Gävle: Lantmateriet, 2007).

Bilker-Koivula, M., Mäkinen, J., Timmen, L., Gitlein, O., Klopping, F., and Falk, R., Repeated absolute gravity measurements in Finland, In Terrestrial Gravimetry: Static and Mobile Measurements (TG-SMM2007), Proceed. Int. Symp. (ed. Peshekhonov V.G.) (St. Petersburg: Elektropribor, 2008) pp. 147–151.

Boedecker, G., and Flury, J. (1995), International Absolute Gravity Basestation Network IAGBN, Catalogue of stations and observations, Report of the IAG International Gravity Commission, IGC-Working Group 2, "World Gravity Standards", available at the Bureau Gravimetrique International, Toulouse.

Boedecker, G., and Fritzer, Th. (1986), International Absolute Gravity Basestation Network, Veröff. Bayer. Komm. für die Internat. Erdmessung der Bayer. Akad. d. Wissensch., Astron.-Geod. Arb. 47, München.

Carter, W.E., Aubrey, D.G., Baker, T., Boucher, C., LeProvost, C., Pugh, D., Peltier, W.R., Zumberge, M., Rapp, R.H., Schutz, R.E., Emery, K.O., and Enfield, D.B. (1989), Geodetic fixing of tide gauge bench marks, Woods Hole Oceanographic Institution Report WHOI-89-31/CRC-89-5, Woods Hole, Mass., U.S.A.

Carter, W. E., Peter, G., Sasagawa, G. S., Klopping, F. J., Berstis, K. A., Hilt, R. L., Nelson, P., Christy, G. L., Niebauer, T. M., Hollander, W., Seeger, H., Richter, B., Wilmes, H., and Lothammer, A. (1994), New gravity meter improves measurements, EOS, Transactions, American Geoph. Union, Vol. 75, No. 08, pp. 90–92.

Ekman, M. (1996), A consistent map of the postglacial uplift of Fennoscandia, Terra Nova 8, 158–165.

Ekman, M., and Mäkinen, J. (1996), *Recent postglacial rebound, gravity change and mantle flow in Fennoscandia*, Geophys. J. Int. *126*, 229–234.

Francis, O., and van Dam, T., Analysis of results of the International Comparison of Absolute Gravimeters in Walferdange (Luxembourg) of November 2003, In: International Comparison of Absolute Gravimeters in Walferdange (Luxembourg) of November 2003, Cahiers du Centre Européen de Géodynamique et de Séismologie 26 (eds. Francis. O., van Dam. T.) (Centre Européen de Géodynamique et de Séismologie: Luxembourg 2006) pp. 1–23.

Francis, O., van Dam, T., Germak, A., Amalvict, M., Bayer, R., Bilker-Koivula, M., Calvo, M., D'Agostino, G.-C., Dell' Acqua, T., Engfeldt, A., Faccia, R., Falk, R., Gitlein, O., Fernandez, Gjevestad, J., Hinderer, J., Jones, Kostelecky, J., Le Moigne, N., Luck, B., Mäkinen, J., McLaughlin, D., Olszak, T., Olsson, P., Pachuta, A., Palinkas, V., Pettersen, B., Pujol, R., Prutkin, I., Quagliotti, D., Reudink, R., Rothleitner, C., Ruess, D., Shen, C., Smith, V., Svitlov, S., Timmen, L., Ulrich, C., Van Camp, M., Walo, J., Wang, L., Wilmes, H., and Xing, L., Results of the European Comparison of Absolute Gravimeters in Walferdange (Luxembourg) of November 2007, In Gravity, Geoid and Earth Observation, IAG Symp. 135 (ed. Mertikas. St.) (Heidelberg: Springer, 2010) pp. 31–36.

Gitlein, O. (2009), Absolutgravimetrische Bestimmung der Fennoskandischen Landhebung mit dem FG5-220, Wissenschaftliche Arbeiten der Fachrichtung Geodäsie und Geoinformatik der Leibniz Universität Hannover 281.

Gitlein, O., Timmen, L., Müller, J., Denker, H., Mäkinen, J., Bilker-Koivula, M., Pettersen, B.R., Lysaker, D.I., Svendsen, J.G.G., Wilmes, H., Falk, R., Reinhold, A., Hoppe, W., Scherneck, H.-G., Engen, B., Omang, O.C.D., Engfeldt, A., Lilje, M., Strykowski, G., and Forsberg, R., Observing Absolute Gravity Acceleration in the Fennoscandian Land Uplift Area, In: Terrestrial Gravimetry: Static and Mobile Measurements (TG-SMM2007), Proceed. Int. Symp. (ed. Peshekhonov. V.G.) (St. Petersburg: Elektropribor, 2008) pp. 175–180.

Hagedoorn, J. M., Wolf, D., and Martinec, Z. (2007), *An estimate of global mean sea-level inferred from tide-gauge measurements using glacial-isostatic models consistent with the relative sea-level record*, Pure Appl. Geophys. *164*, 791–818.

Haas, R., Scherneck, H-G., and Schuh, H., Atmospheric loading corrections in geodetic VLBI and determination of atmospheric loading coefficients, In: Proceedings of the 12th working meeting on European VLBI for Geodesy and Astronomy, Norway (ed. Pettersen. B.R.) (Hønefoss, 1997) pp. 111–121.

Jachens, R.C., The gravity method and interpretive techniques for detecting vertical crustal movements, In: Applications of Geodesy and Geodynamics, Int. Symp., Reports of the Department of Geodetic Science 280 (ed. Mueller. I.I.) (Ohio State University 1978) pp. 153–155.

Johansson, J.M., Davis, J.L., Scherneck, H.-G., Milne, G.A., Vermeer, M., Mitrovica, J.X., Bennett, R.A., Jonsson, B., Elgered, G., Elósegui, P., Koivula, H., Poutanen, M., Rönnäng, B.O., and Sha-piro, I.I. (2002), *Continuous GPS measurements of postglacial adjustment in Fennoscandia, 1. geodetic results*, J. Geophys. Res. *107*, B8, ETG 3, 1–27.

Kaufmann, G., Wu, P., and Li, G. (2000), *Glacial isostatic adjustment in Fennoscandia for a laterally heterogeneous Earth*, Geophys. J. Int. *143*, 262–273.

Klemann, V. (2004), Linear gravity variations for the IfE absolute gravity stations in Fennoscandia predicted by geophysical GIA modelling, Personal communication, Deutsches GeoForschungsZentrum (GFZ), Potsdam.

LAMBECK, K., SMITHER, C., and EKMAN, M. (1998a), *Tests of glacial rebound models for Fennoscandinavia based on instrumented sea- and lake-level records*, Geophys. J. Int. *135*, 375–387.

LAMBECK, K., SMITHER, C., and JOHNSTON, P. (1998b), *Sea level-change, glacial rebound and mantle viscosity for northern Europe*, Geophys. J. Int. *134*, 102–144.

LAMBERT, A., JAMES, T. S., LIARD, J. O., and COUTIER, N., The role and capability of absolute gravity measurements in determining the temporal variations in the earth's gravity field, In Global Gravity Field and Its Temporal Variations, IAG Symp. 116 (eds. Rapp. R.H., Cazenave. A.A., Nerem. R.S.) (Heidelberg: Springer, 1996) pp. 20–29.

LAMBERT, A., COURTIER, N., SASAGAWA, G. S., KLOPPING, F., WINESTER, D., JAMES, T. S., and LIARD, J. O. (2001), *New constraints on laurentide postglacial rebound from absolute gravity measurements*, Geophys. Res. Letters *28*, 2109–2211.

LIDBERG, M., JOHANSSON, J. M., SCHERNECK, H.-G., and DAVIS, J. L. (2007), *An improved and extended GPS-derived 3D velocity field of the glacial isostatic adjustment (GIA) in Fennoscandia*, J. of Geodesy *81*, 213–230.

MÄKINEN, J., ENGFELD, A., HARSSON, B. G., ROUTSALAINEN, H., STRYKOWSKI, G., OJA, T., and WOLF, D., The Fennoscandian Land Uplift Lines 1966–2003, In Gravity, Geoid and Space Missions GGSM2004, IAG Symp. 129 (eds. Jekeli. C., Bastos. L., Fernandes. J.) (Heidelberg: Spinger, 2005) pp. 328–332.

MÄKINEN, J., AMALVICT, M., SHIBUYA, K., and FUKUDA, Y. (2007), *Absolute gravimetry in Antarctica: status and prospects*, J. of Geodyn. *43*, 339–357.

MARTINEC, Z. (2000), *Spectral-finite element approach to three-dimensional viscoelastic relaxation in a spherical earth*, Geophys. J. Int. *142*, 117–141.

Micro-g Solutions Inc (1999) Operator's manual, FG5 absolute gravimeter, Erie, Colorado.

MILNE, G.A., DAVIS, J.L., MITROVICA, J.X., SCHERNECK, H.-G., JOHANNSON, J.M., VERMEER, M., and KOIVULY, H. (2001), *Space-geodetic constraints on glacial isostatic adjustment in Fennoscandia*, Science *291*, 2381–2385.

NAKIBOGLU, S.M., and LAMBECK, K., Secular sea-level change, In Glacial isostasy, sea level and mantle rheology (eds. Sabadini. R., Lambeck. K., Boschi. E.) (Dordrecht: Kluwer Acad. Publ., 1991) pp 237–258.

NIEBAUER T.M., SASAGAVA G.S., FALLER J.E., HILT R., and KLOPPING F. (1995), *A new generation of absolute gravimeters*, Metrologia *32*, 159–180.

SCHERNECK, H.-G., JOHANSSON, J., KOIVULA, H., VAN DAM, T., and DAVIS, J. (2003), *Vertical crustal motion observed in the BIFROST project*, J. of Geodyn. *35*, 425–441.

TAPLEY, B.D., BETTADPUR, S., RIES, J.C., THOMPSON, P.F., and WATKINS, M.M. (2004), *GRACE measurements of mass variability in the Earth system*, Science *305*, 503–505.

TIMMEN, L. (2003), *Precise definition of the effective measurement height of free-fall absolute gravimeters*, Metrologia *40*, 62–65.

TIMMEN, L., GITLEIN, O., MÜLLER, J., DENKER, H., MÄKINEN, J., BILKER, M., PETTERSEN, B. R., LYSAKER, D. I., OMANG, O. C. D., SVENDSEN, J. G. G., WILMES, H., FALK, R., REINHOLD, A., HOPPE, W., SCHERNECK, H.-G., ENGEN, B., HARSSON, B. G., ENGFELDT, A., LILJE, M., STRYKOWSKI, G., and FORSBERG, R., Observing Fennoscandian Gravity Change by Absolute Gravimetry, In: Geodetic Deformation Monitoring: From Geophysical to Engineering Roles, IAG Symp. 131 (eds. Sansò. F., Gil. A.J.) (Heidelberg: Springer, 2006) pp. 193–199.

TIMMEN, L., GITLEIN, O., MÜLLER, J., STRYKOWSKI, G., and FORSBERG, R. (2008), *Absolute gravimetry with the Hannover meters JILAg-3 and FG5-220, and their deployment in a Danish–German cooperation*, zfv—Zeitschrift für Geodäsie, Geoinformation und Landmanagement 3/2008, Jg. *133*, 149–163.

TORGE, W., Gravimetry (Berlin: de-Gruyter, 1989).

TORGE, W. (1993), Gravimetry and tectonics, Publications of the Finnish Geo-detic Insti-tute (Geodesy and Geophysics) *115*, 131–172.

VAN CAMP, M., WILLIAMS, S.D.P., FRANCIS, O. (2005), *Uncertainty of absolute gravity measurements*, J. Geophys. Res. *110*, B05406.

WAHR, J. M., DAZHONG, H., and TRUPIN, A. (1995), *Predictions of vertical uplift caused by changing polar ice volumes on a viscoelastic earth*, Geophys Res Lett *22*(8), 977–980.

WAHR, J., and VELICOGNA, I. (2003), *What might GRACE contribute to studies of post glacial rebound*, Space Science Reviews *108*, 319–330.

WIEHL, M., DIETRICH, R., and LEHMANN, A., How Baltic sea water mass variations mask the postglacial rebound signal in CHAMP and GRACE gravity field solutions, In Earth observation with CHAMP (eds. Reigber. Ch., Lühr. H., Schwintzer. P., Wickert. J.) (Heidelberg: Springer, 2005) pp 181–186.

WILLIAMS, S.D.P., BAKER. T.F., and JEFFRIES, G. (2001), *Absolute gravity measurements at UK tide gauges*, Geophys. Res. Letters *28*, 2317–2329.

WILMES, H., and FALK, R., Bad Homburg—a regional comparison site for absolute gravity meters, In: International Comparison of Absolute Gravimeters in Walferdange (Luxembourg) of November 2003, Cahiers du Centre Européen de Géodynamique et de Séismologie 26 (eds. Francis. O., van Dam. T.) (Luxembourg : Centre Européen de Géodynamique et de Séismologie, 2006) pp. 29–30.

WOLF, D. (1993), *The changing role of the lithosphere in models of glacial isostasy: a historical review*, Global and Planetary Change *8*, 95–106.

WÖPPELMANN, G., MARTIN MIGUES, B., BOUIN, M.-N., and ALTAMIMI, Z. (2007), *Geocentric sea-level trend estimates from GPS analyses at relevant tide gauges world-wide*, Global and Planetary Change *57*, 396–406.

(Received November 17, 2009, revised November 14, 2010, accepted June 12, 2011, Published online September 1, 2011)

Pure Appl. Geophys. 169 (2012), 1343–1356

DOI 10.1007/s00024-011-0398-8

Pure and Applied Geophysics

On the Accuracy of the Calibration of Superconducting Gravimeters Using Absolute and Spring Sensors: a Critical Comparison

Umberto Riccardi,[1,2] Severine Rosat,[1] and Jacques Hinderer[1]

Abstract—Over the past two decades, superconducting gravimeters (SGs) have been a key tool to investigate a number of geophysical processes leading to time-variable gravity changes. As SGs are relative meters, even though they are the most sensitive and stable devices currently available, they need to be accurately calibrated. Each branch of Earth sciences that benefits from high-precision gravity monitoring demands calibration of gravity sensors to accuracy of better than 0.1%. This research deals with a calibration experiment performed at the Strasbourg (France) SG site by means of two FG5 (#206 and #211) absolute gravimeters (AGs) and new-generation spring meters (Scintrex Ltd. Autograv CG-3M and CG5 and Microg-LaCoste gPhone). Our goal is to try to use the newest generation of spring mechanical gravimeters (MGs) for calibrating SGs. We discuss the results in terms of precision and accuracy of the SG calibration by means of different metrological and methodological approaches. With the FG5 #211 we derive scale factors for the SG-GWR C026 located in Strasbourg in agreement with those routinely obtained since 1997 by means of the FG5 #206. This confirms that the estimation of the scale factors is independent of the AG sensor. From a moving-window regression analysis between the synthetic body tides and both the SG and MG gravity records we detect significant fluctuations of the SG scale factors over time due to the instability of the instrumental sensitivity of the MGs. Our main results demonstrate that, owing to the time variability of their sensitivity, the used spring meters, even if well calibrated, cannot be used as a stable reference for SGs. As a result, MGs are not suitable to replace AGs for SG calibration, and we conclude that currently the method using parallel recording with absolute gravity meters is still the most feasible calibration approach for SGs.

Key words: Gravimetry, calibration, superconducting gravimeters, spring gravimeters, absolute gravimeters.

1. *Introduction*

Geodynamical phenomena involving underground mass redistribution and/or change of the Earth's figure affect the gravity field. The magnitude of gravity changes induced by these phenomena can be very small, of the order of 10^{-8}–10^{-9} g (where g is the surface gravity). The long-term stability of superconducting gravimeters (SGs) enables us to collect high-precision gravity observations and investigate a large number of geophysical processes (e.g., Richter *et al.*, 1995a; Hinderer and Crossley, 2000, 2004; Hinderer *et al.*, 2007) inducing surface gravity signals, such as Earth tides, core and wobble modes of the Earth, ocean and atmospheric loading, polar motion, free-core and inner-core nutations, sea-level changes, etc. Furthermore, at the level of precision of SGs, it should be possible to validate Earth tide models. For that purpose, SGs need to be accurately calibrated. So, accurate calibration is still an important issue for obtaining reliable results from high-precision gravity monitoring. Recent advances in theoretical modeling of body and oceanic tides have demanded calibration of the scale factor to accuracy of better than 0.1% (Baker and Bos, 2003). The gravimetric factors for the Dehant, Defraigne, and Wahr (DDW) elastic and inelastic body tide models (Dehant *et al.*, 1999) only differ by 0.12%, so calibrations at such a level of accuracy are needed to validate these models, for instance. This level of accuracy is difficult to reach because of many problems in reproducing an artificial gravity change to be used as a "standard" at such a level of accuracy.

SGs are mostly designed to operate in geophysical observatories as permanent gravity stations; they are not portable and therefore not suitable for measuring on calibration lines. This has forced users to develop in situ calibration methods. Calibration of an SG can be performed by using different approaches that can be categorized into direct and indirect methods.

[1] Institut de Physique du Globe de Strasbourg, IPGS - UMR 7516, CNRS et Université de Strasbourg (EOST), 5 rue René Descartes, 67084 Strasbourg, Cedex, France. E-mail: umbricca@unina.it

[2] Dipartimento di Scienze della Terra, Università "Federico II" di Napoli, L.go S. Marcellino 10, 80138 Naples, Italy.

Direct methods customarily employ devices which allow gravity sensors to be perturbed through an artificial gravity field. This can be done either by moving an external mass (ACHILLI *et al.*, 1995) or equivalently by moving the instrument (inertial acceleration) using an oscillating platform (RICHTER *et al.*, 1995b; VAN RUYMBEKE *et al.*, 1995; FALK *et al.*, 2001); suitable accuracies (10^{-4}) have been achieved with both methods. Unfortunately, direct methods of calibration require some special techniques that sometimes turn out to be problematic, such as the management of heavy reference masses, and that are not feasible at every SG station. Thanks to their feasibility, indirect methods of calibration are frequently preferred over direct ones. Indirect methods are based on monitoring of tides for a long period (2–3 years) followed by the use of well-known tidal amplitudes (e.g., the tidal waves M2 and O1) (MELCHIOR, 1994). The capability to model the ocean tidal loading effect at a suitable level of accuracy is strongly related to the use of the M2 amplitude. In fact, at Continental European stations at mid-latitudes, M2 and O1 have practically equal amplitudes (around 35 μGal) but the oceanic contribution to O1 tidal gravity is rather small. This is why O1 is preferred over M2 tidal wave for calibration purposes. However, some drawbacks of tidal methods remain: first, the requirement for a long-time data series, which makes the method very time consuming; furthermore, the defects in the modeling of gravity changes due to several sources, such as atmospheric air pressure, hydrology, and so on, impact on calibration accuracy.

Currently, the most widely used calibration method is that based on comparison with an absolute gravimeter (AG) (FRANCIS and VAN DAM, 2002). The scale factor of SGs is determined by co-located measurements with AGs (see, e.g., HINDERER *et al.*, 1991; SATO *et al.*, 1996; FRANCIS *et al.*, 1998; AMALVICT *et al.*, 2001; FALK *et al.*, 2001; TAMURA *et al.*, 2001; IMANISHI *et al.*, 2002; FUKUDA *et al.*, 2005). Experience has shown that long data series of up to 7 days are necessary to obtain stable results with accuracy of about 0.1–0.2% (FRANCIS, 1997; FRANCIS *et al.*, 1998) and that days with maximum tidal amplitudes should be used. However, for most stations of the Global Geodynamics Project (GGP) network (CROSSLEY *et al.*, 1999), the accuracy achieved ranges between 0.1% and 0.4% (MEURERS, 2001). This accuracy level may not be sufficient to compare observed tidal factors with those proposed by recent theoretical studies (see, e.g., DEHANT *et al.*, 1999; MATHEWS, 2001).

Use of a well-calibrated relative mechanical spring gravimeter (MG) as a reference for an SG is still under discussion in the gravimetric community, because it needs to be calibrated against some other reference. Moreover, the precision of the calibration is hindered by the effect of error propagation coming from different instrumental sources. A review of tidal gravity observations collected with astatized spring gravimeters in the International Center for Earth Tides (ICET) databank has showed that these spring sensors display calibration uncertainty averaging $\pm 1\%$ (MELCHIOR and FRANCIS, 1998). In principle, such performance would limit the use of spring gravimeters as references for SGs. The ICET databank collects data acquired mainly with LaCoste and Romberg (LR) meters, models G, D, and Earth tides (ET) gravimeters, most of them built during the 1970s and 1980s. The main difficulty in trying to compare the calibration results coming from these gravimeters is that the signal composition of the spring sensors differs due to several reasons (e.g., instrumental noise and response to microseismic noise; instrumental drift, amplitude, and phase response). In addition, LR gravimeters are known to give an abnormal reaction to atmospheric pressure variations due to noncompensated Archimedean forces (e.g., ARNOSO *et al.*, 2001; MEURERS, 2002). In principle, SG calibration could be done in either the time or the frequency domain. In the time domain we obtain the scale factor by following the classical procedure customarily adopted for calibration using AGs, that is to say by least-squares fitting a straight line between the MG observations and the SG output voltage; this can be successfully carried out even on short datasets (several days). On the contrary, frequency-domain calibration can be obtained by comparing the results of tidal analyses applied on the data collected with the two meters; this approach requires long observation periods (>1 month) in order to properly separate the main tidal constituents. Since this method is essentially based on comparative tidal analysis in the main tidal bands 1 and 2 cpd, numerical high-pass filters designed for such analyses are customarily applied to the data, so drift determination and the different air pressure responses of the sensors become less critical.

This paper concerns a calibration experiment performed at the Strasbourg (France) superconducting gravity station (SG GWR–C026) by means of two FG5 absolute gravimeters and three new-generation spring gravimeters: the Scintrex Autograv CG-3M and CG5 and the most recent Microg-LaCoste gPhone. We discuss the results in terms of the precision and accuracy of the SG calibration. In a first part we perform the SG calibration using the FG5 #211 and compare with previous calibrations using the FG5 #206. Then we use the MG record to calibrate the SG. In a third part the time stability of the scale factor is discussed, and finally a tidal analysis is performed on both the gPhone and the SG to check the transfer function of the gPhone in the tidal frequency band.

2. *Calibration by Means of FG5 Absolute Meters*

The calibration of the superconducting gravimeter (SG GWR-C026) at the gravity observatory J9 in Strasbourg (France) by means of parallel absolute gravity measurements dates back to 1997 (Amalvict *et al.*, 2001). The SG-C026 is routinely calibrated about six times per year by means of absolute gravity observations collected with the FG5 #206. Details on the calibration procedures are given in Amalvict *et al.* (2001) and Rosat *et al.* (2009).

During 2008 and 2009, the absolute gravimeter FG5 #211 owned by the Instituto Geográfico Nacional (IGN) of Madrid in Spain has operated at the J9 observatory. Here we report on the determination of the scale factors of the SG-C026 obtained by means of co-located absolute measurements collected with this Spanish meter. Two calibration sessions were performed based on continuous AG records lasting several days: (a) 17–23 December 2008 and (b) 16–23 June 2009, lasting about 6 and 7 days, respectively. We discuss the results in terms of the accuracy achieved in the calibration and compare them with the historical set of scale factors routinely determined at J9 with the French FG5 #206. We follow the procedure described in Rosat *et al.* (2009) to make the calibrations obtained with the FG5 #211 comparable to the historical set of scale factors.

In principle, SG calibration can be done by least-squares fitting a straight line between the AG observations, expressed in gravity units (1 μGal = 10 nm/s^2), to be taken as the dependent variable, and the output voltage from the SG feedback which needs to be calibrated (Fig. 1a, b). The slope coefficient of the fitted line, expressed in gravity units/Volt, provides the scale factor of the SG. The fitting is customarily done assuming that errors only affect the AG data. Accounting for the typical standard deviation of the SG gravity residuals, which is at least two orders lower than the standard deviation of the AG set values, this assumption can be considered realistic. In spite of that, we set up a validation test on this assumption in order to verify that the errors of the SG data do not impact on the fitting procedure. We tested a weighted total least-squares (WTLS) algorithm (Krystek and Anton, 2007) on the calibration data. This approach deals with the well-known problem of fitting a straight line to data with uncertainties in both coordinates. Indeed, no significant difference in the scale factors and their statistics result from the application of the WTLS algorithm; this is due to the very small noise level in the SG gravity record. Therefore, we reduce our study to a problem of simple weighted fitting, only accounting for the errors in the dependent variable.

Assuming that the measurement error of y_n (AG observations) follows a normal distribution, we search for the a and b values that minimize χ^2, the weighted sum of squared residuals,

$$\chi^2 = \sum_{n=1}^{N} \frac{1}{\sigma_n^2}\left[(y_n - ax_n - b)^2\right], \tag{1}$$

where N is the number of data, $\frac{1}{\sigma_n^2}$ is the weight, and σ_n is the standard deviation of each set of absolute gravity measurements. Minimizing χ^2 with respect to the unknown parameters leads to the well-known normal equations. The weighting function in the least-squares procedure plays a crucial role, because the measurement error (σ_n) is far from being homogeneous throughout the observation period. Indeed, several error sources (e.g., laser problems, clock drift,

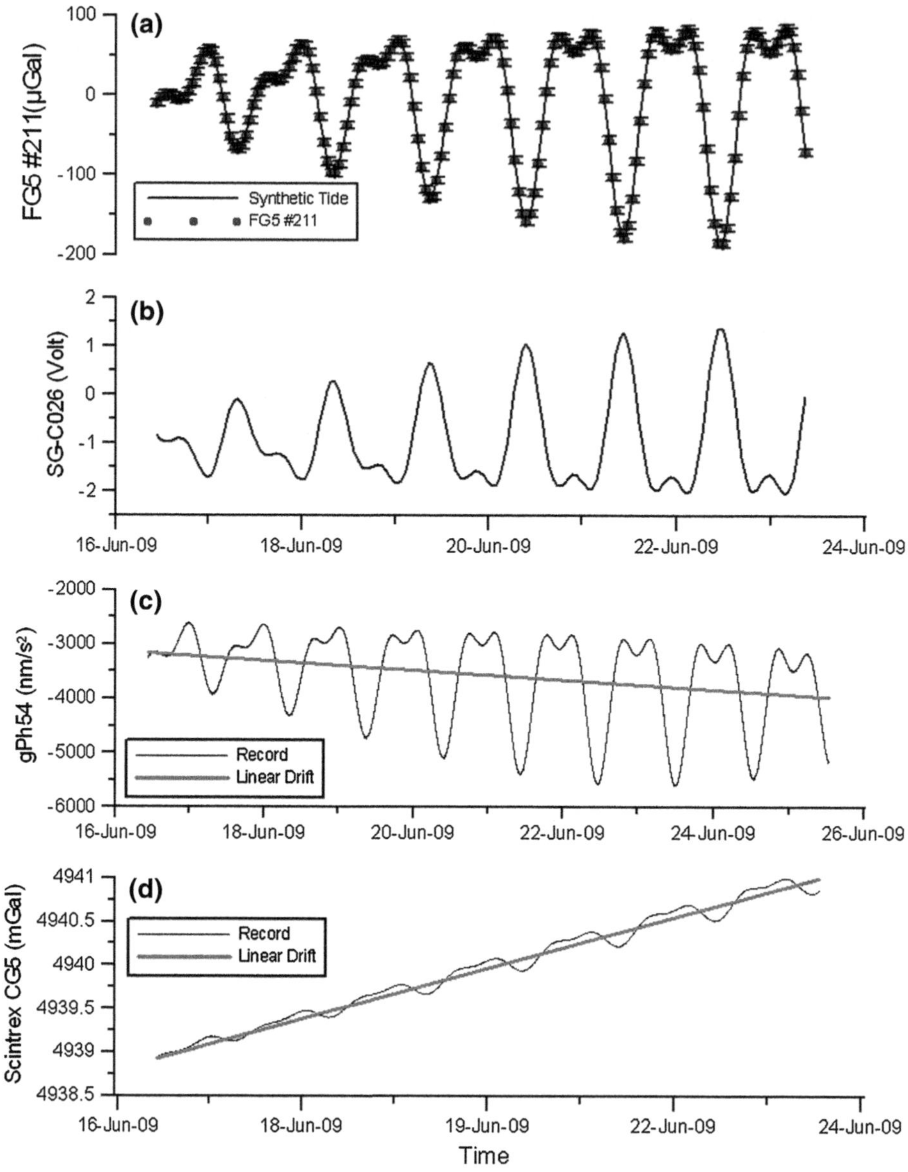

Figure 1
SG-C026 calibration with co-located AG measurements and relative gravity records: **a** hourly absolute measurements collected with the FG5 #211, **b** SG-C026 output voltage, **c** gPh54 gravity record and drift model, and **d** CG5 gravity record and drift model

environmental or microseismic noise) can contribute to the measurement error coming from the drop scattering (Van Camp *et al.*, 2005). Of course the procedure described in Eq. (1) is justified only if the SG and the AG data are perfectly synchronized.

For calibration purpose we use the set of gravity values of the AG resulting from a statistical processing of the drop gravity values; a rejection criterion of outlier drops is applied at 3σ (standard deviations). Disturbances related to earthquakes occurring at the time of the calibration have to be removed from the raw SG records. As a result of the low instrumental drift of the SG (less than 15 nm/s^2 per year) and the short duration of each calibration campaign (6–7 days), the instrumental drift does not need to be removed before the linear regression. Before applying the linear least-squares fitting, we derived the SG values corresponding to the central

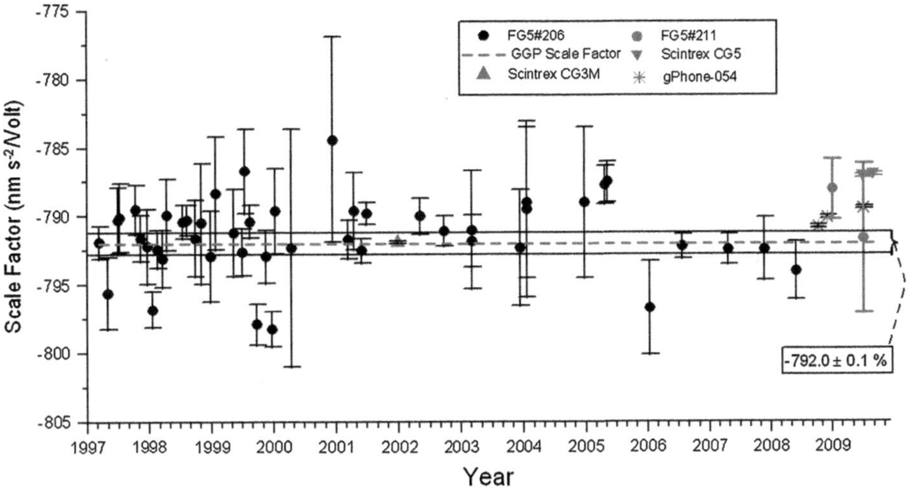

Figure 2
SG-C026 scale factors obtained since 1997 with the FG5 #206 and the most recent ones estimated with the Spanish FG5 #211 and MGs. The 0.1% limit of accuracy (*bold line*) and the scale factor provided in the GGP database (−792 nm/s^2/V, *dotted line*) are also indicated. The *error bars* correspond to 1σ

time of the AG sets from the 1-min SG data. The 1-s SG data were decimated to 1 min using a low-pass filter designed as a standard 504-pole finite impulse response (FIR) filter, having an equivalent half-size length of about 8 min, implemented in the GWR data acquisition system.

These new scale factors, coming from the two calibrations carried out in December 2008 and June 2009, respectively, are plotted in Fig. 2 with the previous calibration factors. The results (-788 ± 2.2 and -792 ± 4.8 nm s^{-2}/V) are in agreement with the scale factors and standard deviations obtained in the past by way of the French FG5 #206.

3. *Calibration by Means of Spring Gravimeters*

This section deals with the calibration of the SG-C026 pursued with the Microg-LaCoste gPhone and the Scintrex Ltd. CG-3M and CG5. These spring sensors represent the two kinds of spring available today: (a) a zero-length metal spring (patented by LaCoste & Romberg and ZLS Corp.) and (b) a monolithic quartz spring sensor (patented by Scintrex Ltd.).

The relative gravimeter gPhone N°54 (gPh54) owned by the Spanish IGN has been operating at J9 over two time intervals (May–December 2008 and May–August 2009). The gPhone is a portable Earth tide gravimeter equipped with 0.1-μGal resolution feedback. It is essentially a LR, model D meter, but with significant upgrades: it has an improved thermal system (a double oven) to increase temperature stability and should have an efficient vacuum seal, making it almost insensitive to buoyancy changes due to atmospheric pressure fluctuations (Microg-LaCoste, 2008).

The CG-3M is an automated gravity meter having a measurement range of over 8,000 mGal and reading resolution of 1 μGal. The sensing element is sealed in a temperature-stabilized vacuum chamber to improve protection from changes in ambient temperature and atmospheric pressure. Internal tilt sensors are more efficient than the conventional bubble levels mounted on the old LR meters, and continuous tilt correction can be pursued (Scintrex Limited, 1995). The newest version CG5 (Scintrex Limited, 2006) comes from a restyling of the CG-3M with reduced meter weight and improved electronics, power supply (battery), and global positioning system (GPS) capability for reliable positioning and timing of the gravity station.

The calibration procedure adopted is the same as described before (Eq. 1), that is to say, we use weighted fitting accounting only for the errors in the MG data. The uncertainties (σ_n) in the MG observations, if not directly available as an acquired channel, as is the case for LR sensors, are quantified through the standard deviations of the decimated residual gravity. Gravity residuals are obtained by subtracting from the record the body tides, the ocean tidal loading using the FES04 model (Letellier, 2004; Lyard *et al.*, 2006; Bos and Scherneck 2009), the atmospheric fluctuations (through a nominal admittance of -3 nm/s^2/hPa), and the drift. Of course, preprocessing aimed at reducing earthquake perturbations, spikes, and gap effects in both SG and MG registrations is applied. The main difference from the calibration approach using the AG data is that drift has to be taken into account. Drift has a strong impact on the calibration accuracy, even more so when using long time series of MG data. In this case the strong and irregular drift of the spring gravimeters has to be carefully modeled or reduced by high-pass filtering, because it is expected to introduce systematic calibration errors. It is well known that drift of the MG sensor is caused by unavoidable creep of the spring, whose length increases under tension; spring elongation on the average of 0.5 ppm/day has been estimated for the quartz sensor at operating temperature of 60°C. This leads to typical linear drift rate of about 500 μGal/day (Scintrex Limited, 1995). Zero-length metal spring sensors usually have smaller drift rates on the order of some tens of μGal/day.

The dataset available for the calibration test consisted of five time windows as listed in Table 1. As an example, an SG calibration test carried out through an 8-day-long parallel record with the gPh54 is depicted in Fig. 1c; the results are listed in Table 1; a very low linear drift of 12 μGal/day has been subtracted from the gPh54 record (Fig. 1c). For the sake of comparison, an analogous SG calibration through a 7-day record collected with the CG5 is depicted in Fig. 1d; in this case, the spring gravimeter displayed a huge drift averaging about 300 μGal/day.

The whole set of SG-C026 scale factors determined since 1997 by using all the gravity sensors available over the different epochs is plotted in Fig. 2. The results obtained with the MGs are in the range of repeatability of the scale factors obtained using the AG measurements. It is quite evident from Fig. 2 that the error bars (standard deviations) associated with the SG calibrations obtained by means of AG and MG measurements are quite different. Actually, the calibration through the relative spring gravimeters appears to be more precise then the one obtained via AG data. Different factors contribute to improve the precision of the calibration achieved with the MGs: the decimation filters used in downsampling the data from 1 to 60 s, the mean scatter of 1-s raw data, which is smaller than that for the drop sets collected with AGs, and subordinately the larger number of minute data available for calibration purposes (60 times larger than the hourly AG ones). However, the greater precision is not necessarily a guarantee of greater accuracy. In fact, if not well-calibrated MG sensors are used, very precise calibrations can be achieved, but the scale factors would systematically deviate from the real values; in other words, we could have very precise but wrong (or poorly accurate) scale factors. This drawback of the use of MGs for SG calibration purpose is more evident when long time series are used.

4. *Time Stability of the SG Calibration Using MGs*

This section deals with two calibrations of the SG-C026 using MG gravity records longer than 1 month. Our intent is, through comparison of the SG gravity record with signals coming from MGs, to focus on the performance of the spring sensors and to highlight the effects that strongly limit the use of such

Table 1

Results of the SG-C026 calibration tests performed with the CG3M, CG5, and gPh54 spring gravimeters

MG	Time span	Length	Scale factor (nm s^{-2}/V)
Scintrex CG-3M	26 October 2001 to 11 February 2002	108 days	-790.5 ± 0.13
Scintrex-CG5	16 June 2009 to 23 June 2009	7 days	-787.1 ± 0.13
gPhone 054	3 September 2008 to 14 October 2008	42 days	-790.8 ± 0.12
gPhone 054	5 November 2008 to 10 December 2008	36 days	-790.1 ± 0.10
gPhone 054	16 June 2009 to 25 June 2009	9 days	-789.4 ± 0.09

gravimeters as a reference for SGs. In Fig. 3, signals collected over more than 3 months with the CG-3M are jointly plotted with the SG-C026 output voltage. As a preliminary model for the drift, we try to fit a straight line to the CG-3M record, but after subtracting just the body tides (Fig. 3c), the gravity signal still reveals a residual trend requiring an additional polynomial reduction. Therefore, a secondary, fifth-degree polynomial is subtracted from the record. Then, we use the CG-3M gravity record corrected for the drift (Fig. 3d) to calibrate the SG output voltage (Fig. 3e). Since a long time interval is used for calibration purposes, a linear drift is subtracted from the superconducting signal as well. The calibration outcomes by using the whole, 108-day gravity record are displayed in Fig. 3f and listed in Table 1. As a long gravity record is available, we decided to analyze the stability of the SG scale factors coming from the intercomparison with the CG-3M over that time span. The dataset was split into several 5-day-long subsets, and the regression analysis was applied on each set. The scale factors resulting from the analysis are plotted in Fig. 4, which features clear time variability, sometimes exceeding 0.1%. In order to understand the source of such fluctuations we used as a reference signal the synthetic body tide computed on the basis of the tidal parameters (amplitude and phase) experimentally determined at Strasbourg; accounting for a standard admittance, the gravity effect due to the atmosphere is even included in the synthetic signal. In other words, we use the synthetic body tides as a calibrator signal for the SG-CO26 output voltage, aiming to obtain a set of synthetic scale factors to be compared with those obtained from the CG-3M (Fig. 4a). It is noteworthy that the synthetic scale factors display very small variability. This result leads to the hypothesis that the time evolution of the scale factors could be due to some instability of the instrumental sensitivity of the MG. Aiming to check the CG-3M sensitivity, we first transformed the gravity signal into electrical units (Volt) through the scale factor provided by the manufacturing company (Scintrex Ltd.), then we calibrated the output voltage with the synthetic body tides; in this way we are able to compute the CG-3M instrumental sensitivity and study its evolution over time (Fig. 4b). Figure 4 clearly shows that the trends of the SG-C026 scale factors and CG-3M sensitivity overlap perfectly. This observation leads us to infer that the observed wandering of the SG-C026 scale factors is exclusively referable to the instability of the sensitivity of the spring sensor. We have no explanation for this instability, but we can exclude that it could be due to some site-dependent effect such as tilting over time; in fact, tilt effects have been reduced in the gravity record according to the signal coming from the biaxial electronic levels. If the meter is tilted by an angle ϕ, we measure a component $g\cos\phi$, which is smaller than the vertical gravity component (g); the reduction from the real vertical g is

$$\Delta g = g\left(-\frac{1}{2}\phi^2\right). \qquad (2)$$

Assuming $g = 9.8$ m/s^2 and expressing ϕ in radians, we obtain a dependency of gravity on tilt changes averaging -4.9 m^{-2}/rad^2 (or equivalently -4.9×10^{-3} nm s^{-2}/μrad^2). If the tilt change is monitored with two orthogonal levels (cross and long levels or [X, Y]), the measured gravity changes in response to tilt can be computed according to the following expression:

$$\Delta g = g\left[1 - \cos\phi_x \cos\phi_y\right], \qquad (3)$$

where ϕ_x and ϕ_y are tilts of the gravity sensor about two horizontal axes (X and Y), converted into angle units through the constants furnished by the manufacturing company.

To state whether the fluctuations of the instrumental sensitivity are a distinguishing feature of the MGs, we repeated the same experiment by using the Spanish gravimeter gPh54; in that case, a 36-day long record was available (Fig. 5). Drift effects were reduced in both the gPh54 and SG-C026 gravity records by fitting a straight line. The linear drift model, as can be argued from the residual gravity (Fig. 5c), seems in both cases to be reasonable. A very small drift rate (<10 μGal/day) was subtracted from the gPh54 record. The same analysis as shown in Fig. 4, consisting of comparing results coming from the calibration obtained from the gPhone with those from the synthetic tides, was done on this dataset (Fig. 6). Again from this experiment, it is clear that the temporal variations observed in the SG

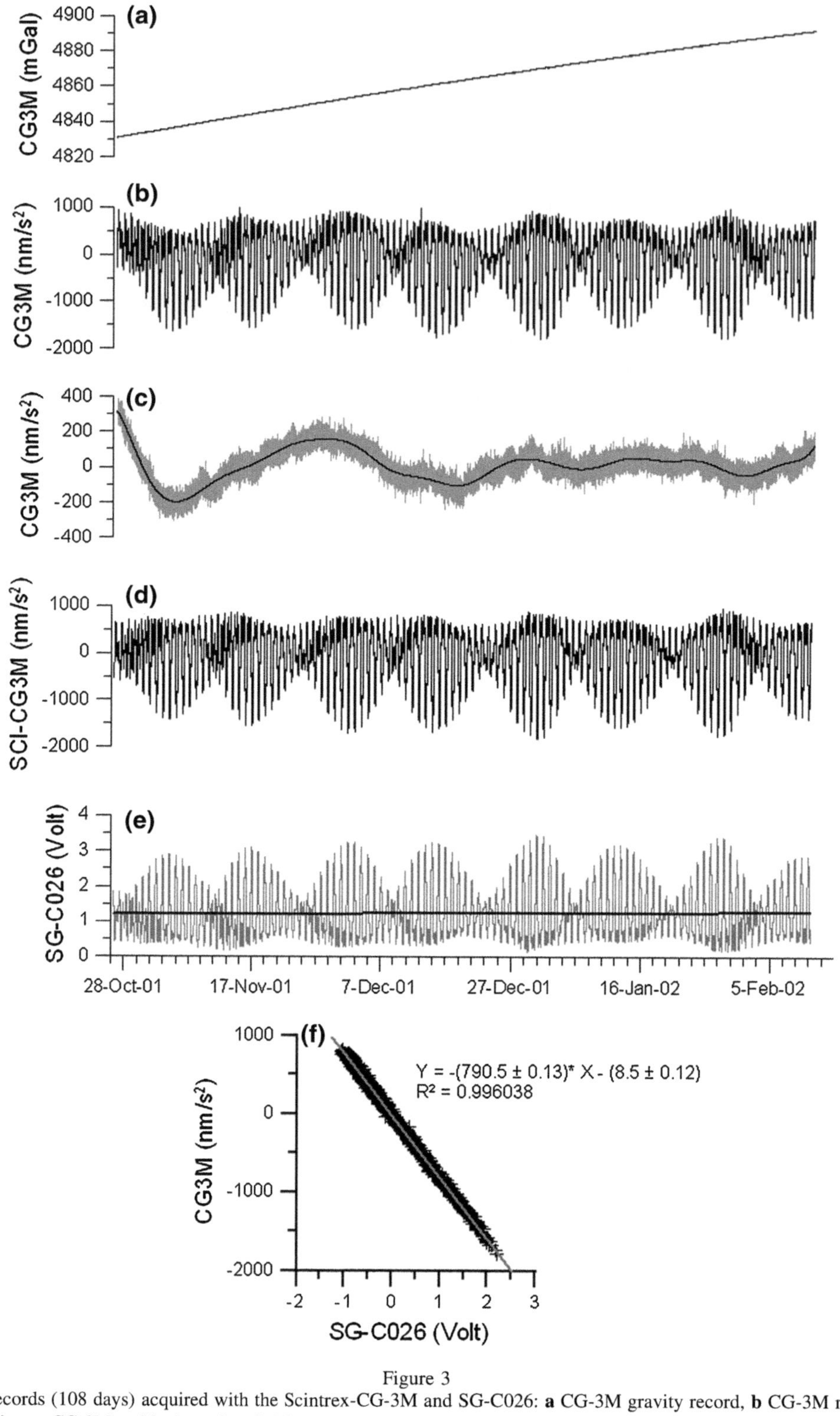

Figure 3

Parallel gravity records (108 days) acquired with the Scintrex-CG-3M and SG-C026: **a** CG-3M gravity record, **b** CG-3M record after a first linear drift reduction, **c** CG-3M residual gravity, **d** CG-3M record after a secondary, polynomial drift correction, **e** SG-C026 output voltage and drift, and **f** least-squares fitting and statistics

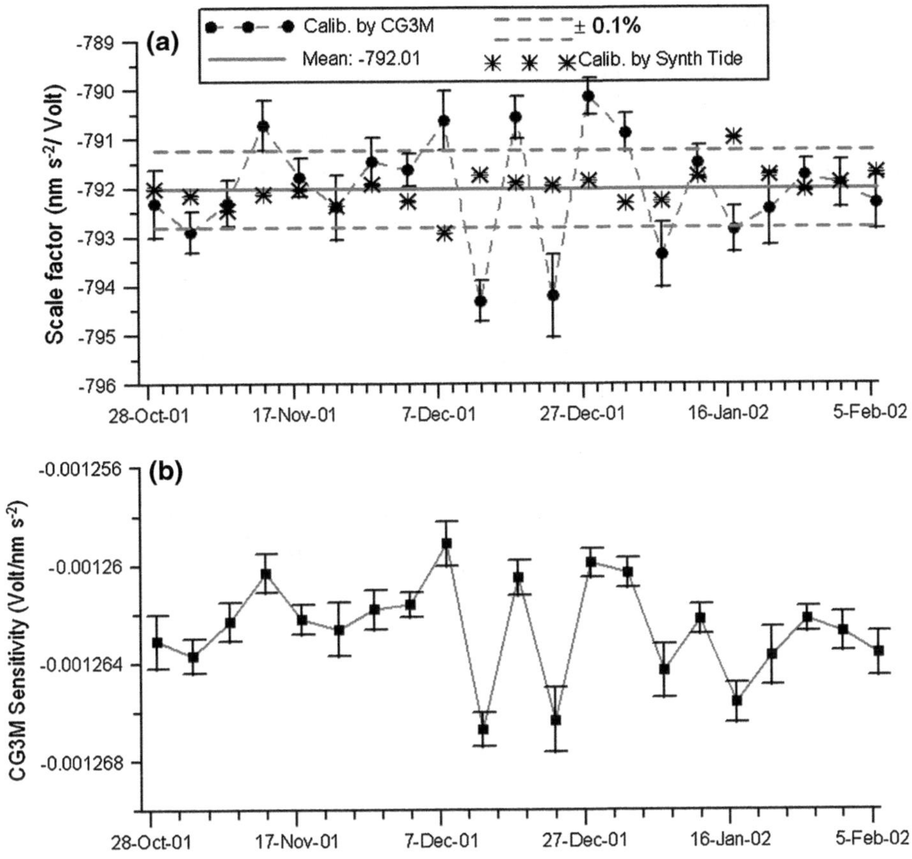

Figure 4
SG-C026 calibration experiment performed with the Scintrex-CG3 M: **a** time fluctuations of the SG-C026 scale factors determined with the CG3 M (*dots*) and with synthetic tides (*stars*), and **b** variability of the CG-3M sensitivity over time

scale factors obtained with the gPh54 are due to fluctuations in the instrumental sensitivity of the MG. Moreover, these results reveal a calibration problem for the gPh54; actually the offset between the average of the SG scale factors and the mean value of the synthetic ones (Fig. 6a) indicates that the output voltage of the gPh54 has been converted into gravity units with an unsuitable scale factor. In that case, unfortunately, we cannot exclude tilt effects during the experiment, because the digital level output was not accounted for due to some problem affecting the sensor. So, the observed change over time of the amplitude response could even be explained by some loss of verticality perturbing the meter.

Given the availability of long gravity records, we further investigated whether the precision of the SG scale factors obtained using the MGs depended on the duration of the record. Such a kind of dependency has already been investigated by Francis *et al.* (1998) for calibration through FG5 gravimeters. In Fig. 7, the variations of the scale factors and the convergence rate of their standard deviations are plotted as a function of the length of the data series used. We split the whole time span (36 days) into different subsets by varying the daily lengths, and the corresponding scale factors were calculated. Each value represents the arithmetic mean of the scale factors computed for each subset. It is remarkable that, even with a short time span (1 day), the deviation from the value obtained from the complete dataset is less than 0.1%, which is considered as the limit for the precision of the calibration in geophysical investigations. As stated before, this is not a guarantee of suitable accuracy but only of good precision, because to achieve the suitable accuracy the MG needs to be calibrated against some other reference. However, an asymptotic behavior has been

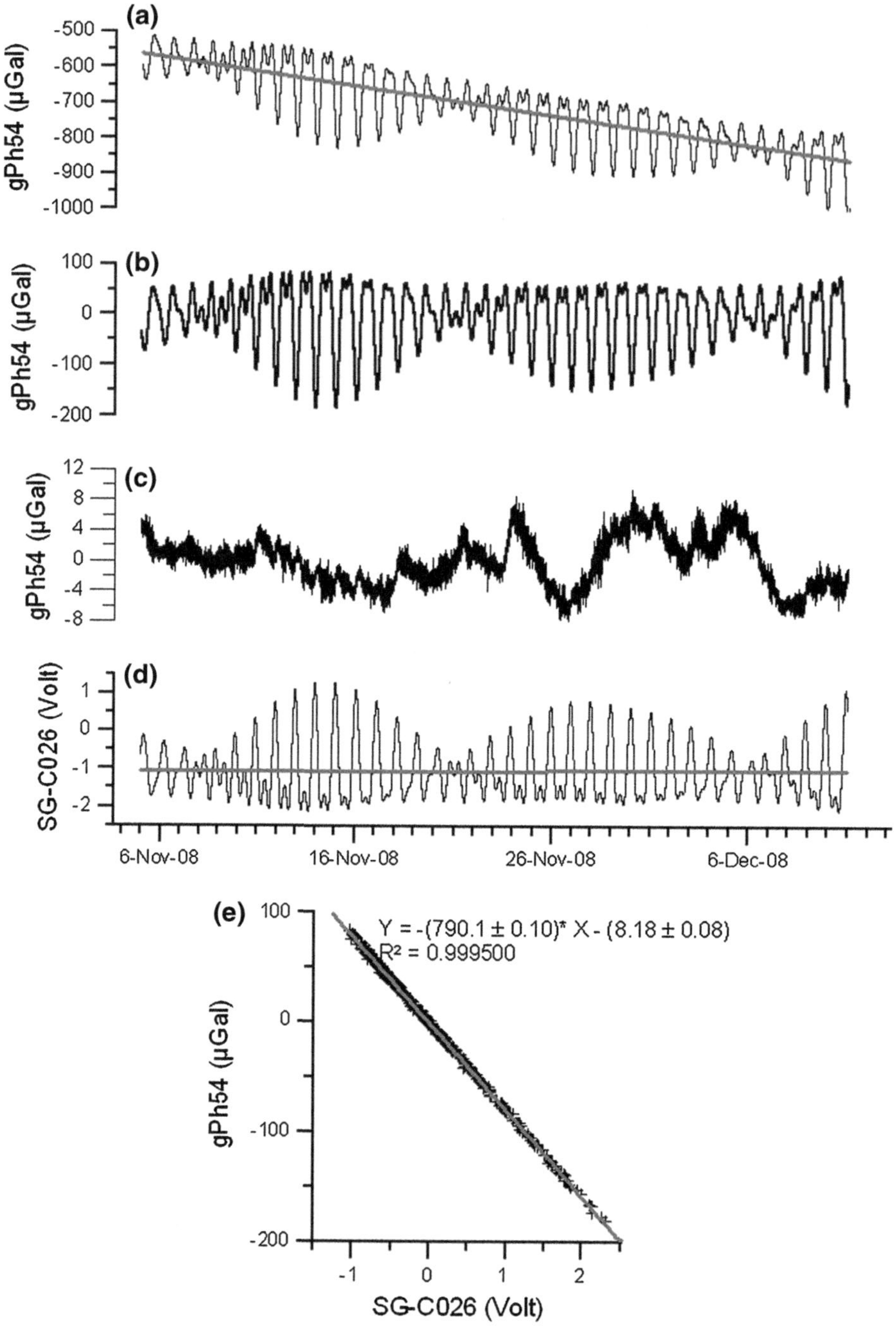

Figure 5
Gravity records (36 days) acquired with the gPhone 054 and with the SG-C026: **a** gPh54 gravity record and modeled drift, **b** drift-corrected gPh54 record, **c** gPh54 residual gravity, **d** SG-C026 output voltage and drift, and **e** least-squares fitting and statistics

observed in the standard deviation of the scale factor (Fig. 7c), which leads to a convergence within 5–6 days, so the scale factor remains stable (Fig. 7a, b). Such asymptotic behavior is still noteworthy because it demonstrates that, when the drift is well reduced, namely when it can be modeled, it does not impact on the precision of the calibration even when using a long time span (>7 days).

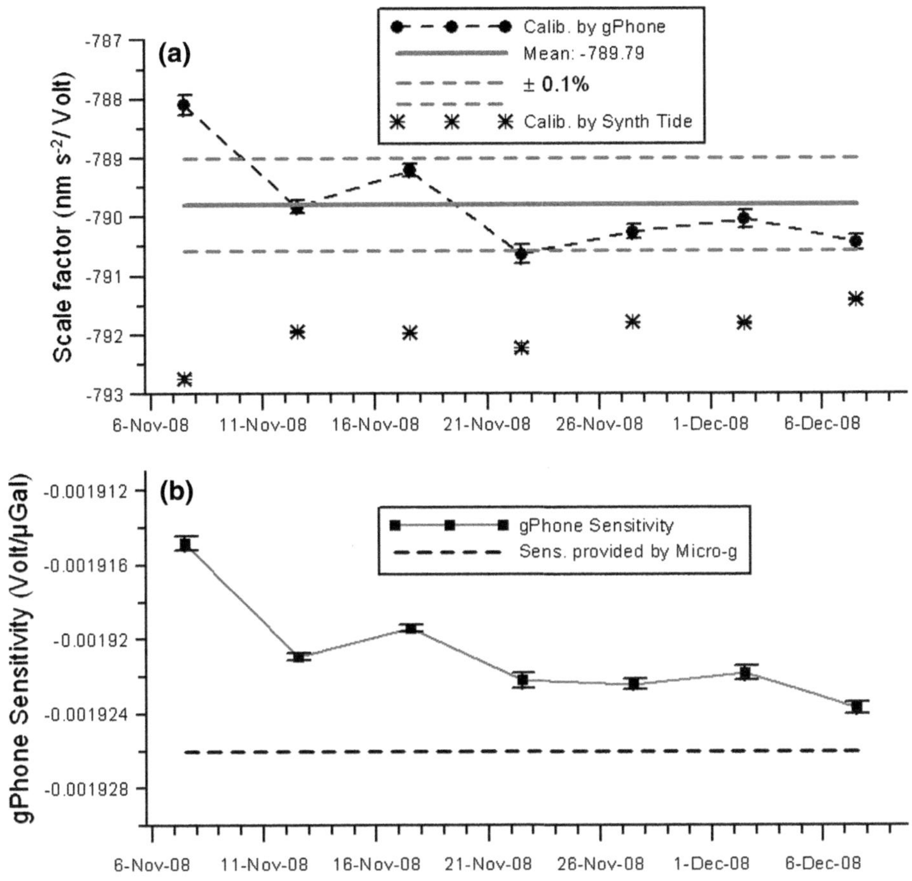

Figure 6

SG-C026 calibration experiment performed with the gPhone 054: **a** time fluctuations of the SG-C026 scale factors determined with the gPh54 (*dots*) and with synthetic tides (*stars*), and **b** variability of the gPh54 sensitivity over time

5. Tidal Analyses

It is indeed possible to achieve the calibration by means of comparative tidal analysis performed on the gravity registrations collected with the SG and the MG using the latter as a reference. This approach was already used by Asch *et al.* (1986) to calibrate some Askania gravimeters through intercomparison with an LR Earth tide gravimeter. Such analysis is in principle a calibration in the frequency domain, because the SG amplitude response can be checked for each separated wave in the tidal bands, according to Rayleigh's criterion. In addition, if the MG is well calibrated in phase, rough knowledge of the SG phase delay, if not already available, can also be obtained. Actually, an error of 0.02° in the M2 phase, which is the typical standard deviation achievable in tidal analysis of an MG record, will lead to accuracy of about 2.5 s for the phase delay of the SG. This is quite a rough determination considering the typical accuracy of 0.01 s achievable by injecting electric signals into the control electronics of the SG (Van Camp *et al.*, 2000).

Table 2 lists for the main tidal waves the outcomes of an analysis performed on a 105-day-long gravity record collected with SG-C026 and gPh54. The analysis was performed on hourly gravity observations decimated from 1-min data, spanning from 19 May 2009 until 31 August 2009. Standard preprocessing aimed at reducing spikes, steps, and gaps in the registrations was applied prior to launching ETERNA 3.4 tidal analysis (Wenzel, 1996); the Pertzev59 high-pass filter with 51 coefficients was used in the analyses. The SG-C026 output

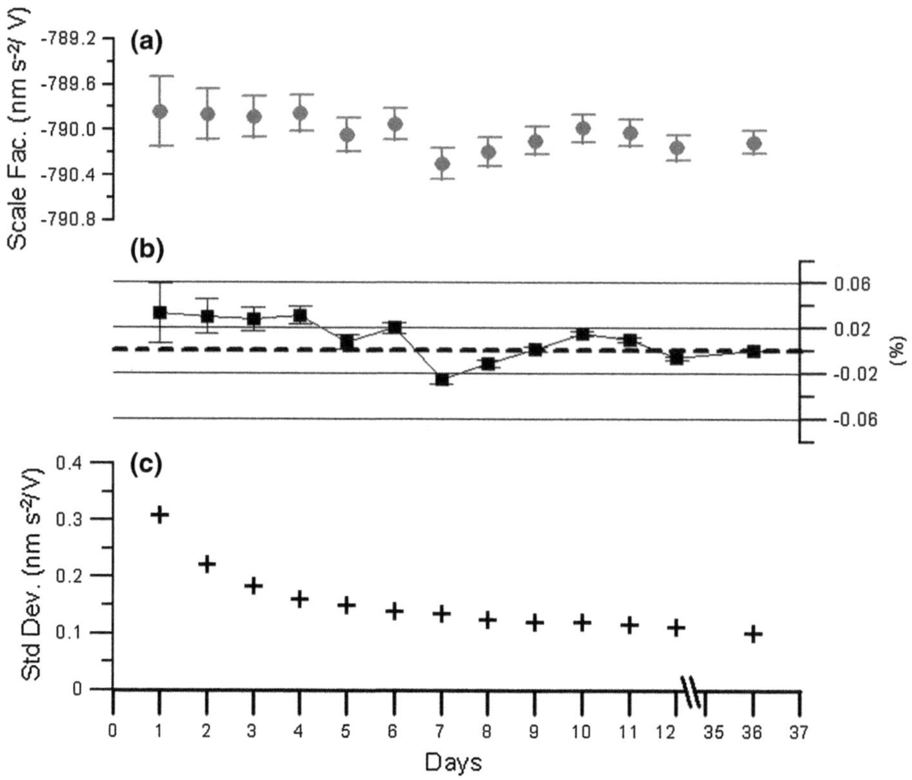

Figure 7
Evolution of the SG-C026 calibration results as a function of the length of the reference data: **a** scale factors, **b** percentage difference with respect to the reference scale factor, and **c** standard deviations of the scale factors. The reference scale factor is the value estimated by using the whole 36-day-long dataset (-790.1 ± 0.1 nm s^{-2}/V)

voltage was calibrated with the usual scale factor used in the GGP database, while the gPh54 signal was converted into gravity units using the calibration factor provided by the manufacturing company. The calibration was pursued through analysis of the amplitude ratios (SG-C026/gPh54). The results for the main tidal waves coherently indicate that the amplitude response of the SG-C026 is slightly larger (0.3%) than that of the gPh054; such a discrepancy is exactly the one we found with the moving-window regression analysis (Fig. 6b). Therefore, we can conclude that the tidal analysis confirms a calibration defect for the gPh54. No significant differences in phase can be envisaged from the tidal analysis. The SG-C026 was already calibrated in phase during 1999 by injecting reference signals into the feedback loop (Van Camp *et al.*, 2000), so these results lead us to state that the gPh54 is well calibrated in phase.

6. Conclusions

The main outcome of this calibration experiment is that co-located absolute gravity measurements are currently still the most feasible calibration approach for SGs. The agreement of the scale factors obtained by means of the Spanish FG5 #211 with the long historical series of the factors obtained since 1997 using the French FG5 #206 indicates that change of the absolute instrument does not change the calibration results, or alternatively, the estimation of the scale factors is independent of the instrument used. The AG is ideal for this purpose because it provides an absolute acceleration scale without any need for calibration against some other reference. Last but not least, being an indirect method, it does not perturb the meter as is the case when, for instance, oscillating platforms are used.

Table 2

Results of tidal analysis carried out on high-pass-filtered hourly gravity observations collected with the SG-C026 and the gPh54 for the main tidal constituents

Wave	gPhone-054				SG-C026				Comparison					
	δ	SD	Phase (°)	SD (°)	δ	SD	Phase (°)	SD (°)	δ_SG/gPh	SD	Diff. phase (°)	SD (°)	Phase lag (s)	SD (s)
Ql	1.144	0.002	−0.05	0.15	1.1486	0.0009	−0.25	0.07	1.004	0.0019	−0.200	0166	−53.52	44.3
O1	1.1461	0.0005	0.12	0.033	1.1496	0.0002	0.09	0.015	1.003	0.0005	−0.03	0.036	−7.73	9.3
PSK1	1.1342	0.0003	0.33	0.22	1.1369	0.0001	0.22	0.01	1.002	0.0003	−0.110	0.220	−26.36	52.8
N2	1.167	0.0015	2.64	0.07	1.1718	0.0008	2.58	0.037	1.004	0.0015	−0.060	0.079	−7.59	10.0
M2	1.1857	0.0003	2.14	0.016	1.1884	0.0002	2.154	0.007	1.003	0.0003	0.014	0.0170	1.74	2.2
S2K2	1.1869	0.0007	0.89	0.04	1.1912	0.0004	0.74	0.02	1.004	0.0007	−0.150	0.045	−17.98	5.4

On the contrary, MGs need to be accurately calibrated at a suitable level (0.1% in amplitude, 0.01 s in phase) to be used as a reference for SGs. We tried to address the issue of whether well-calibrated MGs can replace AGs in the calibration of the SG. We used some new-generation spring gravimeters to calibrate the SG, and the results were in agreement with the scale factors obtained using the AGs. Looking at the errors in the scale factors, we deduce that the calibrations via MG registrations are more precise then those obtained via AG data. We demonstrate that the greater precision achieved with the spring gravimeters is not necessarily a guarantee of greater accuracy, but it is just a consequence of some concurrent factors (decimation filters, less noisy raw observations, and larger amount of data available for calibration). Indeed, if not well-calibrated MG sensors are used, very precise calibration can be achieved, but the scale factors will systematically deviate from the real values. Accounting for the amplitude response of the spring gravimeters that we used in this study, we can state that it is risky to use even well-calibrated MGs as reference for SG calibration because they could exhibit some instability over time. The instrumental sensitivity variations could be due to various instrumental defects. Therefore, MGs cannot be used as a stable reference for SGs. On the contrary, we think that the spring gravimeters, to be used as permanent gravity station, could take advantage from measuring in well-calibrated superconducting stations. In fact recording in a superconducting gravity observatory could provide a complete definition of the MGs transfer function in the tidal band, in terms of phase and amplitude. Moreover the instrumental drift affecting the MGs requires careful modeling. It is possible to get rid of the drift problems by deriving the calibration from a comparative tidal analysis carried out on a monthly gravity record, but even in this case the MG needs to be accurately calibrated.

Acknowledgments

The authors thank the IGN of Madrid for providing the gPhone 054 and the FG5 #211 for this study and Mr S. Sainz-Maza Aparicio and Mr. P.A. Vaquero-Fernandez for their support in the installation of the meter. The authors are grateful to Walter Zürn and an anonymous referee for their valuable comments that significantly improved this paper.

References

Achilli, V., Baldi, P., Casula, G., Errani, M., Focardi, S., Guerzoni, M., Palmonari, F. and Raguni, G. (1995), *A calibration system for superconducting gravimeters*, Bull. Geod. *69*, 73–80.

Amalvict, M., Hinderer, J., Boy, J-P. and Gegout, P. (2001), *A three year comparison between a superconducting gravimeter (GWR C026) and an absolute gravimeter (FG5#206) in Strasbourg (France)*, J. Geod. Soc. Japan. *47*, 334–340.

Arnoso, J., Vieira, R., Velez, E.J., Van Ruymbeke, M. and Venedikov, A.P. (2001), *Studies of tides and instrumental performance of three gravimeters at Cueva de los Verdes (Lanzarote, Spain)*, J. Geod. Soc. Japan. *47*, 1, 70–75.

Asch, G., Jahr, T., Jentzsch, G., Kiviniemi, A., Kääriäinen, J., Plag, H.-P. and Thiel, W. (1986), Loading tides along the 'Blue Road' Geotraverse in Fennoscandia, Prc. 10th Int. Sympos. On Earth tides, Madrid, *1985*, 707–717.

Baker, T.F. and Bos, M.S. (2003), *Validating Earth and ocean tide models using tidal gravity measurements*, Geophys. J. Int. *152*, 468–485.

Bos, M.S., and Scherneck, H.-G., (2009) Website: http://www.oso.chalmers.se/~loading/. Accessed 2009

Crossley, D., Hinderer, J., Casula, G., Francis, O., Hsu, H.T., Imanishi, Y., Jentzsch, G., Kääriäinen, J., Merriam, J., Meurers, B., Neumeyer, J., Richter, B., Shibuya, K., Sato, T. and Van Dam, T. (1999), *Network of superconducting gravimeters benefits a number of disciplines*, EOS *80* (11), 121/125–126.

Dehant, V., Defraigne, P. and Wahr, J.M. (1999), *Tides for a convective Earth*, J. Geophys. Res. *104*, 1035–1058.

Falk, R., Harnish, M., Harnish, G., Nowak, I., Richter, B. and Wolf, P. (2001), Calibration of the superconducting gravimeter SG193, C023, CD029 and CD030, J. Geod. Soc. Japan. *47*, 22–27.

Francis, O. Calibration of the C021 Superconducting Gravimeter in Membach (Belgium) using 47 days of absolute gravity measurements, In International Association of Geodesy Symposia, (Springer-Verlag, Berlin, 1997) 117, pp. 212–219.

Francis, O., Niebauer, T.M., Sasagawa, G., Klopping, F. and Gschwind, J. (1998), *Calibration of a superconducting gravimeter by comparison with an absolute gravimeter FG5 in Boulder*, Geophys. Res. Lett. *25*, 1075–1078.

Francis, O. and Van Dam, T. (2002), *Evaluation of the precision of using absolute gravimeters to calibrate superconducting gravimeters*, Metrologia. *39*, 485–488.

Fukuda, Y., Iwano, S., Ikeda, H., Hiraoka, Y. and Doi, K. (2005), *Calibration of the superconducting gravimeter CT#043 with an absolute gravimeter FG5#210 at Syowa Station, Antarctica*, Polar Geosciences. *18*, 41–48.

Hinderer, J. and Crossley, D. (2000), *Time variations in gravity and inferences on the Earth's structure and dynamics*, Surv. Geophys. *21*, 1–45.

Hinderer, J. and Crossley, D. (2004), *Scientific achievements from the first phase (1997–2003) of the Global Geodynamics Project using a worldwide network of superconducting gravimeters*, J. Geodynamics. *38*, 237–262.

Hinderer, J., Crossley, D. and Warburton, R.J., Gravimetric methods: Superconducting gravity meters. In Treatise on Geophysics (ed. Schubert G.) (Elsevier, New York, 2007) pp. 65–122.

Hinderer, J., Florsch, N., Mäkinen, J., Legros, H. and Faller, J.E. (1991), *On the calibration of a superconducting gravimeter using absolute gravity measurements*, Geophys. J. Int. *106*, 491–497.

Imanishi, Y., Higashi, T. and Fukuda, Y. (2002), *Calibration of the superconducting gravimeter T011 by parallel observation with the absolute gravimeter FG5 #210: a Bayesian approach*, Geophys. J. Int. *151*, 867–878.

Krystek, M. and Anton, M. (2007), *A weighted total least-squares algorithm for fitting a straight line*, Meas. Sci. Technol. *18*, 3438–3442. doi:10.1088/0957-0233/18/11/025.

Letellier, T. (2004), Etude des ondes de marée sur les plateux continentaux. Thèse doctorale, Université de Toulouse III, Ecole Doctorale des Sciences de l'Univers, de l'Environnement et de l'Espace, 237 pp.

Lyard, F., Lefévre, F., Letellier, T. and Francis, O. (2006), *Modelling the global ocean tides: modern insights from FES2004*, Ocean Dynamics. *56* (5–6), 394–415.

Mathews, P.M. (2001), *Consistent modeling of the effects of the diurnal tidal potential*, J. Geod. Soc. Japan. *47*, 219–224.

Melchior, P., (1994), *A new data bank for tidal gravity measurements (DB92)*, Phys. Earth Planet. Int. *82*, 125–155.

Melchior, P. and Francis, O. (1998), Proper Usage of the ICET Data Bank: Du bon usage de la Banque de Données ICET, Comparison with theoretical applications on Earth Models, Proc. XIII Int. Symp. on Earth Tides, Ed. B. Ducarme and P. Pâquet, July 22–25, 1997, Bruxelles, 377–396.

Meurers, B. (2001), Superconducting gravimetry in geophysical research today, J. Geod. Soc. Japan. 47, 300–307.

Meurers, B. (2002), *Aspects of gravimeter calibration by time domain comparison of gravity records*, Bull. d'Inform. Marèes Terrestres. *135*, 10643–10650.

Microg-LaCoste. (2008), gPhone/P.E.T Hardware Manual V1. 16 pp.

Richter, B., Wenzel, H.-G., Zürn, W. and Klopping, F. (1995a) *From Chandler wobble to free oscillations: comparison of cryogenic gravimeters and other instruments in a wide period range*, Phys. Earth Planet. Int. *91*, 131–148.

Richter, B.,Wilmes, H. and Nowak, I. (1995b) *The Frankfurt calibration system for relative gravimeters*, Metrologia. *32*, 217–223.

Rosat, S., Boy, J-P., Ferhat, G., Hinderer, J., Amalvict, M., Gegout, P. and Luck, B. (2009), *Analysis of a 10-year (1997–2007) record of time-varying gravity in Strasbourg using absolute and superconducting gravimeters: New results on the calibration and comparison with GPS height changes and hydrology*, J. Geodynamics. *48*, 360–365.

Sato, T., Tamura, Y., Okubo, S. and Yoshida, S. (1996), *Calibration of scale factor of superconducting gravimeter at Esashi using an absolute gravimeter FG5*, J. Geod. Soc. Japan. *42*, 225–232.

Scintrex Limited (1995) CG-3/3 M Autograv automated gravity meter operator manual.

Scintrex Limited (2006), CG5 Autograv automated gravity meter operator manual (Rev. 4). pp. 308.

Tamura, Y., Sato, T., Fukuda, Y. and Higashi, T. (2001), *Scale factor calibration of a superconducting gravimeter at Esashi Station, Japan, using absolute gravity measurements*, J. Geodesy. *71*, doi:10.1007/s00190-004-0415-0.

Van Camp, M., Wenzel, H.-G., Schott, P., Vauterin, P. and Francis, O. (2000), *Accurate transfer function determination for superconducting gravimeters*, Geophys. Res. Lett. *27* (1), 37–40.

Van Camp, M., Williams, S.D.P. and Francis, O. (2005), *Uncertainty of absolute gravity measurements*, J. Geophys. Res. *110*, B05406, doi:10.1029/2004JB003497.

van Ruymbeke, M., Somerhausen, A. and Grammatica, N. (1995) *Calibration of gravimeters with the VRR 8601 calibration platform*, Metrologia. *32*, 209–216.

Wenzel, H.-G. (1996), *The nanogal software: Earth tide data processing package ETERNA 3.30, Bull. d'Inform.* Marèes Terrestres. *124*, 9425–9439.

(Received February 26, 2010, revised November 11, 2010, accepted August 12, 2011, Published online August 31, 2011)

Pure Appl. Geophys. 169 (2012), 1357–1372
© 2011 Springer Basel AG
DOI 10.1007/s00024-011-0399-7

Pure and Applied Geophysics

Separation of the Geodetic Consequences of Past and Present Ice-Mass Change: Influence of Topography with Application to Svalbard (Norway)

A. Mémin,[1] J. Hinderer,[1] and Y. Rogister[1]

Abstract—Polar regions such as Greenland, Svalbard and Antarctica are deforming today because of both the present-day ice-mass (PDIM) change of glaciers and the glacial isostatic adjustment (GIA) following the Pleistocene deglaciation. Observations handled in these areas contain both the contributions from the PDIM change and GIA. This study aims at separating them by considering two specific gravity variation-to-vertical displacement ratios. We first review the case of the viscoelastic rebound (GIA) subsequent to the Pleistocene deglaciation leading to a ratio C^v. The outcome of previous studies is that C^v is approximately equal to -0.15 μGal/mm and almost independent of the deglaciation history, ice geometry and viscosity profile of the mantle. Similarly we consider the elastic deformation resulting from PDIM change which leads to a second ratio $C^{e,N}$. Several studies have shown that $C^{e,N} \approx -0.26\,\mu$Gal/mm if one assumes that the changing glaciers are thin layers over the surface of a spherical Earth model. In this case, we show that the separation between the contributions from PDIM change and GIA is unique if both gravity and height changes observations are available at the same station. Next, we focus on $C^{e,N}$ and show that according to the deglaciation/glaciation context and from colocated gravity variation and ground vertical velocity measurements one can deduce a range of possible values for $C^{e,N}$. Studying the influence of the topography on $C^{e,N}$ we first show that it tends to positive values if most of surrounding ice-mass changes above the altitude of the observation site and to values lower than -0.26 μGal/mm if changes are below. We next apply our general formalism to the case of the past and PDIM changes in Svalbard, Norway. We compute the ratio $C^{e,N}$ at the geodetic observatory at Ny-Ålesund and show the influence of the topography of the surrounding glaciers on the measured gravity and uplift rates. We show that if the ice-mass change is spatially uniform, $C^{e,N}$ does not depend on the speed of ice-mass change, and hence the separation of the contributions from PDIM changes and GIA can still be done univocally. However, if the ice-mass change is not spatially uniform, $C^{e,N}$ depends on both the speed of ice-mass change and the volume of ice-change rate.

Key words: Gravity variation-to-vertical displacement ratio, present-day ice thinning, glacial isostatic adjustment, topography, Svalbard.

1. Introduction

In response to climatic changes, ice-masses vary. Currently a reduction of the volume of ice is observed using different methods. For example, the analysis of space gravimetric observations from GRACE (Gravity Recovery and Climate Experiment) suggests ice-mass loss over Alaska (e.g. Luthcke *et al.* 2008), Greenland and West Antarctica (e.g. Barletta *et al.* 2008; Wouters *et al.* 2008; Slobbe *et al.* 2009; Horwath and Dietrich 2009; Velicogna 2009). Comparison of digital elevation models deduced from satellite altimetry or photogrammetry have shown ice thinning in Svalbard (e.g. Kohler *et al.* 2007; Kääb 2008; Nuth *et al.* 2010). Ice thinning has also been shown in Alaska by using satellite imagery (Berthier *et al.* 2010).

The solid Earth elastically deforms because of this present-day ice-mass (PDIM) change. Moreover, most of the regions where the ice-mass presently decreases is also subject to the glacial isostatic adjustment (GIA), which is the viscous relaxation that follows the Pleistocene deglaciation. Of course, the observations (ground gravimetry and precise positioning) do not separate the two effects (e.g. Bevis *et al.* 2010). Usually, models of deglaciation histories are used to compute the GIA, which is subtracted from the observations. The volume of ice loss can then be deduced from the residuals. The contribution of the GIA to the observations is sometimes estimated using the vertical displacement to gravity conversion factor of Wahr *et al.* (1995). It appears difficult to separate the two contributions by using observations only.

We study the separation between the contributions from past and present deglaciations to the observations by using two gravity variation-to-vertical displacement ratios, for both the viscous (C^v) and

[1] IPGS/UMR 7516, Université de Strasbourg/EOST, CNRS, 5 rue René Descartes, 67084 Strasbourg Cedex, France. E-mail: anthony.memin@unistra.fr

elastic ($C^{e,N}$) deformations. In Sect. 2, we first list the relations between the observed and theoretical vertical displacement and gravity rates. Then we collect ratios of gravity rate to vertical velocity found in the literature (Sects. 2.1 and 2.2) and relate them to the observed parameters (Sect. 2.3). In Sect. 3 we focus on $C^{e,N}$. We study the influence of the glaciation/deglaciation context (Sect. 3.1) and the effect of the topography, for which we derive a general expression (Sects. 3.2 and 3.3). We investigate in the last section the influence of the topography and spatial distributions of ice-mass variations on the separation of the geodetic consequences of past and present-day ice-mass change over the Svalbard archipelago, Norway (Sect. 4). After a brief geographical description of Svalbard in Sect. 4.1, we introduce, in Sect. 4.2, the spatial distribution of the glaciers and the different vertical profiles of ice-mass change used to model the gravity variations and ground velocity (Sect. 4.3). We discuss the results from models and observations in Sect. 4.4.

2. *Viscoelastic and Elastic Gravity and Uplift Rates*

Let δu^v and δg^v be respectively the time variations of the vertical displacement and gravity rate due to the GIA at the Earth's surface. The uplift and gravity rates induced by the PDIM change are respectively denoted by δu^e and $\delta g^e + \delta g^N$. The first term δg^e is due to the elastic deformation of the ground, the second term is the Newtonian attraction of the varying mass. In areas subject to both the GIA and PDIM change, the observed vertical velocity δu^{obs} and gravity rates δg^{obs} are given by

$$\delta u^{\text{obs}} = \delta u^e + \delta u^v, \tag{1}$$

$$\delta g^{\text{obs}} = \delta g^e + \delta g^N + \delta g^v. \tag{2}$$

2.1. *Gravity Variation-to-Vertical Displacement Ratio for Viscoelastic Deformation*

Using different deglaciation histories, ice geometries and viscosity profiles for the mantle of a viscoelastic Maxwell Earth, Wahr *et al.* (1995) found the following relation between the viscous gravity and uplift rates:

$$C^v = \frac{\delta g^v}{\delta u^v} \approx -0.15\,\mu\text{Gal/mm}. \tag{3}$$

This ratio was studied by Fang and Hager (2001) who confirmed that it is independent of the radial viscosity profile of the Earth, which is due to the nearly incompressible viscous response of a Maxwell Earth. Using the ICE-3G history of Tushingham and Peltier (1991) to model the GIA in Antarctica, James and Ivins (1998) numerically found a ratio of -0.16 μGal/mm, close to ratio (3). Actually, the viscous response of the Earth involves both changes in the height of the surface and mantle mass redistribution. The motion of the surface involves a variation of the gravity measured by an instrument moving with the surface. This variation is the so-called free-air gradient, which is $-2\ g_0/a \approx -0.31\ \mu$Gal/mm, where a is the mean Earth radius and $g_0 = 9.81$ m/s^2 is the surface gravity. The second effect can be approximated by the Bouguer plate formula leading to $2\pi G\rho_m$ for the gravity variation-to-vertical displacement ratio, where G is the gravitational constant and ρ_m is the density of the plate. Taking $\rho_m = 3{,}350$ kg/m^3 as an average density for the upper mantle, we have $2\pi G\ \rho_m = 0.14\ \mu$Gal/mm and $\delta g^v/\delta u^v \approx -0.17\ \mu$Gal/mm, as found for example in Ekman and Mäkinen (1996), James and Ivins (1998) and Le Meur and Huybrechts (2001). This ratio is close to the one first derived by Wahr *et al.* (1995), although it is derived from a simpler modeling.

The ratio $C^v = -0.15\ \mu$Gal/mm given by Wahr *et al.* (1995) was also obtained by Larson and van Dam (2000) in North America from uplift and gravity observations, whereas Lambert *et al.* (2006) obtained $-0.18 \pm 0.03\ \mu$Gal/mm from similar observations. In Fennoscandia, Mäkinen *et al.* (2005) obtained a ratio in the range $[-0.18 \pm 0.06,\ -0.16 \pm 0.04]\ \mu$Gal/mm. Some studies (e.g. Wolf *et al.* 1997; Lambert *et al.* 2001; Mäkinen *et al.* 2007; Amalvict *et al.* 2009) used a ratio in the range $[-0.17,\ -0.15]\ \mu$Gal/mm to predict the gravity rate from a modelled displacement rate.

2.2. *Gravity Variation-to-Vertical Displacement Ratio for Elastic Deformation*

Using a spectral approach, de Linage *et al.* (2007) computed the ratio

$$C^e = \frac{\delta g^e}{\delta u^e} \quad (4)$$

between the rates of gravity and vertical displacement for an elastic deformation. According to the compressibility of the uppermost layer of the Earth model, its value is either ≈ -0.20 μGal/mm if the layer is incompressible or ≈ -0.24 μGal/mm if it is compressible.

These values do not take into account the direct attraction of the changing mass expressed by δg^N. Considering that the mass changes occur at the surface of the spherical and compressible Earth model, de LINAGE *et al.* (2007) found that outside the area where the load occurs

$$C^{e,N} = C^e + C^N \approx -0.26\,\mu\text{Gal/mm}, \quad (5)$$

where

$$C^N = \frac{\delta g^N}{\delta u^e}. \quad (6)$$

They obtained this value by averaging over the harmonic degrees 2–50 the ratio between the nth spectral components of the transfer functions of the gravity variation and vertical displacement. For the harmonic degrees 6 and 3,000, this ratio takes the values -0.2881 and -0.2467 μGal/mm, respectively.

JAMES and IVINS (1998) obtained $C^{e,N} = -0.27$ μGal/mm, which is close to the value (5) found by de LINAGE *et al.* (2007) and which corresponds to 85% of the free-air gradient. As for the viscous ratio, $C^{e,N}$ is used to estimate secular gravity variation from modelled uplift rate such as that due to the PDIM change in Antarctica (e.g. MÄKINEN *et al.* 2007; AMALVICT *et al.* 2009).

2.3. *Separation Between the Geodetic Consequences of GIA and PDIM Change*

For colocated gravimetric and geodetic stations, relations (1), (2), (3) and (5) provide

$$\delta u^e = \frac{\delta g^{\text{obs}} - C^v \delta u^{\text{obs}}}{C^{e,N} - C^v}, \quad (7)$$

$$\delta u^v = -\frac{\delta g^{\text{obs}} - C^{e,N} \delta u^{\text{obs}}}{C^{e,N} - C^v}, \quad (8)$$

which, in turn, give δg^v and $\delta g^e + \delta g^N$ by using (3) and (5):

$$\delta g^e + \delta g^N = C^{e,N} \frac{\delta g^{\text{obs}} - C^v \delta u^{\text{obs}}}{C^{e,N} - C^v}, \quad (9)$$

$$\delta g^v = -C^v \frac{\delta g^{\text{obs}} - C^{e,N} \delta u^{\text{obs}}}{C^{e,N} - C^v}. \quad (10)$$

Therefore, the contributions δg^v and δu^v of the GIA and $\delta g^e + \delta g^N$ and δu^e of the PDIM change to the observed gravity rate and vertical velocity of the ground can be uniquely solved and directly estimated from gravity and geodetic observations and theoretical ratios (3) and (5).

So far, we have considered that the remote unloaded/loaded area is located beneath the horizontal plane passing through the observation point (JAMES and IVINS 1998; LE MEUR and HUYBRECHTS 2001; MÄKINEN *et al.* 2007). However, since the Newtonian part of the gravity depends on the relative position of the observation point and location where mass changes occur (MERRIAM 1992; BOY *et al.* 2002; MÉMIN *et al.* 2009), it depends on the topography of the loaded area. In the next section, we examine the influences of the geophysical context, topography and load variation on $C^{e,N}$.

3. *Study of $C^{e,N}$*

3.1. *$C^{e,N}$ in Areas Subject to both Past and Present-Day Ice-Mass Changes*

The expressions (7)–(10) are valid if $C^{e,N}$ is known, which seems to be the case when the topography is neglected (Sect. 2.2), and if $C^{e,N} \neq C^v$. It is, indeed, impossible to discriminate processes producing the same gravity variation-to-vertical displacement ratio.

To put constraints on $C^{e,N}$, we consider the ratio $\delta u^v/\delta u^e$. Let us first assume that δu^e and δu^v have the same sign, so $\delta u^v/\delta u^e$ is positive. Therefore,

$$\frac{\delta u^v}{\delta u^e} = -\frac{\delta g^{\text{obs}} - C^{e,N} \delta u^{\text{obs}}}{\delta g^{\text{obs}} - C^v \delta u^{\text{obs}}} \geq 0, \quad (11)$$

and

$$C^{e,N} \begin{cases} \geq \frac{\delta g^{\text{obs}}}{\delta u^{\text{obs}}}, & \text{if } \delta g^{\text{obs}} - C^v \delta u^{\text{obs}} > 0 \\ \leq \frac{\delta g^{\text{obs}}}{\delta u^{\text{obs}}}, & \text{if } \delta g^{\text{obs}} - C^v \delta u^{\text{obs}} < 0 \end{cases}. \quad (12)$$

If, moreover, $\frac{\delta u^v}{\delta u^e} < 1$, then $C^{e,N}$ satisfies the following conditions:

$$C^{e,N}\begin{cases} \leq 2\frac{\delta g^{\text{obs}}}{\delta u^{\text{obs}}} - C^v, & \text{if } \delta g^{\text{obs}} - C^v\delta u^{\text{obs}} > 0 \\ \geq 2\frac{\delta g^{\text{obs}}}{\delta u^{\text{obs}}} - C^v, & \text{if } \delta g^{\text{obs}} - C^v\delta u^{\text{obs}} < 0 \end{cases}. \tag{13}$$

Second, we assume that δu^e and δu^v have opposite signs. Therefore,

$$C^{e,N}\begin{cases} \leq \frac{\delta g^{\text{obs}}}{\delta u^{\text{obs}}}, & \text{if } \delta g^{\text{obs}} - C^v\delta u^{\text{obs}} > 0 \\ \geq \frac{\delta g^{\text{obs}}}{\delta u^{\text{obs}}}, & \text{if } \delta g^{\text{obs}} - C^v\delta u^{\text{obs}} < 0 \end{cases}. \tag{14}$$

If, moreover, $\delta u^v/\delta u^e > -1$, then

$$C^{e,N}\begin{cases} > C^v, & \text{if } \delta g^{\text{obs}} - C^v\delta u^{\text{obs}} > 0 \\ < C^v, & \text{if } \delta g^{\text{obs}} - C^v\delta u^{\text{obs}} < 0 \end{cases}. \tag{15}$$

If $\delta u^v/\delta u^e = -1$, then $C^{e,N} = C^v$ and expressions (7) and (8) are no longer valid. The case $|\delta u^v/\delta u^e| > 1$ concerns either local ice-mass change which induces effects lower than that induced by the Pleistocene deglaciation or ice-mass change which is too remote to have sufficiently large effects. The different inequalities are shown in Fig. 1.

In conclusion, knowing the observed gravity and vertical displacement rates, as well as the glaciation and deglaciation context, we can provide a range of values for $C^{e,N}$.

In Sects. 3.2 and 3.3, we take now the topography into account for calculating $C^{e,N}$.

3.2. *Expression for $C^{e,N}$*

According to Eq. 5,

$$C^{e,N} = \frac{\delta g^e + \delta g^N}{\delta u^e}. \tag{16}$$

In this section we derive an expression for $C^{e,N}$ that explicitly contains the distance between the observation site and the location where the loading is applied. The derived expression is used in Sect. 3.3 to study the influence of the topography on $C^{e,N}$.

In spherical coordinates r, θ, ϕ, the Newtonian (δg^N) and elastic (δg^e) gravity variations and the vertical displacement rate (δu^e) are respectively given by

$$\delta g^N(\mathbf{r}) = \iiint_{\upsilon} \rho(\mathbf{r}')G_{g^n}(\mathbf{r},\mathbf{r}')r'^2\mathrm{d}r'\sin\theta'\mathrm{d}\theta'\mathrm{d}\phi' \tag{17}$$

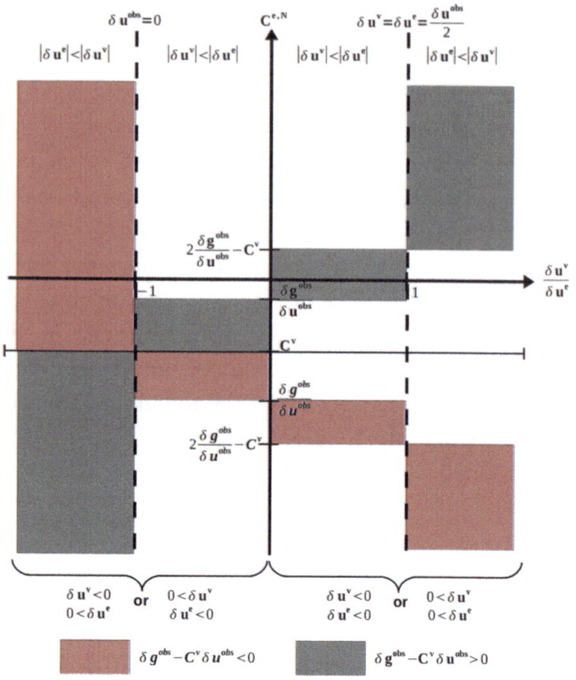

Figure 1
Summary of the different cases, corresponding to various geophysical processes, discussed in the text for the ratio $C^{e,N}$

$$\delta g^e(\mathbf{r}) = \iint_{\Omega} \rho(\mathbf{r}')\delta h(\mathbf{r}')G_{g^e}(\mathbf{r}-\mathbf{r}')r'^2\sin\theta'\mathrm{d}\theta'\mathrm{d}\phi' \tag{18}$$

$$\delta u^e(\mathbf{r}) = \iint_{\Omega} \rho(\mathbf{r}')\delta h(\mathbf{r}')G_{u^e}(\mathbf{r}-\mathbf{r}')r'^2\sin\theta'\mathrm{d}\theta'\mathrm{d}\phi', \tag{19}$$

where $\mathbf{r}'$ is the position of an element of ice-height variation δh and density ρ. The surface and volume of the ice load are, respectively, Ω and υ. Green functions for the vertical displacement and elastic gravity variations are respectively (Farrell 1972):

$$G_{u^e}(\mathbf{r}-\mathbf{r}') = \frac{G}{a g_0}\sum_{n=1}^{\infty} h'_n P_n(\cos\psi), \tag{20}$$

$$G_{g^e}(\mathbf{r}-\mathbf{r}') = -\frac{G}{a^2}\sum_{n=1}^{\infty}[2h'_n - (n+1)k'_n]P_n(\cos\psi). \tag{21}$$

G is the Newtonian constant of gravitation and g_0, the gravity at the surface of the spherical Earth model of mean radius a. h'_n and k'_n are the load Love numbers of degree n for the displacement and variation of the gravity potential respectively. P_n is the Legendre polynomial of degree n. ψ is the angular distance between the observation point and loading point. The Green function for the Newtonian gravity variation is (MERRIAM 1992; BOY *et al.* 2002):

$$G_{g^n}(\mathbf{r},\mathbf{r}') = G\frac{r - r'\cos\psi}{(r^2 + r'^2 - 2rr'\cos\psi)^{3/2}}. \quad (22)$$

Angular distance ψ is given by

$$\cos\psi = \cos\theta\cos\theta' + \sin\theta\sin\theta'\cos(\phi - \phi'). \quad (23)$$

Inserting Eqs. 17–22 in Eq. 16 and assuming a constant density for the load, we obtain

$$C^{e,N}(\mathbf{r}) = \frac{\rho}{\delta u^e(\mathbf{r})}\iint_\Omega \left[\delta h(\mathbf{r}')G_{g^e}(\mathbf{r}-\mathbf{r}')r'^2 + \int_{r'}^{r'+\delta h(\mathbf{r}')} G_{g^n}(\mathbf{r},\mathbf{r}'_1)r_1'^2 dr'_1\right] \sin\theta' \mathrm{d}\theta' \mathrm{d}\phi', \quad (24)$$

where the coordinates of the point at $\mathbf{r}'_1$ are $r' + \mathrm{d}z', \theta', \phi'$. One can rewrite Eq. 24

$$C^{e,N}(\mathbf{r}) = \frac{\rho}{\delta u^e(\mathbf{r})}\iint_\Omega \left[\delta h(\mathbf{r}')G_{g^e}(\mathbf{r}-\mathbf{r}')r'^2 + \int_0^{\delta h(\mathbf{r}')} G_{g^n}(\mathbf{r},\mathbf{r}'+z'\mathbf{e_r})(r'+z')^2 \mathrm{d}z'\right] \sin\theta' \mathrm{d}\theta' \mathrm{d}\phi'. \quad (25)$$

If we denote by z^{load} and z^{obs} respectively the altitudes of the loading point and observation point, then $r' = a + z^{\text{load}} \gg z'$ and $r = a + z^{\text{obs}}$. Thus, we obtain

$$C^{e,N}(\mathbf{r}) = \frac{\rho}{\delta u^e(\mathbf{r})}\iint_\Omega \left[\delta h(\mathbf{r}')G_{g^e}(\mathbf{r}-\mathbf{r}')r'^2 + \int_0^{\delta h(\mathbf{r}')} G_{g^n}(\theta-\theta',\phi-\phi',z^{\text{obs}}-z^{\text{load}}-z')r'^2 \mathrm{d}z'\right] \times \sin\theta' \mathrm{d}\theta' \mathrm{d}\phi' \quad (26)$$

or

$$C^{e,N}(\mathbf{r}) = \frac{\rho}{\delta u^e(\mathbf{r})}\iint_\Omega \left[\delta h(\mathbf{r}')G_{g^e}(\mathbf{r}-\mathbf{r}') + \int_{z^{\text{load}}}^{z^{\text{load}}+\delta h(\mathbf{r}')} G_{g^n}(\theta-\theta',\phi-\phi',z^{\text{obs}}-z')\mathrm{d}z'\right] \times r'^2 \sin\theta' \mathrm{d}\theta' \mathrm{d}\phi'. \quad (27)$$

We introduce the pseudo Green function of $C^{e,N}$ that we name $G_{C^{e,N}}(\mathbf{r},\mathbf{r}')$. Actually, $G_{C^{e,N}}$ is a function of $\mathbf{r}-\mathbf{r}'$:

$$G_{C^{e,N}}(\mathbf{r}-\mathbf{r}') = \frac{1}{\delta h(\mathbf{r}')G_{u^e}(\mathbf{r}-\mathbf{r}')}\left[\delta h(\mathbf{r}')G_{g^e}(\mathbf{r}-\mathbf{r}') + \int_{z^{\text{load}}}^{z^{\text{load}}+\delta h(\mathbf{r}')} G_{g^n}(\theta-\theta',\phi-\phi',z^{\text{obs}}-z')\mathrm{d}z'\right] \quad (28)$$

We call it a pseudo Green function because it is the ratio produced by a mass point of height δh, in other terms, it is a unit-mass point response scaled to δh. Using Eq. 28, Eq. 16 transforms to

$$C^{e,N}(\mathbf{r}) = \frac{\rho}{\delta u^e(\mathbf{r})}\iint_\Omega \delta h(\mathbf{r}')G_{u^e}(\mathbf{r}-\mathbf{r}')G_{C^{e,N}}(\mathbf{r}-\mathbf{r}') \times r'^2 \sin\theta' \mathrm{d}\theta' \mathrm{d}\phi'. \quad (29)$$

We now write

$$\alpha(\mathbf{r},\mathbf{r}') = \delta h(\mathbf{r}')\frac{G_{u^e}(\mathbf{r}-\mathbf{r}')}{\delta u^e(\mathbf{r})}, \quad (30)$$

and introduce it in Eq. 29. We finally obtain

$$C^{e,N}(\mathbf{r}) = \rho\iint_\Omega \alpha(\mathbf{r},\mathbf{r}')G_{C^{e,N}}(\mathbf{r}-\mathbf{r}')r'^2 \sin\theta' \mathrm{d}\theta' \mathrm{d}\phi'. \quad (31)$$

Thus, as predicted, $C^{e,N}$ depends on the distance between the observation and loading points and, consequently, on the surface of the load. It also depends on the ice-height variations at each loading point. Function α is a weighting factor that indicates how $C^{e,N}$ for a specific loading point contributes to the total $C^{e,N}$ ratio induced by the whole load. Its behaviour is shown in Fig. 2, where $\alpha\delta u^e/\delta h$ is plotted. For example, loading points located either 200 km or 1 km from the observation point have the same weight if the height variation of the farthest loading point is almost 1,000 times larger than that of the closest. Consequently, the contribution of the $C^{e,N}$ ratio of one specific loading point is strongly influenced by the ice-height variation if it is larger than that of other further away loading points.

3.3. *Study of $G_{C^{e,N}}$*

Using the Green functions for the deformation of a symmetric, non-rotating, elastically isotropic Earth model (FARRELL 1972), we compute $G_{C^{e,N}}$ for δh = 1 m/year of water (ρ = 1,000 kg m^{-3}). The observation points are at a distance ranging from 1 to 1,111 km from the loading point. To take into account the influence of the topography, namely the relative elevation between the load and the observation point, we compute $G_{C^{e,N}}$ for a load located at several altitudes ($z^{\text{load}} \in [0; 2{,}000]$ m). The altitude of the observation point is z^{obs} = 1,000 m. The difference between the altitudes of the observation point and load is $\Delta z = z^{\text{load}} - z^{\text{obs}}$. Plots are shown in Fig. 3.

First, we show that $G_{C^{e,N}} \sim -0.26$ μGal/mm for Δz = 0 m. This is valid for a point load acting at any distance from the observation site. Consequently, for any load, we have $C^{e,N} \sim -0.26$ μGal/mm, as found by de LINAGE *et al.* (2007).

Next, $G_{C^{e,N}}$ varies very differently closer to the observation point according to Δz. The strong increase toward positive numbers for $\Delta z \geq 0$ is because mass changes are above the observation point. Increasing (resp. decreasing) the mass decreases (resp. increases) the gravity and leads to a negative (resp. positive) gravity variation. Because of an increased (resp. a decreased) mass, the ground subsides (resp. uplifts) and a negative (resp. positive) displacement variation can be observed. Consequently, the resulting ratio is positive. If $\Delta z \leq 0$, the mass changes beneath the observation site and $G_{C^{e,N}}$ strongly decreases for both increasing and decreasing masses. Indeed, the gravity variation and vertical velocity have opposite signs in these cases. Otherwise, with the distance the ratios converge to that obtained for Δz = 0, namely about -0.26 μGal/mm.

We repeat the same computation for $\delta h = \{2, 5, 10, 50, 100\}$ m/year and denoting $G^1_{C^{e,N}}(\mathbf{r} - \mathbf{r}')$ the function $G_{C^{e,N}}(\mathbf{r} - \mathbf{r}')$ for δh = 1 m/year, we plot

$$\Delta G_{C^{e,N}}(\mathbf{r} - \mathbf{r}') = \text{Log}|G^1_{C^{e,N}}(\mathbf{r} - \mathbf{r}') - G_{C^{e,N}}(\mathbf{r} - \mathbf{r}')| \quad (32)$$

on Figs. 4 and 5 for several relative altitudes. We study the influence of the load variation rate δh. We obtain:

- $\Delta G_{C^{e,N}}$ increases with the distance to the load for any Δz, it is about 0.01 μGal/mm at 50 km from the loading point for $|\delta h|$ = 100 m/year,

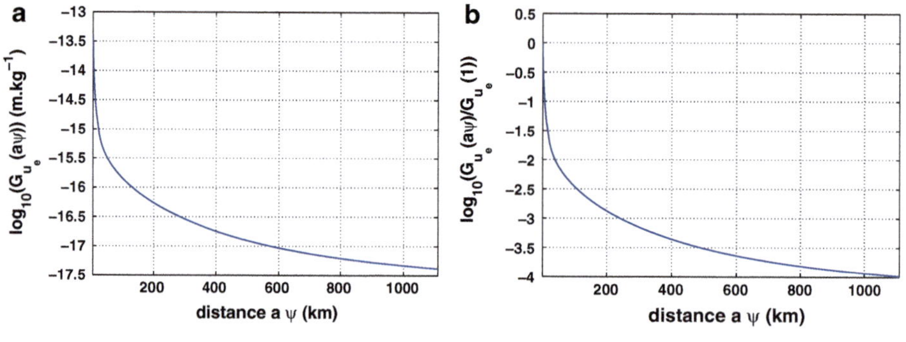

Figure 2
Weighting factor $\alpha\delta u^e/\delta h$ plotted as a function of the distance of the observation point from the load (*left*) and divided by that computed for an observation point located at 1 km from the load (*right*)

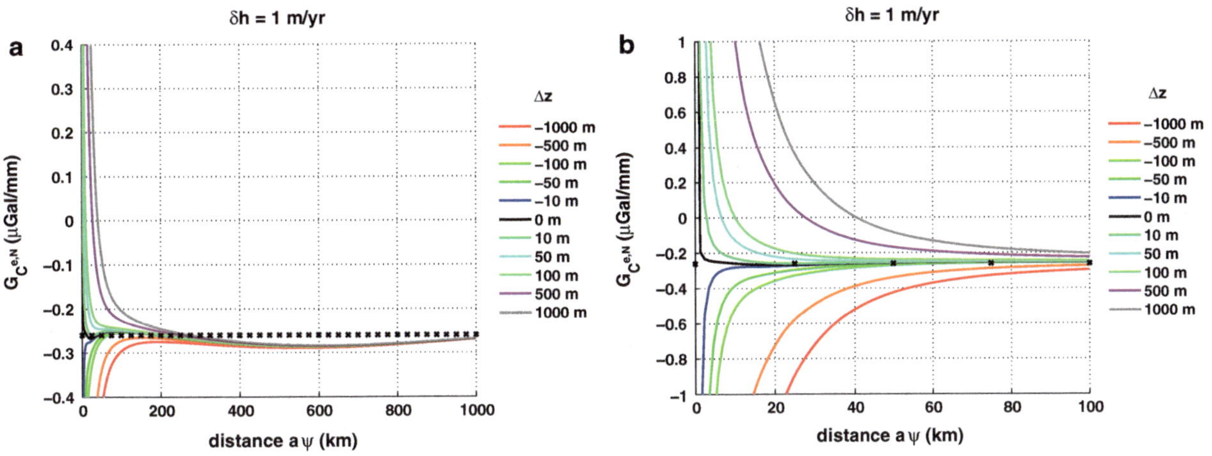

Figure 3
$G_{C^{e,N}}$ as a function of the distance to the observation point from the load for $\delta h = 1$ m/year and several relative altitudes $\Delta z = z^{\mathrm{load}} - z^{\mathrm{obs}}$

- $\Delta G_{C^{e,N}} \leq 0.01$ μGal/mm for $|\delta h| \leq 10$ m/year from 11 km from the loading point,
- $\Delta G_{C^{e,N}} \leq 0.01$ μGal/mm for $|\delta h| \leq 5$ m/year from 6.5 km from the loading point,
- $\Delta G_{C^{e,N}} \leq 0.08$ μGal/mm for $|\delta h| \leq 5$ m/year from 2 km from the loading point.

If we have a digital elevation model and if load variations are farther than 2 km away from the observation point, we can estimate $G_{C^{e,N}}$ independently of the load variations and Δz if height variations are lower than 5 m/year. In that case, the accuracy is better than 0.08 μGal/mm. The accuracy would be better than 0.02 μGal/mm if $|\delta h| \leq 2$ m/year. For distances farther than 6.5 or 11 km, the load variation has to be lower than 5 or 10 m/year respectively to have $G_{C^{e,N}}$ with 0.01 μGal/mm accuracy.

However, the topography of the load is not sufficient to completely determine the total $C^{e,N}$ ratio induced by a load acting on any area that would allow the unique determination of the contributions from the GIA and the PDIM change. Indeed, the height variations of the load appear in the coefficient α (Eq. 30) and need to be known. So, in Sect. 4, we review the specific case of Svalbard, which has already been studied (Sato *et al.* 2006; Mémin *et al.* 2011), and focus on what can be extracted from the $C^{e,N}$ ratio.

4. Past and Present-Day Ice-Mass Changes in Svalbard

4.1. Location of Svalbard and Geodetic Observations

4.1.1 Svalbard Archipelago and Ny-Ålesund Observatory

The Arctic archipelago of Svalbard is located north of Norway between 76°N and 81°N of latitude and 11°E and 26°E of longitude. It is covered by about 36,000 km^2 of ice which represents 60 % of the total area. Most of the ice surface is thinning (Kohler *et al.* 2007; Dowdeswell *et al.* 2008; Kääb 2008; Moholdt *et al.* 2009; Nuth *et al.* 2010), which induces deformation and gravity variations. Svalbard is also subject to the GIA following the last deglaciation (e.g. Tushingham and Peltier 1991). At the Geodetic Observatory of Ny-Ålesund (11.855°E, 78.929°N, 43 m), Very Long Baseline Interferometry (VLBI), GPS or Doppler Orbitography and Radiopositioning Integrated by Satellite (DORIS) data have been collected for up to 18 years. The gravity variation is measured since 1999 with a superconducting gravimeter, which is part of the Global Geodynamics Project (Crossley *et al.* 1999), and the absolute gravity has been measured six times with FG5 absolute gravimeters: in 1998, 2000, 2001, 2002, 2004, and 2007.

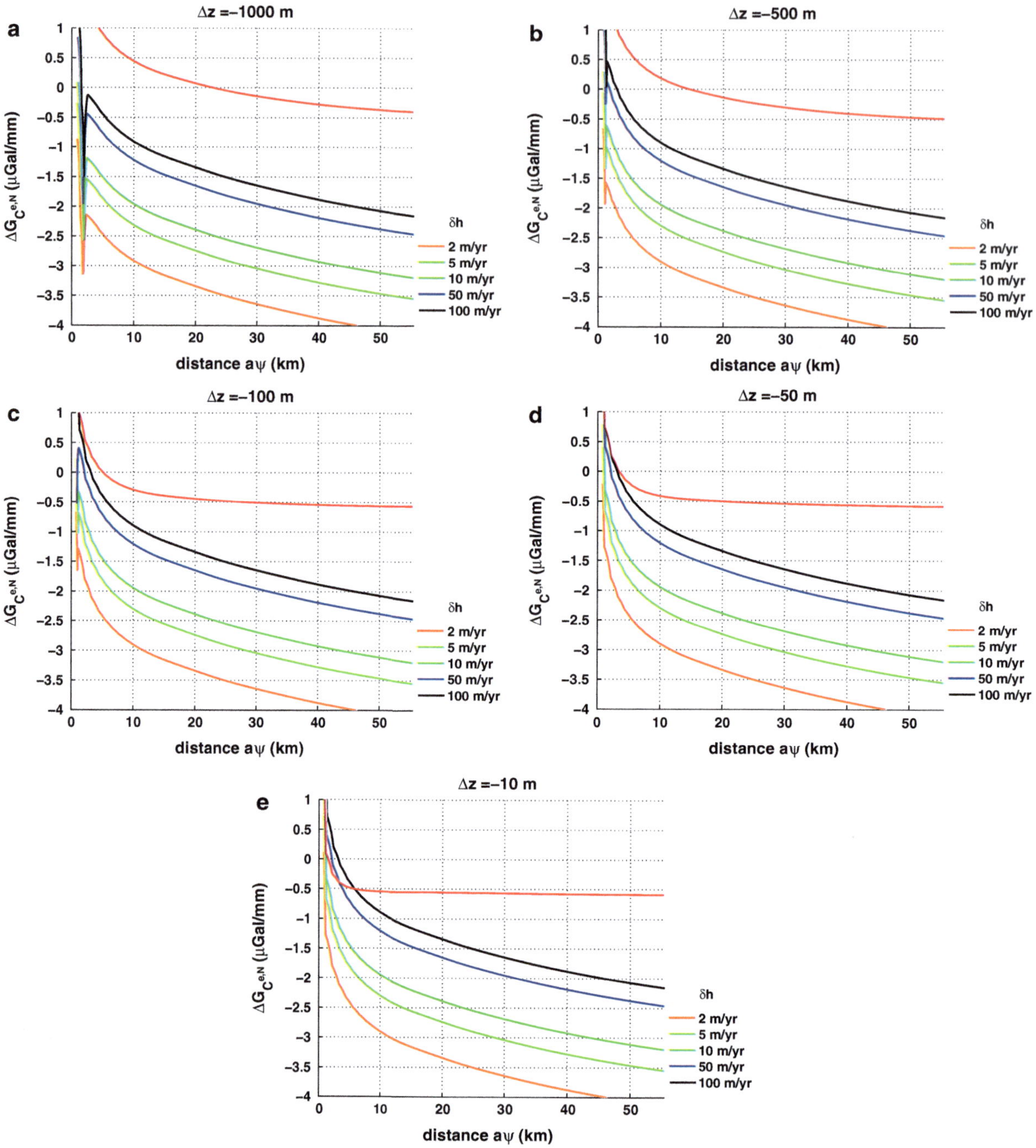

Figure 4
$G_{C^{e,N}}(\mathbf{r}-\mathbf{r}')$ computed for $\delta h = 1$ m/year (*red curve*) and $\Delta G_{C^{e,N}}$ (*other color curves*) as a function of the distance of the observation point from the load for five relative altitudes ($\Delta z = z^{\mathrm{load}} - z^{\mathrm{obs}} < 0$). $\Delta G_{C^{e,N}}$ is computed for five load variations ($\delta h = \{2, 5, 10, 50, 100\}$ m/year)

The Digital Chart of the World (DCW, http://www.maproom.psu.edu/dcw/) provides a realistic geographical distribution of the glaciers. More accurate location and altitude of the glaciers near Ny-Ålesund are given by the Digital Elevation Model (DEM) from the SPIRIT project (Korona *et al.*

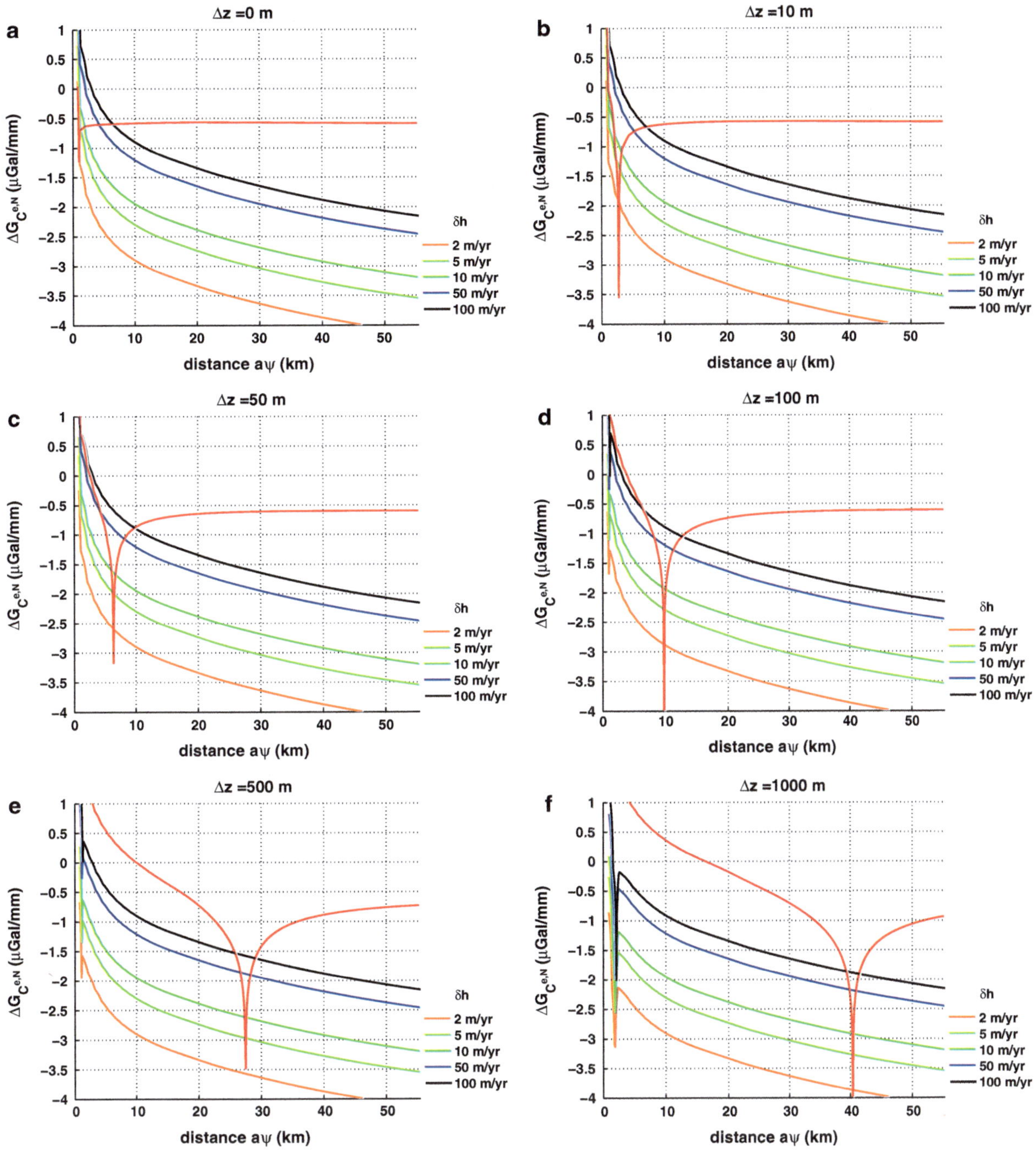

Figure 5
Same as Fig. 4 for six other relative altitudes ($\Delta z = z^{\text{load}} - z^{\text{obs}} \geq 0$)

2009). For all the other glaciers, the topography is provided by the GTOPO30 DEM (http://edc.usgs.gov/products/elevation/gtopo30/gtopo30.html). The total surface of ice is divided into seven basins as shown in Fig. 6.

4.1.2 *Observations at Ny-Ålesund*

Sato *et al.* (2006) obtained from VLBI, GPS and gravity observations an uplift rate of 5.2 ± 0.6 mm/year and a gravity rate of −2.5 ± 0.9 μGal/year at

Ny-Ålesund. These rates have been revised by Mémin *et al.* (2011) using longer datasets. They use six absolute gravity measurements for the period 1998–2007, instead of four for Sato *et al.* (2006) for the period 1998–2002, and find a lower value of -1.02 ± 0.48 μGal/year. They propose an uplift rate of 5.64 ± 1.57 mm/year using velocity observations found in the literature. A direct modeling of their observations with a uniform ice-mass loss rate of 75 cm/year allows to fit the observations. This rate is consistent with the one proposed by Sato *et al.* (2006). However, the corresponding volume of ice loss, ~ 25 km^3/year, is larger than the one derived from the analysis of the GRACE data (Mémin *et al.* 2011) which is ranging between 5 and 18 km^3/year. The GRACE derived volume of ice loss is in agreement with glaciological studies estimating ice loss to be between 4 and 14.2 km^3/year. They associate a part of the discrepancy between ice losses derived from ground and space observations to be due to the difference of sensitivity of both methods. Indeed GRACE measurements are mostly sensitive to the total loss of mass while ground gravity measurements are sensitive to local effects. To reduce the discrepancy, they propose to take into account the altitude dependency of ice-mass change in the modeling of PDIM change effects. Using ground observations, we evaluate non uniform ice-mass change scenarios by focusing on the $C^{e,N}$ ratio.

4.2. *Ice-Mass Change Distribution in Svalbard*

4.2.1 *Glaciers Distribution in Svalbard with Respect to Ny-Ålesund Observatory*

Figure 7 shows the distribution of area of basin 1, as a function of altitude and distance to Ny-Ålesund observatory, relative to the total area of the seven basins. All the glaciers in this basin, which is the smallest, are located above Ny-Ålesund. Moreover, a large part ($\sim 25\%$) of the ice in this region is located between 2 and 5 km from the station and between 40

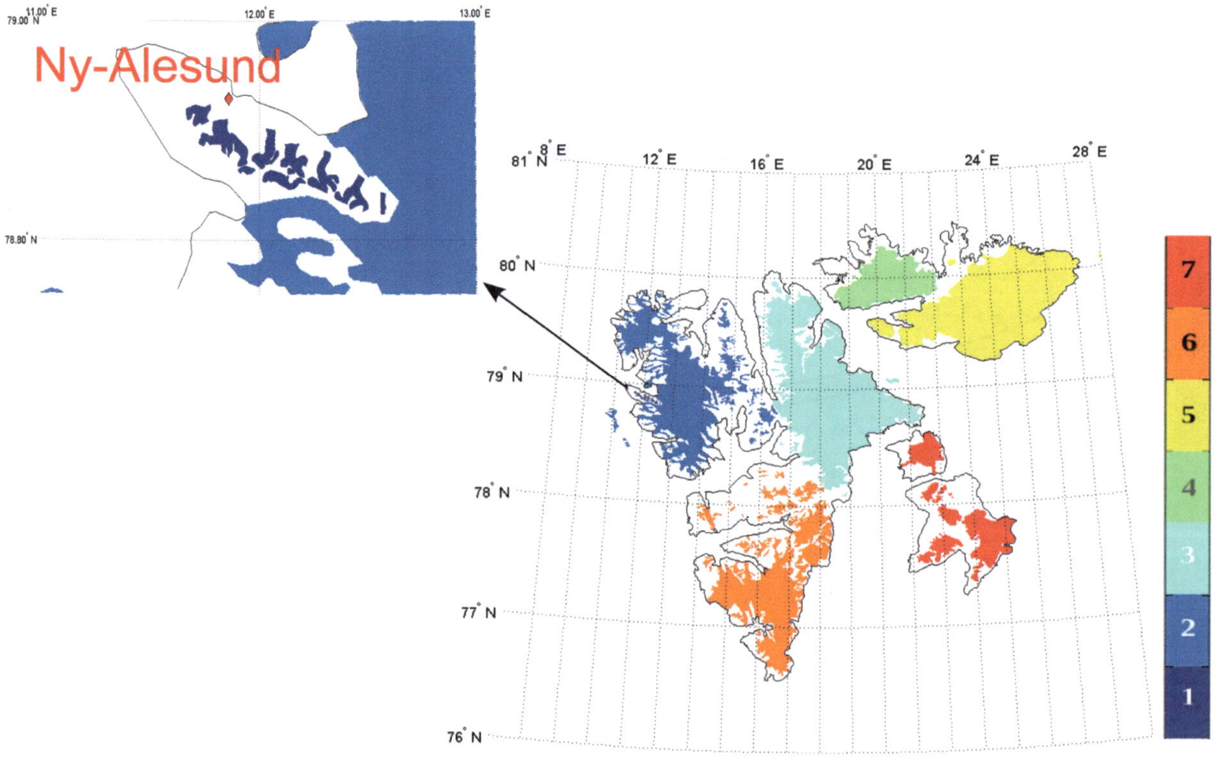

Figure 6
Ice-covered area in Svalbard from the Digital Chard of the World. The total surface is divided into seven basins, whose numbers appear in the *color bar* on the *right* of the map

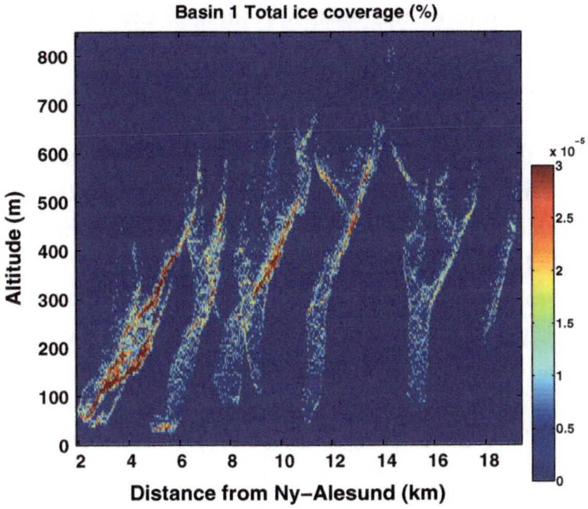

Figure 7
Distribution of ice-covered area (%) for basin 1, scaled to the total area of ice coverage, as a function of both the distance from Ny-Ålesund and altitude of the load

and 500 m of altitude. Figure 8 shows the distribution of the ice in each basin as a function of the distance to the station. Basin 2 extends from 10 to 110 km, basin 3 from 100 to 210 km, basin 4 from 165 to 240 km, basin 5 from 175 to 350 km, basin 6 from 115 to 280 km, and basin 7 from 205 to 320 km.

4.2.2 Ice-Mass Change Profiles

Figure 9 shows the rate of the thickness variation, d*h*/d*t*, of the ice load for the seven basins as a function of the altitude. We assume that, for a given altitude, d*h*/d*t* is the same over an entire basin. The profiles for basins 3, 4, 6 and 7 are based on the study by Nuth *et al.* (2010). The profile for basin 5 is based on the study by Moholdt *et al.* (2009). For both basins 1 and 2, we consider two different profiles, labelled 1a, 1b, 2a and 2b. Profiles 1a and 1b are obtained from the height change rates provided by Kohler *et al.* (2007). Profile 2a is also given by Nuth *et al.* (2010). Finally, we derive profile 2b by using an average of the thinning rates provided by Kierulf *et al.* (2009).

4.3. Gravity and Uplift Rates at Ny-Ålesund

4.3.1 Computation of Gravity and Uplift Rates

The geodetic effects of ice-mass change are numerically calculated with account for the geographic distribution and topographic height of the glaciers. The total gravity rates or vertical velocity at the observation station are obtained using Eqs. 17–19 of Sect. 2 assuming that the load has a uniform density $\rho = 1{,}000\ \mathrm{kg/m^3}$.

The computed vertical motion and gravity rate at Ny-Ålesund for five models of ice-mass change are listed in Table 1. In models 1, 2, and 5, we use profiles 3–7. In models 1 and 2, we respectively use the couple of profiles 1a–2a and 1b–2b. In models 3 and 4, we assume a uniform thinning rate of 1 m/year. In model 4, we do not take into account the topography of the glaciers. Model 5 is the same as model 2 but the profile 2b is multiplied by 2. This increases the thinning or thickening rates and changes the ice loss. These five models will allow us to study the influence of (1) the topography of the glaciers on the geodetic consequences of ice-mass change and (2)

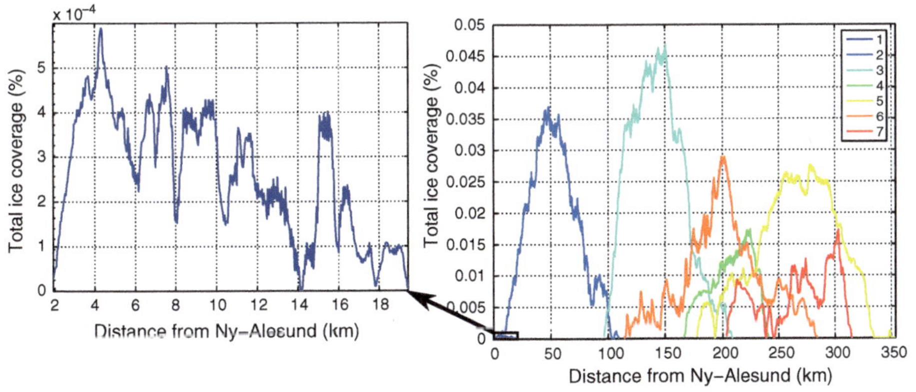

Figure 8
Distribution of ice-covered area (%) for each basin of Fig. 6 as a function of the distance from Ny-Ålesund

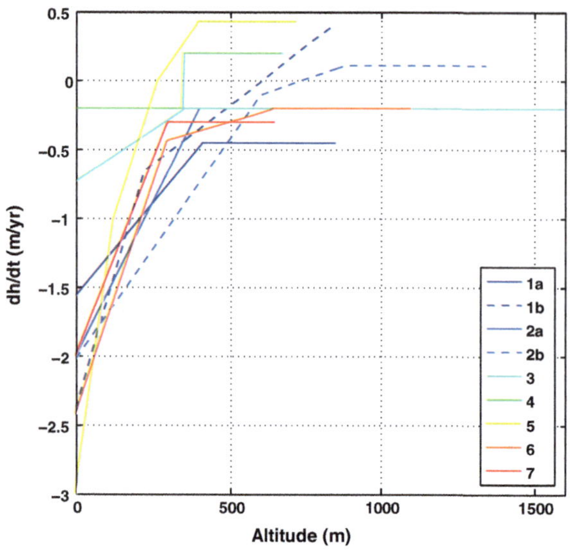

Figure 9
Rate of ice height change dh/dt as a function of the altitude for each basin of Fig. 6. *Curves 1a* and *1b* are used for basin 1 and 2a and 2b are used for basin 2

the geographical distribution of ice-mass change over a given glacier.

4.3.2 Study of Different Ice-Mass Change Scenarios

Comparison of models 3 and 4 in Table 1 shows that the topography plays a role only for the closest glaciers, in basins 1 and 2. The influence on the gravity rate of the mass variation of the glaciers in basins 3–7, which are more than 100 km away from the station, is so small that their topography does not need to be taken into account.

Figure 10 shows the computed gravity rates as a function of the uplift rates at Ny-Ålesund for the five models of ice-mass changes. It also shows the GIA effect (solid black line), corresponding to a slope of approximately −0.15 μGal/mm ratio (Wahr *et al.* 1995), and the elastic PDIM change effect without any topography (black dashed line), which corresponds to the −0.26 μGal/mm ratio theoretically found by de Linage *et al.* (2007). We assume that these two lines cross at the point which corresponds to the GIA effects computed by Sato *et al.* (2006) namely −1.88 mm/year for the vertical velocity and −0.31 μGal/year for the annual gravity rate.

We see in Fig. 10 that neither model 1 nor model 2 can explain observations within their error bars while models 3, 4 and 5 could. However, model 4 is discarded since no topography is taken into account. If we consider a model similar to model 3, but with a uniform ice-mass change rate of −0.85 m/year, we obtain 3.60 mm/year for the vertical velocity and −0.37 μGal/year for the annual gravity rate. These values are very close to the ones found for model 5 (3.52 mm/year and −0.39 μGal/year), see Fig. 10. But, the annual ice losses are very different for the

Table 1

At Ny-Ålesund, vertical velocity δu^e (mm/year) and total gravity rate $\delta g^{e+N} = \delta g^e + \delta g^N$ (μGal/year) due to ice-mass change

Model		1		2		3		4	5	
dh/dt profiles		1a, 2a, 3–7		1b, 2b, 3–7		−1 m/year			1b, 2 × 2b, 3–7	
Volume of ice loss (km^3/year)		−11.64		−12.22		−35.48			−15.67	
Basin	Area (km^2)	δu^e	δg^{e+N}	δu^e	δg^{e+N}	δu^e	δg^{e+N}	δg^{e+N*}	δu^e	δg^{e+N}
1	40	0.175	0.198	0.166	0.172	0.191	0.226	−0.046	0.166	0.172
2	6,411	1.21	−0.175	1.43	−0.219	2.37	−0.231	−0.626	2.86	−0.439
3	9,453	0.208	−0.050	0.208	−0.050	0.844	−0.201	−0.216	0.208	−0.050
4	2,543	−0.0622	0.0016	−0.0622	0.0016	0.127	−0.033	−0.0339	−0.0622	0.0016
5	8,419	0.0232	−0.00667	0.0232	−0.00667	0.289	−0.0785	−0.0795	0.0232	−0.00667
6	5,875	0.211	−0.055	0.211	−0.055	0.320	−0.0827	−0.0847	0.211	−0.055
7	2,741	0.0576	−0.0158	0.0576	−0.0158	0.0928	−0.0253	−0.0256	0.0576	−0.0158
Total	35,482	1.88	−0.10	2.09	−0.17	4.23	−0.43	−1.11	3.52	−0.39
Total with GIA		3.76	−0.41	3.97	−0.48	6.11	−0.74	−1.42	5.4	−0.7
$\delta g^{e+N}/\delta u^e$ (μGal/mm)		−0.06		−0.08			−0.10	−0.26	−0.11	

δg^{e+N*} is the gravity variation computed without taking into account the topography of the changing glaciers. According to Sato *et al.* (2006), the uplift and gravity variation associated to the GIA are, respectively, 1.88 mm/year and −0.31 μGal/year two models: it is approximately 30 km^3/year for the former and 15 km^3/year for the latter. Even if the model 3 scaled to −0.85 m/year could explain ground observations, it does not explain GRACE satellite gravimetric measurements. Model 5 is more appropriate to explain both the ground and space observations.

4.4. $C^{e,N}$ at Ny-Ålesund, Svalbard

4.4.1 Results of Modelling

The influence of the topography in basins 1 and 2 is significant: if we take it into account, the gravity rate to vertical velocity ratio is −0.10 μGal/mm, while we obtained −0.26 μGal/mm in Sect. 2, where the topography was neglected. We have checked that other uniform thinning rates lead to the same $C^{e,N}$ (black dotted dashed line on Fig. 10). This is a direct consequence of the results of Sect. 3 for $|\delta h| \leq 1$ m/year. Therefore, if one considers a uniform loading, the problem of Sect. 2 is still uniquely solved even if the topography is taken into account.

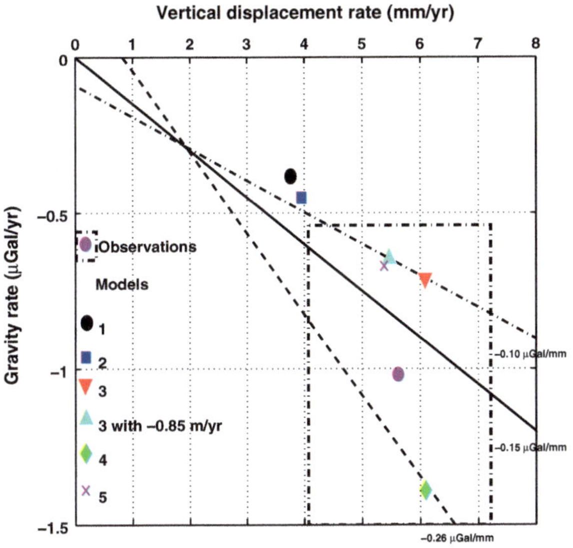

Figure 10
Computed gravity rate as a function of the uplift rate at Ny-Ålesund. The *solid black line* gives the GIA effect. Its slope is approximately −0.15 μGal/mm (Wahr *et al.* 1995). The *black dashed line* corresponds to the −0.26 μGal/mm ratio theoretically found by de Linage *et al.* (2007) for the elastic PDIM change effect without any topography. The *black dotted dashed line* corresponds to the −0.10 μGal/mm ratio found for different values of ice-thinning rate with the model 3. All *these lines* cross at the point which corresponds to the GIA effects computed by Sato *et al.* (2006). The *black symbols* are the computed rates for the model 1 (*circle*), 2 (*square*), 3 (*inverted triangle*), 4 (*diamond*) and 5 (*cross*). The *triangle* corresponds to the model 3 with a uniform ice thinning of 0.85 m/year. Observations (*magenta circle*) are from Mémin *et al.* (2011)

Comparison of models 1, 2 and 5 shows that $\delta g^e + \delta g^N$ and δu^e at Ny-Ålesund, as well as $C^{e,N}$, depend on the spatial distribution of ice-mass change over basins 1 and 2. For models 1, 2, 3 and 5, the absolute value of $C^{e,N}$ is 2–4 times smaller than for model 4, in which the topography is neglected.

Figure 11 shows $C^{e,N}$ due to ice-mass change in each basin separately. For basins 1 and 2, the ratio is clearly dependent on the load while for basins 3 to 7, the ratio is close to −0.26 μGal/mm in agreement with results of Sect. 3. When the topography is not taken into account, the ratio is close to −0.26 μGal/mm for all the basins as proposed by de Linage *et al.* (2007).

4.4.2 $C^{e,N}$ Estimated from Observations

The gravity and vertical displacement variations at Ny-Ålesund are, respectively, −1.02 μGal/year and 5.64 mm/year (Sect. 1) leading to $\delta g^{obs}/\delta u^{obs} = -0.18$ μGal/mm and $\delta g^{obs} - C^v\ \delta u^{obs} < 0$.

Knowing that most of ice in Svalbard is thinning and that Ny-Ålesund is subject to the uplift due to the Pleistocene deglaciation, then according to Sect. 1, $C^{e,N}$ should be lower than −0.18 μGal/mm. Besides, if displacement variations due to PDIM are larger than those due to GIA, then $C^{e,N}$ should be higher than $2\delta g^{obs}/\delta u^{obs} - C^v = -0.21$ μGal/mm. When $C^{e,N}$ decreases from −0.18 μGal/mm, δu^v increases while δu^e decreases and for $C^{e,N} = -0.21$ μGal/mm, $\delta u^v = \delta u^e = \delta u^{obs}/2$.

The fact that $C^{e,N}$ can be different from −0.26 μGal/mm shows that ice-mass change occurs at different altitudes. Moreover, if $C^{e,N}$ is higher than −0.26 μGal/mm, this means that most of ice-mass surrounding Ny-Ålesund changes above the altitude of the observation site (Sect. 3). If $C^{e,N} = -0.26$ μGal/mm, GIA effects would be larger than PDIM change. The expected uplift rate induced by GIA would be, in this case, about 3.96 mm/year which is more than twice that modeled by Sato *et al.* (2006). Indeed, as seen in Sect. 3, they obtained

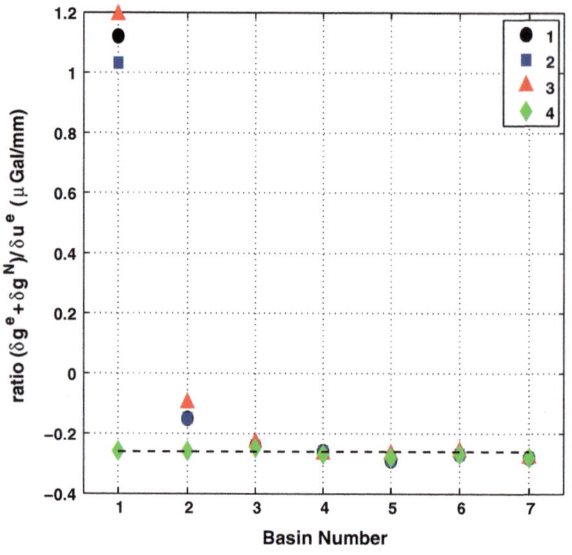

Figure 11
$C^{e,N}$ separately computed for each basin of Fig. 6, using the profiles of ice-mass change of Fig. 9 or a uniform thinning rate of 1 m/year. The *black dashed line* is -0.26 μGal/mm

$u^v = 1.88$ mm/year which leads to $u^e = 3.72$ mm/year and $C^{e,N} = -0.20$ μGal/mm. Using $C^{e,N} = -0.26$ μGal/mm leads to $u^e = 1.63$ mm/year.

From observations and the glaciation/deglaciation context we suggest that $C^{e,N}$ ranges between -0.21 and -0.18 μGal/mm whereas the best model provides -0.10 μGal/mm for the same context. The discrepancy is likely due to measurement accuracies which remain an important issue for the separation of geodetic consequences of GIA and PDIM change.

5. *Conclusion*

By modeling the elastic and viscoelastic deformations of the earth, one can compute the gravity variation-to-vertical displacement ratios C^v and $C^{e,N}$ that are defined for the GIA and PDIM change processes, respectively. They allow for a unique separation of the two effects, which are simultaneously observed by using geodetic and gravimetric techniques, provided one assumes the ice-mass change is uniform in a thin layer over the surface of the spherical model.

In this paper we have focused on $C^{e,N}$ and shown that according to the glaciation/deglaciation context and from the measurement of gravity variation and ground vertical velocity one can deduce a range of possible values for the $C^{e,N}$ ratio. Introducing the pseudo Green function $G_{C^{e,N}}$ we have shown that $C^{e,N}$ not only depends on the topography but also on the height variation of the ice load. Studying $G_{C^{e,N}}$ for the influence of the topography, we have shown that $C^{e,N}$ tends to positive values if most of surrounding ice-mass changes above the altitude of the observation site and to values lower than -0.26 μGal/mm if it changes below. We have also shown that $G_{C^{e,N}}$ can be known independently from the ice-height variation using a DEM with a 0.01 μGal/mm accuracy provided the ice load is located at least 6.5 km from the observation site and its variations are lower than 5 m/year. However, in general, for short distances and large ice-height variations, the determination of $C^{e,N}$ from a DEM only is not possible.

Using a particular example in Svalbard we have pointed out that different changes of ice volume and different load distributions can give similar vertical displacements and gravity variations or similar gravity variation-to-vertical displacement-rate ratio at a single observation station. This is clearly important in regions involving glaciers where the thinning rate varies from one glacier to the other. We have shown that $C^{e,N}$ does not depend on the rate of ice-mass change if it is spatially uniform. In this case, $C^{e,N}$ is larger than when the topography is neglected, and -0.26 μGal/mm can be considered to be a lower limit for the effect of PDIM change. If the ice-mass change is not spatially uniform, $C^{e,N}$ depends on the rate of change of the closest ice-covered area.

Acknowledgments

A. Mémin acknowledges financial support from the Centre National d'Études Spatiales. We thank H. Steffen and two anonymous reviewers for their comments on the first drafts of the paper.

References

Amalvict M., Willis P., Wöppelmann G., Ivins E. R., Bouin M.-N., Testut L., and Hinderer J., 2009. *Isostatic stability of the East Antarctic station Dumont d'Urville from long-term geodetic observations and geophysical models*, Polar Research, doi: 10.1111/j.1751-8369.2008.00091.x.

Barletta V. R., Sabadini R., and Bordoni A., 2008. *Isolating the PGR signal in the GRACE data: impact on mass balance estimates in Antarctica and Greenland*, Geophys. J. Int., *172*, 18–30.

Berthier E., Schiefer E., Clarke G. K. C., Menounos B., and Rémy F., 2010. *Contribution of Alaskan glaciers to sea-level rise derived from satellite imagery*, Nature Geoscience, *3*, 92–95, doi: 10.1038/NGEO737.

Bevis M., Kendrick E., Smalley R., Dalziel I., Caccamise D., Sasgen I., Helsen M., Taylor F. W., Zhou H., Brown A., Raleigh D., Willis M., Wilson T., and Konfal S., 2010. *Geodetic measurements of vertical crustal velocity in West Antarctica and the implications for ice mass balance*, Geochem. Geophys. Geosyst., *10* (10), Q10005, doi:10.129/2009GC002642.

Boy J.-P., Gegout P. and Hinderer J., 2002. *Reduction of surface gravity data from global atmospheric pressure loading*, Geophys. J. Int., *149*, 534–545.

Crossley D., Hinderer J., Casula G., Francis O., Hsu H.-T., Imanishi Y., Jentzsh G., Kääriäinen J., Merriam J., Meurers B., Neumeyer J., Richter B., Shibuya K., Sato T., and van Dam T., 1999. *Network of superconducting gravimeters benefits a number of disciplines*, EOS. Am. Geophys. Union, *80* (11), 125–126.

Dowdeswell J. A., Benham T. J., Strozzi T., and Hagen J. O., 2008. *Iceberg calving flux and mass balance of the Austfonna ice cap on Nordaustlandet, Svalbard*, J. Geophys. Res., *113*, F03022, doi:10.1029/2007JF000905.

Ekman M. and Mäkinen J., 1996. *Recent postglacial rebound, gravity change and mantle flow in Fennoscandia*, Geophys. J. Int., *126*, 229–234.

Fang M. and Hager B. H., 2001. *Vertical deformation and absolute gravity*, Geophys. J. Int., *146*, 539–548.

Farrell W. E., 1972. *Deformation of the Earth by surface loads*, Rev. Geophys. Space Phys., *10*, 761–797.

Horwath M. and Dietrich R., 2009. *Signal and error in mass change inferences from GRACE: the case of Antarctica*, Geophys. J. Int., *177*, 849–864.

James T. S. and Ivins E. R., 1998. *Predictions of Antarctic crustal motions driven by present-day ice sheet evolution and by isostatic memory of the last glacial maximum*, J. Geophys. Res., *103* (B3), 4993–5017.

Kääb A., 2008. *Glacier volume changes using ASTER satellite stereo and ICESat GLAS laser altimetry. A test study on Edgeøya, Eastern Svalbard*, IEEE Transactions on Geoscience and Remote Sensing, *46*, 10, 2823–2830.

Kierulf H. P., Plag H.-P., and Kohler J., 2009. *Surface deformation induced by present-day ice melting in Svalbard*, Geophys. J. Int., *179*, 1–13.

Kohler J., James T. D., Murray T., Nuth C., Brandt O., Barrand N. E., Aas H. F., and Luckman A., 2007. *Acceleration in thinning rate on western Svalbard glaciers*, Geophys. Res. Lett., *34*, L18502, doi:10.1029/2007GL030681.

Korona J., Berthier E., Bernard M., Remy F., and Thouvenot E., 2009. *SPIRIT. SPOT 5 stereoscopic survey of Polar Ice: Reference Images and Topographies during the fourth International Polar Year (2007–2009)*, ISPRS Journal of Photogrammetry and Remote Sens., *64*(2), 204–212.

Lambert A., Courtier N., Sasagawa G. S., Klopping F., Winester D., James T. S., and Liard J. O., 2001. *New constraints on Laurentide postglacial rebound from absolute gravity measurements*, Geophys. Res. Lett., *28*(10), 2109–2112.

Lambert A., Courtier N., and James T. S., 2006. *Long-term monitoring by absolute gravimetry: tides to postglacial rebound*, J. Geodyn., *43*, 339–357, doi:10.1016/j.jog.2006.08.002.

Larson K. M. and van Dam T., 2000. *Measuring postglacial rebound with GPS and absolute gravity*, Geophys. Res. Lett., *27* (23), 3925–3928.

Le Meur E. and Huybrechts P., 2001. *A model computation of the temporal changes of surface gravity and geoidal signal induced by the evolving Greenland ice sheet*, Geophys. J. Int., *145*, 835–849.

de Linage C., Hinderer J., and Rogister Y., 2007. *A search for the ratio between gravity variation and vertical displacement due to a surface load*, Geophys. J. Int., *171*, 986–994, doi:10.1111/j.1365-246X.2007.03613.x.

Luthcke S. B., Arendt A. A., Rowlands D. D., McCarthy J. J., and Larsen C. F., 2008. *Recent glaciers mass changes in the Gulf of Alaska region from GRACE mascon solutions*, J. Glaciol., *54* (188), 2109–2112.

Mäkinen J., Engfeldt A., Harsson B. G., Ruotsalainen H., Strykowski G., Oja T., and Wolf D., 2005. *The Fennoscandian land uplift gravity lines 1966–2003*. In: Jekeli C., Bastos L., Fernandes J. (Eds), Gravity, Geoid and Space Missions–GGSM2004. IAG International Symposium. Porto, Portugal, August 30–September 3, 2004, IAG Symposia 129, Springer, pp. 328–332.

Mäkinen J., Amalvict M., Shibuya K., and Fukuda Y, 2007. *Absolute gravimetry in Antarctica: status and prospects*, J. Geodyn., *41*, 307–317, doi:10.1016/j.jog.2005.08.032.

Mémin A., Rogister Y., Hinderer J., Llubes M., Berthier E. and Boy J.-P., 2009. *Ground deformation and gravity variations modelled from present-day ice thinning in the vicinity of glaciers*, J. Geodyn., *48* (3–5), 195–203, doi:10.1016/j.jog.2009.09.006.

Mémin A., Rogister Y., Hinderer J., Omang O. C. and Luck B., 2011. *Secular gravity variation at Svalbard (Norway) from ground observations and GRACE satellite data*, Geophys. J. Int., *184*(3), 1119–1130, doi:10.1111/j.1365-246X.2010.04922.x.

Merriam J. B., 1992. *Atmospheric pressure and gravity*, Geophys. J. Int., *109*, 488–500.

Moholdt G., Hagen J. O., Eiken T., and Schuler T. V., 2009. *Geometric changes and mass balance of the Austfonna ice cap, Svalbard*, The Cryosphere Discussions.

Nuth C., Moholdt G., Kohler J., Hagen J. O., and Kääb A., 2010. *Svalbard glacier elevation changes and contribution to sea level rise*, J. Geophys. Res., *115*, F01008,doi:10.1029/2008JF001223.

Sato T., Okuno J., Hinderer J., MacMillan D. S., Plag H. P., Francis O., Falk R., and Fukuda Y., 2006. *A geophysical interpretation of the secular displacement and gravity rates observed at , Svalbard in the Arctic–effects of post-glacial rebound and present-day ice melting*, Geophys. J. Int., *165*, 729–743.

Slobbe D. C., Ditmar P. and Lindenbergh R. C., 2009. *Estimating the rates of mass change, ice volume change and snow volume change in Greenland from ICESat and GRACE data*, Geophys. J. Int., *176*, 95–106.

Tushingham A.M. and Peltier W.R., 1991. *ICE-3G: a new global model of Late Pleistocene deglaciation based upon geophysical predictions of postglacial relative sea level change*, J. Geophys. Res., *96*, 4497–4523.

Velicogna I., 2009. *Increasing rates of ice mass loss from the Greenland and Antarctic ice sheets revealed by GRACE*, Geophys. Res. Lett., *36*, L19503, doi:10.1029/2009GL040222.

WAHR J., DAZHONG H., and TRUPIN A., 1995. *Predictions of vertical uplift caused by changing polar ice volumes on a viscoelastic earth*, Geophys. Res. Lett., *22* (8), 977–980.

WOLF D., BARTHELMES F., and SIGMUNDSSON F., 1997. *Predictions of deformation and gravity variation caused by recent change of Vatnajökull ice cap, Iceland*, Compt. Rend. J. Luxemb. Geodyn., *82*, 36–42.

WOUTERS B., CHAMBERS D., and SCHRAMA E. J. O., 2008. *GRACE observes small-scales mass loss in Greenland*, Geophys. Res. Lett., *35*, L20501, doi:10.1029/2008GL034816.

(Received May 13, 2010, revised January 3, 2011, accepted June 21, 2011, Published online September 3, 2011)

Pure Appl. Geophys. 169 (2012), 1373–1390

DOI 10.1007/s00024-011-0416-x

Pure and Applied Geophysics

Retrieval of Large-Scale Hydrological Signals in Africa from GRACE Time-Variable Gravity Fields

Jean-Paul Boy,[1,2] Jacques Hinderer,[1] and Caroline de Linage[3]

Abstract—Since its launch in April 2002, the Gravity Recovery and Climate Experiment (GRACE) mission is recording the Earth's time-variable gravity field with temporal and spatial resolutions of typically 7–30 days and a few hundreds of kilometers, allowing the monitoring of continental water storage variations from both continental and river-basin scales. We investigate here large scale hydrological variations in Africa using different GRACE spherical harmonic solutions, using different processing strategies (constrained and unconstrained solutions). We compare our GRACE estimates to different global hydrology models, with different land-surface schemes and also precipitation forcing. We validate GRACE observations through two different techniques: first by studying desert areas, providing an estimate of the precision. Then we compare GRACE recovered mass variations of main lakes to volume changes derived from radar altimetry measurements. We also study the differences between different publicly available precipitation datasets from both space measurements and ground rain gauges, and their impact on soil-moisture estimates.

Key words: Time-variable gravity, global change from geodesy, hydrology, Africa.

1. Introduction

The global circulation of surface geophysical fluids (atmosphere, oceans, continental water storage, ice sheets, etc.) induces global mass variations at the Earth's surface, and therefore, time-variable gravity field variations, as well as crustal deformation. Since its launch in April 2002, the GRACE (Gravity Recovery and Climate Experiment) mission records the Earth's time variable gravity field with spatial resolution of a few hundreds of kilometers and typical temporal sampling of 7–30 days (Tapley *et al.*, 2004). Continental hydrological variations in various large scale river basins have been successfully extracted from GRACE data (see, for example, Ramillien *et al.*, 2005; Syed *et al.*, 2005; Crowley *et al.*, 2006, 2008; Strassberg *et al.*, 2007).

We focus in this paper on the analysis of GRACE recovered hydrological variations over Africa. The Congo and Zambezi River basins as well as Lakes Chad and Victoria have already been studied using GRACE spherical harmonic gravity field solutions by, respectively, Crowley *et al.* (2006), Winsemius *et al.* (2006), Ramillien and Boronina (2008) and Awange *et al.* (2008). We extend these previous studies to all major African river basins, using the most recent GRACE reprocessed gravity field as well as by comparing with many different global hydrological models and different precipitation datasets.

Most of previous studies have consisted in comparison of GRACE recovered hydrological signals with different hydrology models at both continental and river basin scales. However, with the help of other datasets (such as precipitation, fresh water fluxes from atmospheric models, river discharge gauges), it is possible to solve the water mass balance equation. For example Syed *et al.* (2005), Winsemius *et al.* (2006) and Schmidt *et al.* (2008) used this approach to study continental water storage variations for the Amazon and Mississippi basins, the Zambezi basin and major river basins in South America (Amazon, Orinoco and Tocantins). In our study, we investigate annual and long-term large-scale hydrological variations over the entire African continent, using four different GRACE spherical harmonic solutions and various remote-sensing precipitation datasets.

[1] EOST-IPGS (UMR 7516 CNRS-UdS), 5 rue Rene Descartes, 67084 Strasbourg, France. E-mail: jeanpaul.boy@unistra.fr; jacques.hinderer@unistra.fr;
[2] Planetary Geodynamics Laboratory, Code 698, NASA Goddard Space Flight Center, Greenbelt, MD 20771, USA.
[3] Department of Earth System Science, University of California, Irvine, CA 92697, USA. E-mail: caroline.delinage@uci.edu

This paper is also linked to the current ongoing GHYRAF (Gravity and Hydrology in Africa) project (Hinderer *et al.*, 2009) devoted to the multidisciplinary observations (ground and space geodesy, hydrology and meteorology) of the water storage variations mainly in Niger and Benin, from the Sahara arid to the equatorial monsoon areas. The studied regions are located in the Niger River and Lake Chad basins.

In Sect. 2 we investigate seasonal variations and trends from both GRACE recovered continental water storage and hydrology models on the continental scale. Section 3 is devoted to the investigation of the major river basins in Africa, and the comparison between the different GRACE solutions and hydrology models. Using the Sahara and Libyan deserts, we also estimate the precision of the retrieved equivalent water height from GRACE. As another way to validate GRACE observations, we compare in Sect. 4, mass changes of major lakes and reservoirs in Africa from GRACE to volume estimates from radar altimetry. In Sect. 5, we investigate the differences between various precipitation datasets (from both space measurement and ground rain gauges) and compare them to our GRACE results. Finally, comments and conclusive remarks are given in Sect. 6.

2. *GRACE Recovered Equivalent Water Height*

2.1. *GRACE Gravity Field Solutions*

We analyze spherical harmonic solutions provided by four different centers, 10-day Release-2 constrained solutions from CNES/GRGS (Lemoine *et al.*, 2007; Bruinsma *et al.*, 2010), monthly unconstrained

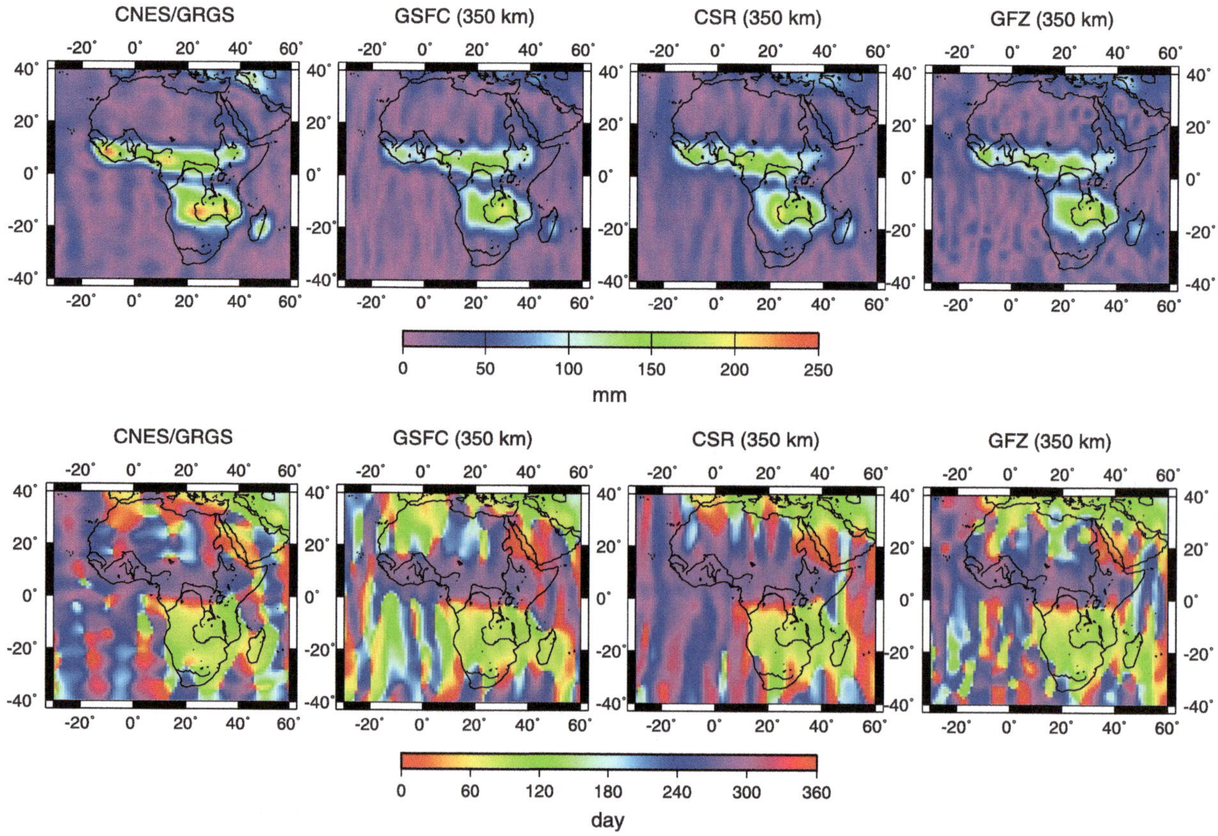

Figure 1
Amplitude (*top*) and phase (*bottom*) of the annual recovered equivalent water height. All unconstrained solutions (CSR, GSFC and GFZ) have been smoothed using a 350-km half width Gaussian filter

solutions obtained from only inter-satellite range rate data from NASA Goddard Space Flight Center (GSFC) (Luthcke *et al.*, 2006) and the monthly Release-4 unconstrained solutions from University of Texas Center for Space Research (CSR) and GeoForschungs-Zentrum (GFZ) (Flechner, 2007). CSR, GFZ, GSFC and CNES/GRGS cover, respectively, the 2002/04–2009/11, 2002/08–2009/10, 2003/04–2009/07 and 2002/08–2009/04 periods.

In addition to different processing strategies, these solutions use different forward modelling for handling atmospheric and oceanic induced effects. Atmospheric contributions to the time-variable gravity field are modelled using 6-hourly 3-D density profiles (Boy and Chao, 2005) from ECMWF (European Centre for Medium-Range Weather Forecasts) operational fields for CNES/GRGS, CSR and GFZ, whereas 3-hourly ECMWF surface pressure (Rabier *et al.*, 2000) are used for the GSFC solution.

The effects of the high-frequency oceanic circulation are modelled using OMCT (Ocean Model for Circulation and Tides) (Thomas, 2002) baroclinic model for CSR and GFZ, whereas the HUGO-m (Hydrodynamic Unstructured Grid Ocean model) (Carrère and Lyard, 2003) barotropic model is used for CNES/GRGS and GSFC solutions. Ocean tides are modelled using FES2004 (Lyard *et al.*, 2006), except for GSFC which uses GOT4.7 (Ray, 1999).

Because of their different processing schemes, the post-processing of the time-variable gravity field differs: Stokes coefficients have been truncated to degree 60 (50 for CNES/GRGS) and a Gaussian smoothing (of different radii) has been applied to the monthly unconstrained solutions. Because of its

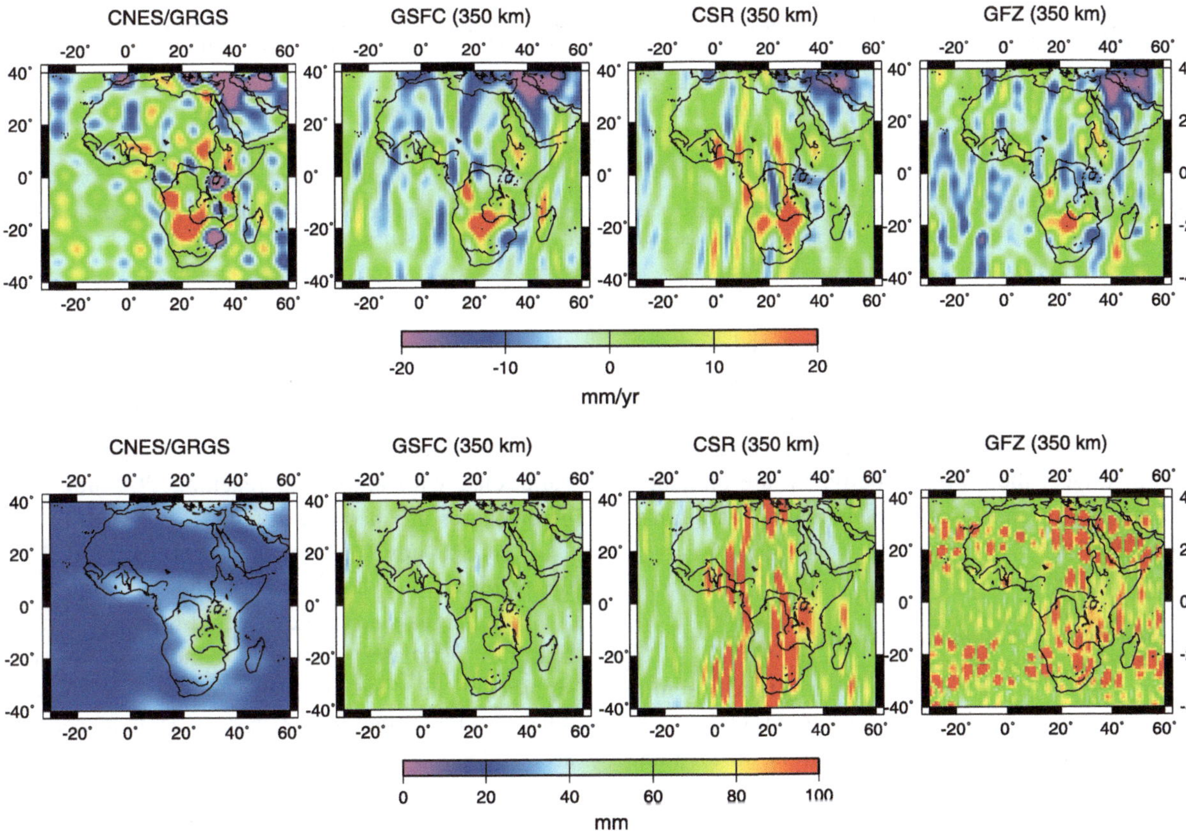

Figure 2
Trends (*top*) and RMS of residuals (*bottom*) of the recovered equivalent water height. All unconstrained solutions (CSR, GSFC and GFZ) have been smoothed using a 350-km half width Gaussian filter

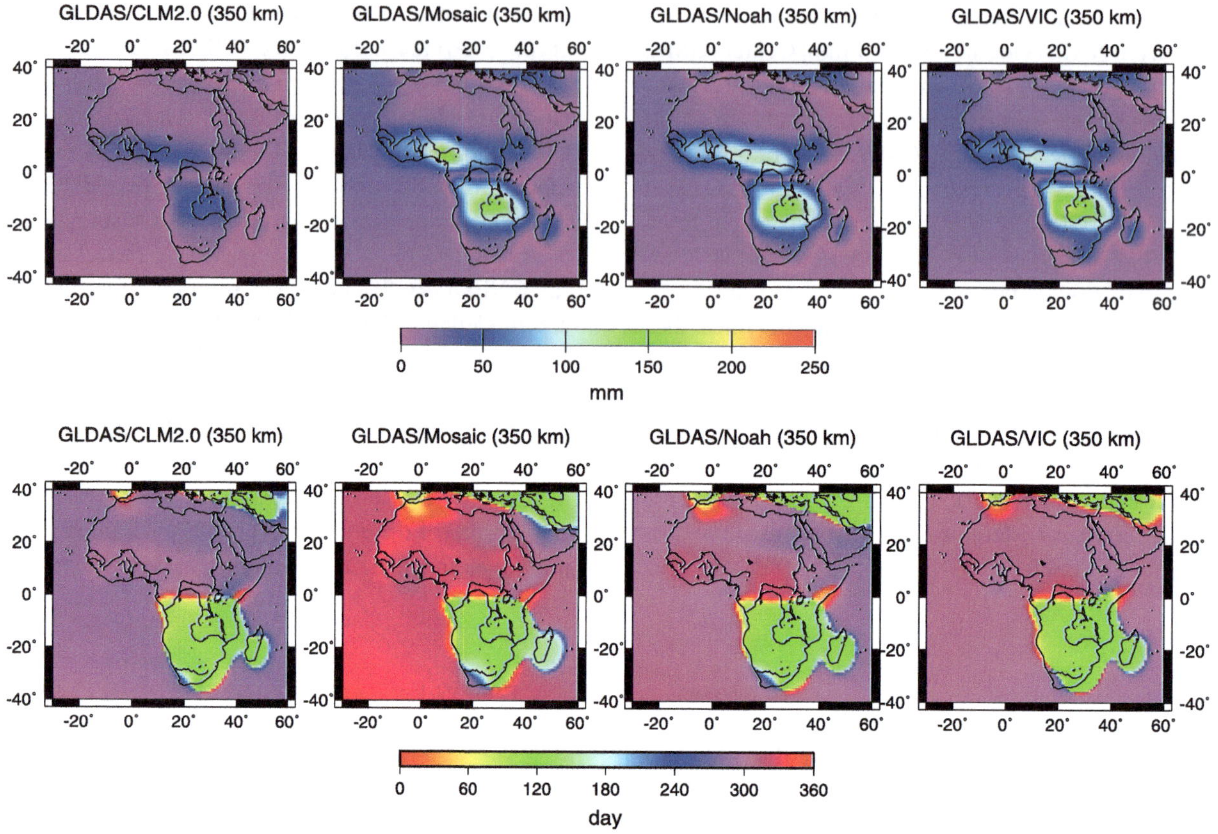

Figure 3
Amplitude (*top*) and phase (*bottom*) of seasonal continental water storage variations according to GLDAS models. A 350-km half width Gaussian filter has been applied to monthly timeseries

constraints, such filtering is not required for the CNES/GRGS solutions. We do not apply any destriping or rescaling to all GRACE solutions.

We assume that all mass variations occur at the Earth's surface; then Stokes coefficient are uniquely converted into equivalent water height variations (CHAO, 2005).

2.2. Seasonal Variations and Trends

We adjust by least square fitting annual, semi-annual, 161-days (aliasing of S2 ocean tide) (RAY and LUTHCKE, 2006) terms as well as a trend to all four time-variable gravity field solutions. Figure 1 shows the amplitude and phase (with respect to the 1st of January) of seasonal variations. Because of the post-processing of unconstrained solutions (350-km half width Gaussian filtering), CNES/GRGS solution shows higher amplitudes as well as smaller wavelength content. Amplitudes over Sahara desert and the oceans give an idea of the noise level in each solution. Equatorial regions show large seasonal variations, with amplitude reaching 200 mm. Both hemispheres are also out of phase.

Figure 2 shows the trends as well as the RMS of the residuals.

Trend signals are more affected by noise than the seasonal estimates. However, one can notice a significant increase of water storage in southern Africa (Okawango River basin) and a decrease around Lake Victoria (AWANGE *et al.*, 2008).

Unconstrained monthly solutions show higher residuals than the 10-day constrained solutions from CNES/GRGS. As GSFC solutions are obtained from K-band range rate (KBRR) data only, they exhibit less striping than CSR and GFZ solutions, as also shown by

Luthcke *et al.* (2006). If a 500-km (instead of 350-km) half width Gaussian filter is applied to the unconstrained solutions, the RMS of the residuals are much closer to the ones from CNES/GRGS solution.

Figures 3 and 4 give, respectively, annual amplitude and phase of seasonal variations and trends and RMS of residuals (after adjusting annual, semi-annual and trend) of the four GLDAS (Rodell *et al.*, 2004) models using different land surface schemes: CLM 2.0 (Bonan, 1998), Mosaic (Koster and Suarez, 1996), Noah (Chen *et al.*, 1996) and VIC (Liang *et al.*, 1994). All models have been smoothed using a 350-km half width Gaussian filter.

Except for CLM2.0 showing smaller amplitudes, the hydrological models agree well with seasonal GRACE recovered water storage variations, both in amplitude and phase. On the other hand, the four models show significant different trend signals, none of them being in agreement with any GRACE observation. Only Mosaic and Noah models show negative trends in the vicinity of Lake Victoria. If GRACE trends are mainly due to water decrease cause by lake level changes (Awange *et al.*, 2008), these variations cannot be modelled by GLDAS hydrology model as it does not include surface water (as well as groundwater). Only Mosaic and, to a smaller extent, VIC shows some positive trend in the Okawango basin, but much smaller than in GRACE data. In addition, the high-frequency content (expressed here in terms of RMS of residuals) of the four models are also significantly different.

3. *Hydrological Variations in the Major River Basins*

In this section, we investigate continental water storage for each individual basin from both GRACE observations and global hydrology models.

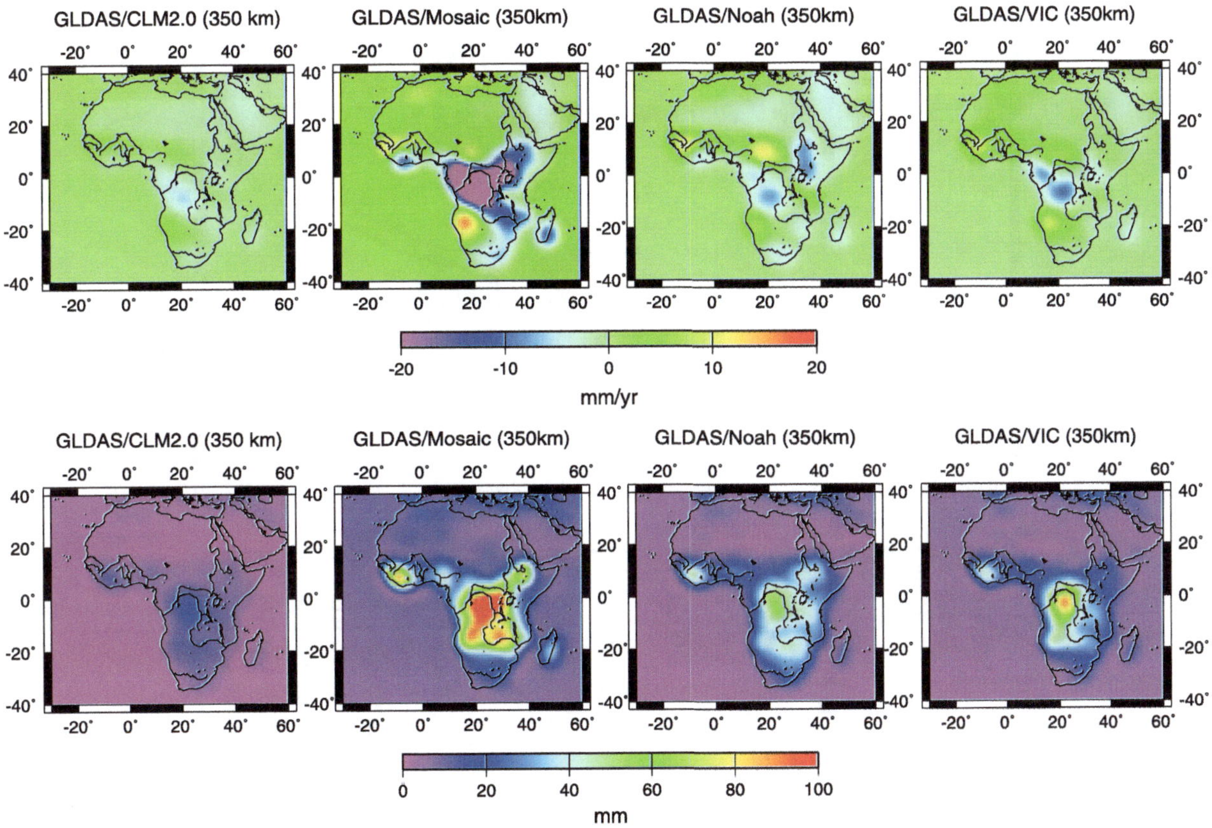

Figure 4
Trends (*top*) and RMS of residuals (*bottom*) of continental water storage variations according to GLDAS models. A 350-km half width Gaussian filter has been applied to monthly timeseries

3.1. River Basin Definition

Figure 5 shows the different major river basins in Africa (Jenness *et al.*, 2008). Because of their size and their different hydrological behavior, Nile, Niger and Congo Rivers and Lake Chad basins are divided into two (three for Congo) sub-basins. We also consider the desert areas of Sahara and Libya (which also include some parts of Egypt and Sudan). Because of their relatively small spatial extent, some minor rivers have been included into the Senegal and Volta regions.

3.2. Null Test

The quantification of the errors of the retrieved mass water storage variations is quite challenging. Several methods have been proposed, including a null

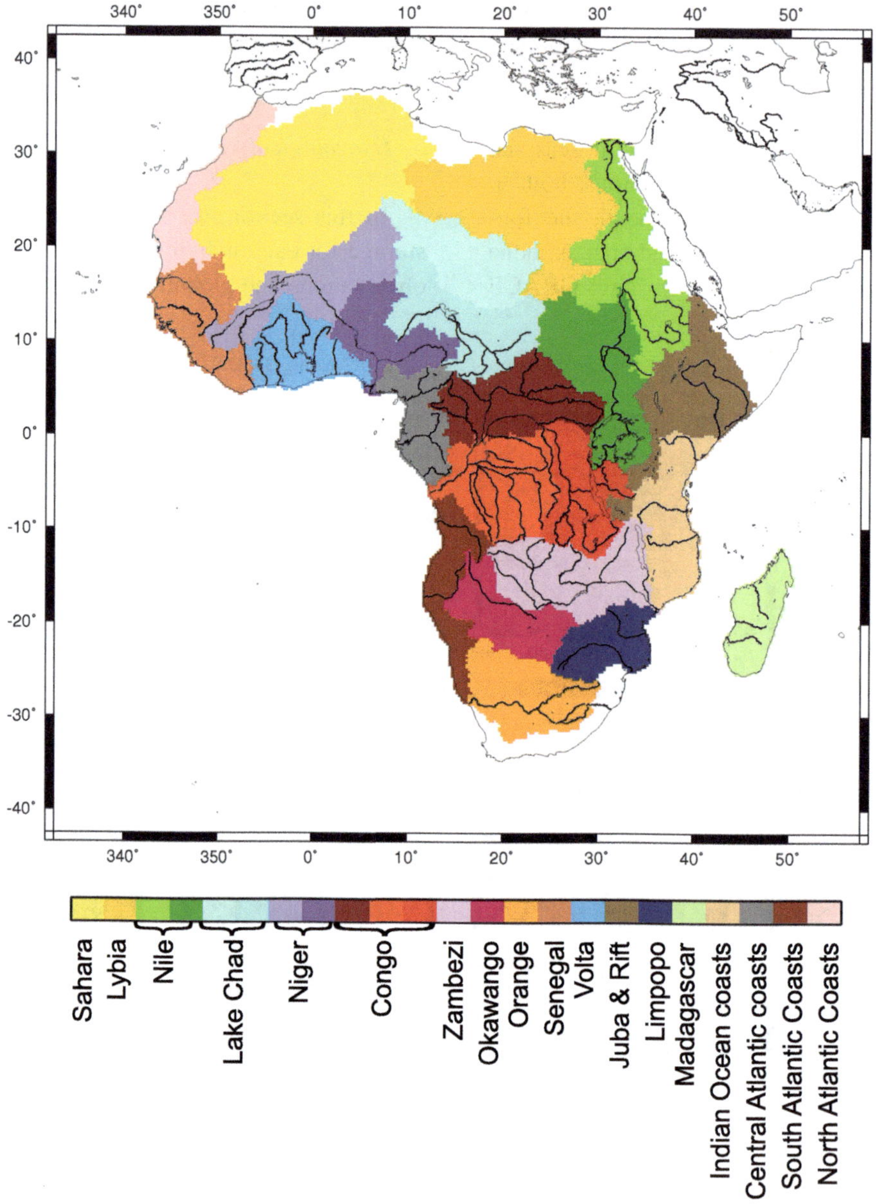

Figure 5
River basin definition (from Jenness *et al.*, 2008)

test experiment. Because of its large scale, the Sahara desert appears to be one of the possible candidates. Table 1 gives the RMS (root mean square) as well as the annual amplitude of each GRACE solution for the two large desert areas shown on Fig. 5 (Sahara and Libya regions). Two different GSFC solutions are provided: v05 corresponds to a classical solution where hydrology is not forward modelled. On the other hand, v06 solution includes the forward modelling of hydrology using GLDAS/Noah model (Rodell *et al.*, 2004). This solution should minimize leakage errors of surrounding areas.

Table 1

RMS and annual amplitude in mm of GRACE recovered equivalent water height over Sahara and Libya regions (see Fig. 5)

	Sahara		Libya	
	RMS	Amplitude	RMS	Amplitude
CSR	17.79	11.92	13.77	12.14
GFZ	13.55	2.91	13.22	8.42
GSFC (v05)	15.54	4.94	18.40	13.83
GSFC (v06)	16.21	7.16	14.65	4.09
CNES/GRGS	14.16	6.79	15.56	9.74

A 500-km half width Gaussian filter has been applied to the unconstrained solutions

If RMS values do not significantly differ (from 13 to 18 mm) for each GRACE solution (with or without forward modelling hydrology) and the two areas, this is not the case for the annual variations. For example, annual amplitudes are 2.91 and 11.92 mm in Sahara for GFZ and respectively CSR solutions. The forward modelling of hydrology (GSFC v06) does not reduce the RMS and annual amplitude in Sahara versus the classical solution (GSFC v05) which contains all hydrological signals. This is not the case for Libya, where both RMS (from 7.16 to 6.79 mm) and annual

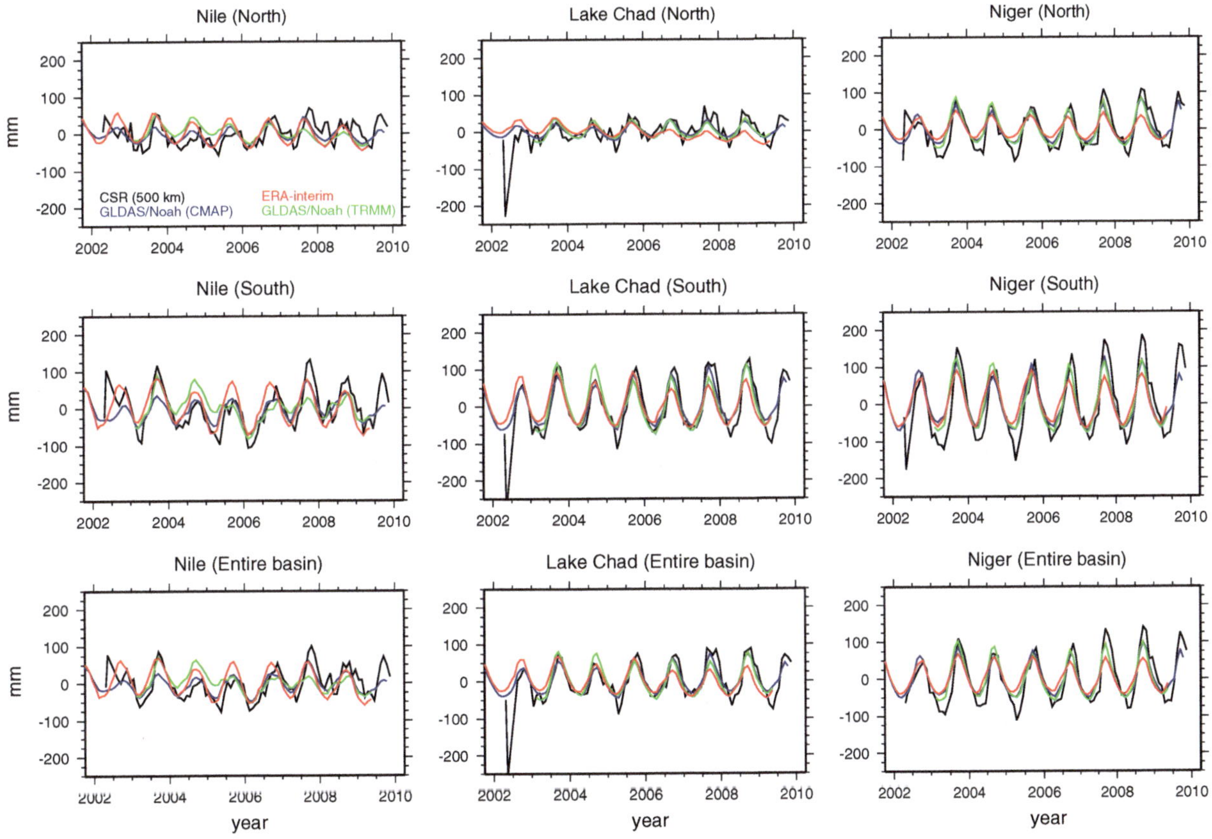

Figure 6

Comparison between CSR solutions (500 km Gaussian smoothing applied), two GLDAS/Noah (forced respectively forced by CMAP and TRMM precipitation) and ECMWF Reanalysis (ERA-interim) for Nile, Lake Chad and Niger basins amplitude (from 13.83 to 4.09 mm) are significantly reduced when hydrology is forward modelled. It is difficult to compare, in terms of RMS, the monthly unconstrained solutions (CSR, GFZ and GSFC) to the constrained solution (CNES/GRGS) because of the different time sampling (monthly versus 10 days respectively). However, the CNES/GRGS annual amplitudes are generally lower than for most of the unconstrained solutions. On average, we can consider that the error of monthly recovered continental water storage from GRACE is about 10–20 mm of equivalent water height.

3.3. *Comparison of GRACE and Global Hydrology Models*

We compare here GRACE recovered water storage in the different basins from the four spherical harmonic solutions to different high-resolution (0.25° and 0.7°) global hydrology models. Figures 6, 7 and 8 show the comparison between CSR monthly solutions and three global hydrology models, after applying to all datasets a 500-km half width Gaussian filter:

- the classical 3-h, 0.25° GLDAS/Noah (Rodell *et al.*, 2004) forced by CMAP (Xie *et al.*, 2003) precipitation field,
- the 3-h, 0.25° GLDAS/Noah forced by TRMM (Huffman *et al.*, 2007) precipitation field,
- the new ECMWF reanalysis: 6-h, 0.7° ERA-interim (Simmons *et al.*, 2007; Uppala *et al.*, 2007).

Except for some higher noise levels in 2002 and early 2003, GRACE solutions generally agree with the different hydrology models. The main differences occur at sub-basin scales (see, for example, North Congo and Madagascar in Fig. 8).

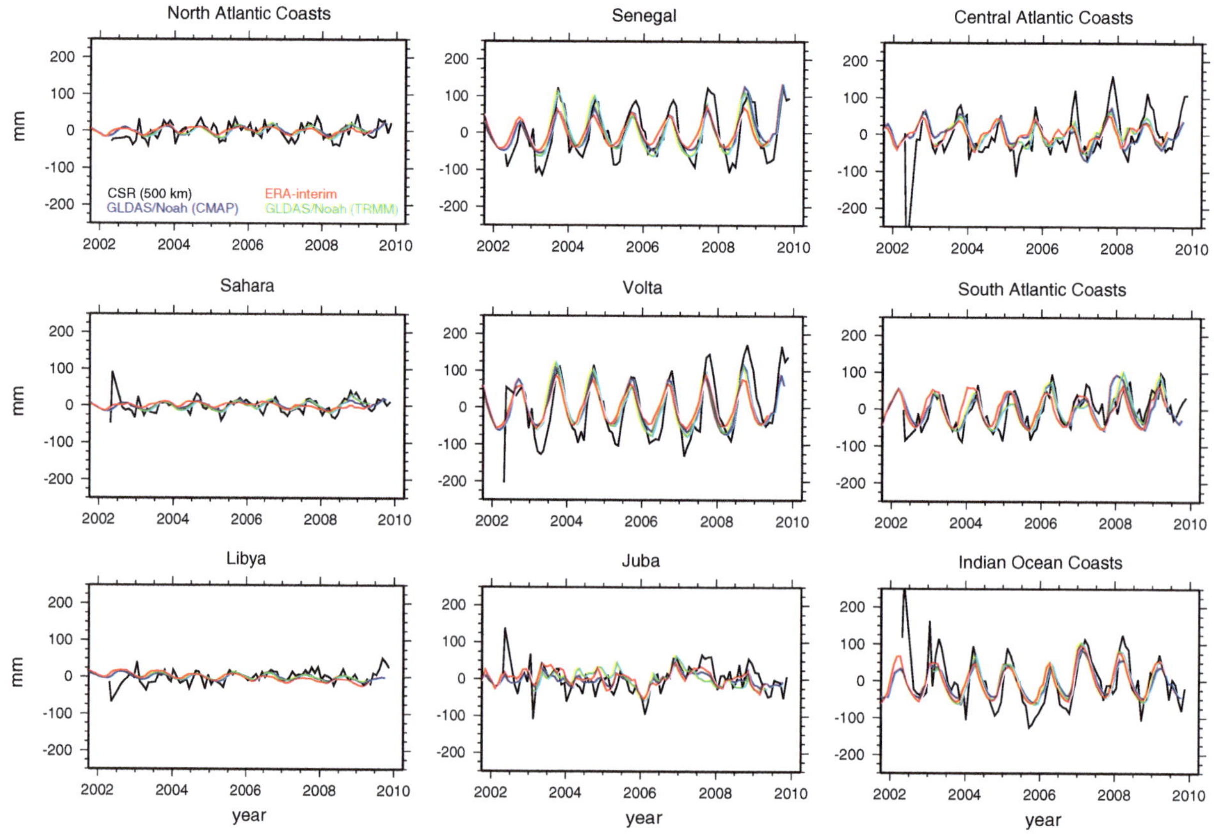

Figure 7
Same as Fig. 6, but for coastal regions, desert areas (Sahara and Libya) and Senegal, Volta and Juba River basins

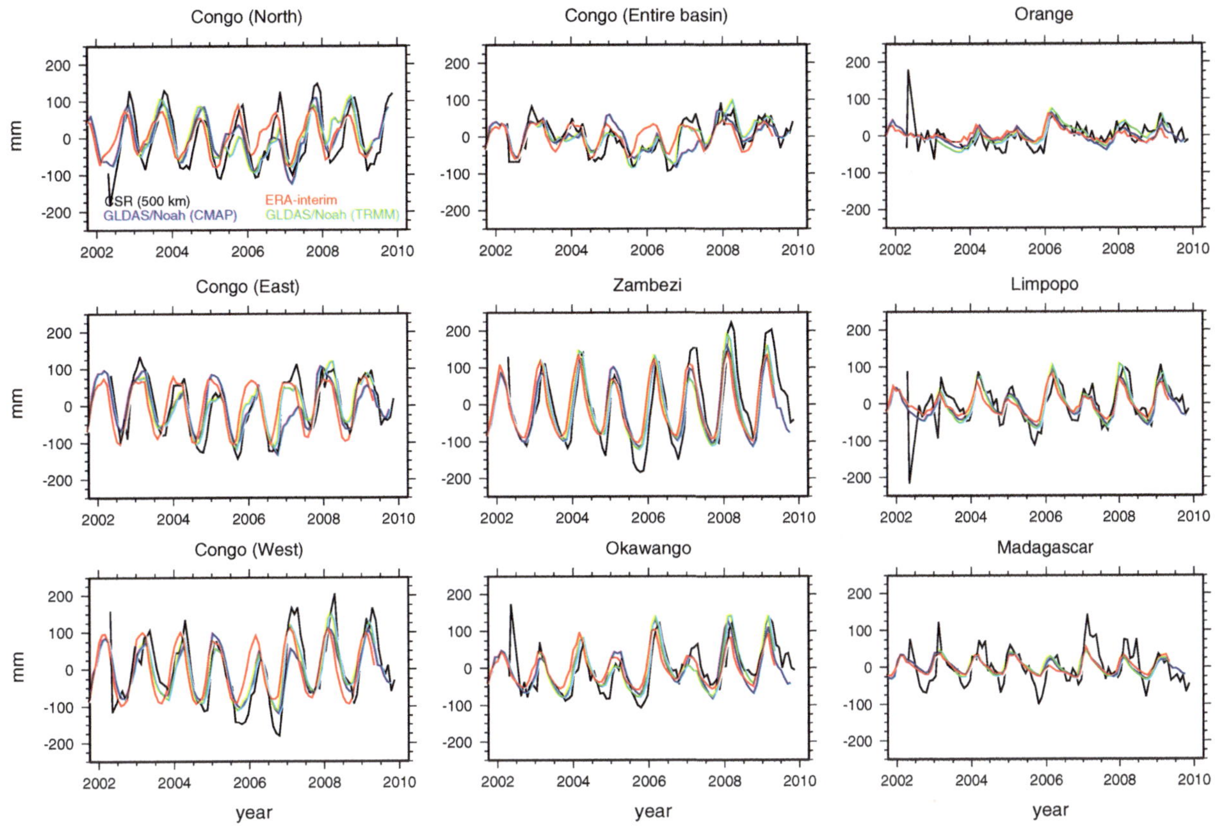

Figure 8
Same as Fig. 6, but for Congo, Zambezi, Okawango, Orange and Limpopo River basins and Madagascar

Table 2 gives the correlation coefficient for seven different hydrological basins (Sahara, Libya, Nile, Lake Chad, Niger, Congo and Zambezi basins) between GRACE solutions (monthly unconstrained from CSR, GFZ and GSFC, and 10-day contrained from CNES/GRGS) and GLDAS/Noah models (forced by CMAP or TRMM precipitation products) as well as ERA interim for a 70 month period spanning from July 2003 to May 2009. Because of the lack of GRACE data, we exclude January 2004.

Despite the higher temporal sampling, and therefore, some high-frequency noise, there are no major differences between CNES/GRGS and GSFC (the better unconstrained solutions in terms of agreement with hydrology models) solutions in terms of correlation with hydrology models.

Although GLDAS models do not include surface water and groundwater, which can have a significant contribution to gravity variations (see, for example, Han *et al.*, 2009), GRACE is in better agreement with GLDAS/Noah model forced by TRMM precipitation (compared to CMAP), for Lake Chad, Niger, Congo and Zambezi River basins.

For Sahara and Libya regions and to a smaller extend for Lake Chad basin, the correlations between GRACE solutions and global hydrology models are small due to the low signal-to-noise ratio compared to the other basins (see Figs. 9, 10 and 11). In general, the ERA-interim hydrology model show smaller correlations with GRACE than the two GLDAS/Noah models. This is probably due to the smaller spatial resolution of the model (0.7° compared to 0.25°) and its simpler land-surface model. This is especially true for the desert areas (Sahara and Libya regions).

There are two main reasons for differences in water content in hydrology models: differences in the forcing data, especially precipitation, and differences

Table 2

Correlation between GRACE solutions and GLDAS/Noah models and ERA-interim for the 2003/07-2009/05 period (2004/01 excluded). CNES/GRGS solutions are not filtered

	Sahara	Libya	Nile	Chad	Niger	Congo	Zambezi
CSR (350 km)							
GLDAS/Noah/CMAP	0.682	0.561	0.775	0.893	0.922	0.679	0.895
GLDAS/Noah/TRMM	0.734	0.449	0.516	0.820	0.901	0.849	0.929
ERA interim	0.168	0.119	0.595	0.266	0.608	0.684	0.835
CSR (500 km)							
GLDAS/Noah/CMAP	0.729	0.556	0.781	0.918	0.924	0.678	0.904
GLDAS/Noah/TRMM	0.778	0.451	0.509	0.844	0.903	0.861	0.934
ERA interim	0.174	0.092	0.590	0.292	0.606	0.678	0.846
GFZ (350 km)							
GLDAS/Noah/CMAP	0.233	0.310	0.822	0.768	0.949	0.690	0.886
GLDAS/Noah/TRMM	0.201	0.281	0.490	0.814	0.939	0.801	0.915
ERA interim	−0.038	0.108	0.589	0.319	0.650	0.548	0.807
GFZ (500 km)							
GLDAS/Noah/CMAP	0.341	0.372	0.832	0.819	0.948	0.723	0.903
GLDAS/Noah/TRMM	0.314	0.362	0.540	0.851	0.945	0.822	0.923
ERA interim	0.023	0.115	0.600	0.343	0.655	0.535	0.826
GSFC (350 km)							
GLDAS/Noah/CMAP	−0.231	0.564	0.816	0.875	0.909	0.690	0.896
GLDAS/Noah/TRMM	−0.240	0.636	0.500	0.887	0.903	0.806	0.930
ERA interim	0.094	0.196	0.655	0.401	0.620	0.639	0.828
GSFC (500 km)							
GLDAS/Noah/CMAP	−0.143	0.611	0.813	0.899	0.926	0.694	0.913
GLDAS/Noah/TRMM	−0.158	0.666	0.520	0.908	0.921	0.820	0.938
ERA interim	0.134	0.203	0.654	0.411	0.634	0.635	0.848
CNES/GRGS							
GLDAS/Noah/CMAP	0.389	0.199	0.768	0.866	0.940	0.662	0.898
GLDAS/Noah/TRMM	0.235	0.240	0.501	0.868	0.912	0.856	0.910
ERA interim	0.114	−0.040	0.717	0.282	0.696	0.664	0.857

in the land surface models. These three models differ in the precipitation forcing, CMAP and TRMM for GLDAS/Noah and the precipitation derived from ERA-interim atmospheric model, but also from the land model, Noah (Chen *et al.*, 1996) for GLDAS and TESSEL (Balsamo *et al.*, 2009) for ERA-interim. Although both CMAP and TRMM are calibrated with rain gauge data, the differences between the two GLDAS models illustrate the differences in precipitation forcing.

In addition to the 1/4-degree GLDAS/Noah model, GLDAS also releases four 3-hour 1-degree models, all forced by CMAP, but with four different land surface models: CLM 2.0, Mosaic, Noah and VIC. Figures 9, 10 and 11 show the comparison between GSFC monthly solutions (Luthcke *et al.*, 2006) and the four GLDAS 1-degree models after applying a 500 km Gaussian filter to all datasets.

As already known, continental hydrology variations are under-estimated in the GLDAS/CLM2.0 model. However, there are significant discrepancies between the three other models (Mosaic, Noah and VIC), especially in equatorial areas (see Congo basin and sub-basins in Fig. 11).

4. Retrieval of Major Lake Mass Variations from GRACE

One way to validate GRACE solutions is to compare mass estimates with known mass variations. As a result of several decades of radar altimetry, water level variations of major lakes and reservoirs are monitored with a precision of a few centimeters. These induced volume variations should be equivalent to the mass variations as retrieved by GRACE, if thermal expansion is neglected.

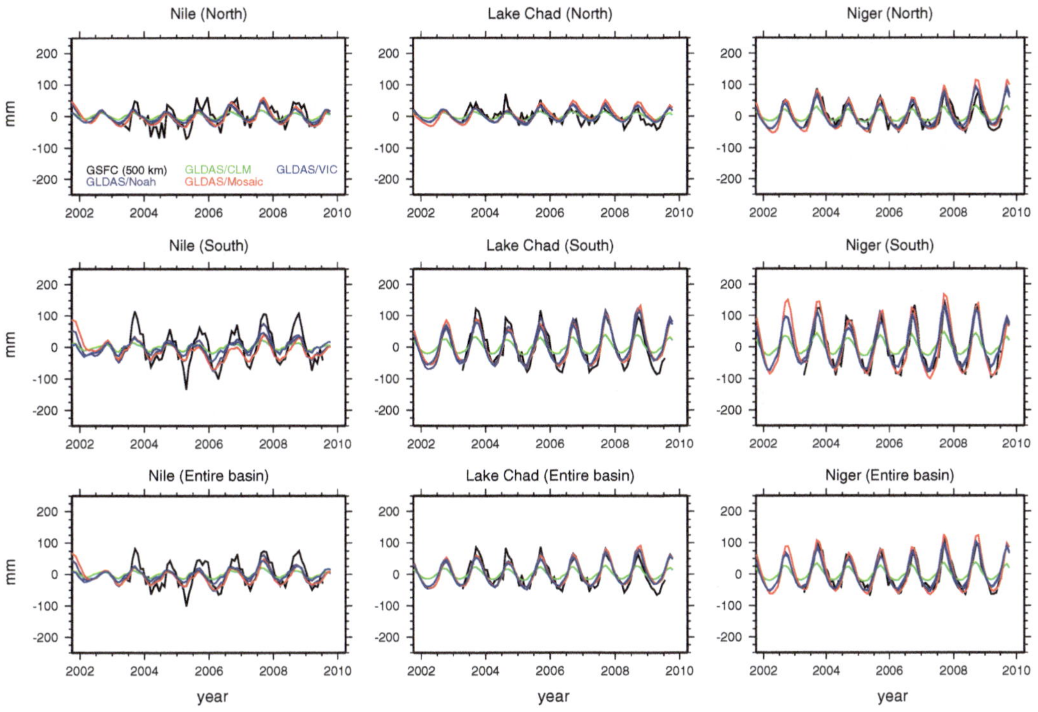

Figure 9
Comparison between GSFC solutions (500-km Gaussian smoothing applied) and the four GLDAS 1-degree models for Nile, Lake Chad and Niger basins

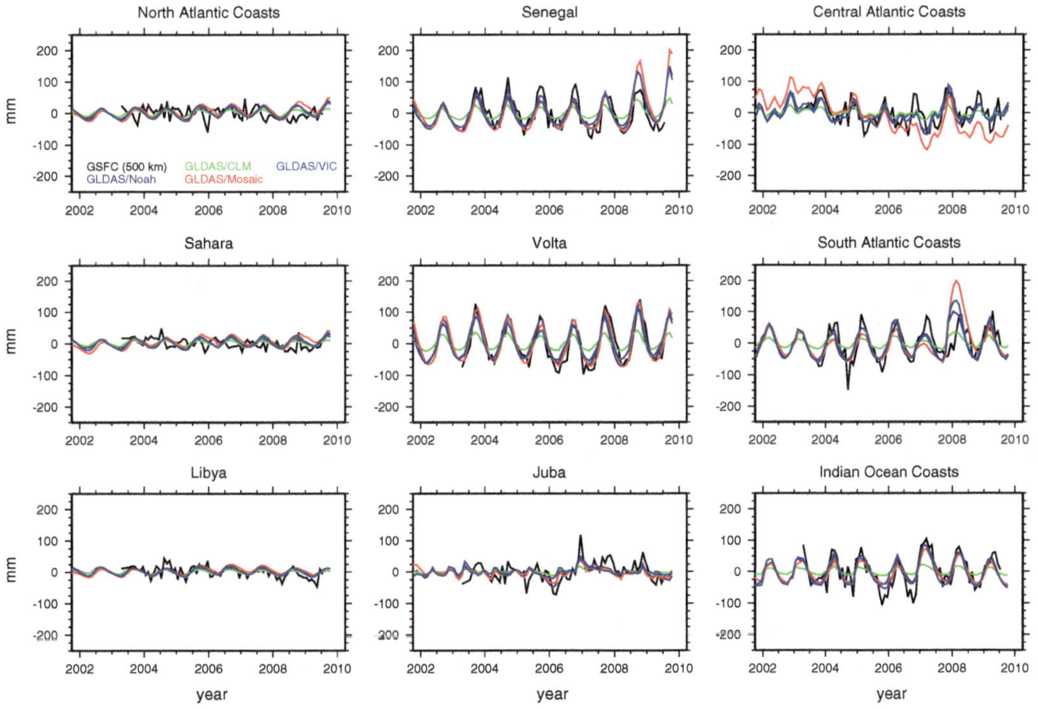

Figure 10
Same as Fig. 9, but for coastal regions, desert areas (Sahara and Libya) and Senegal, Volta and Juba River basins

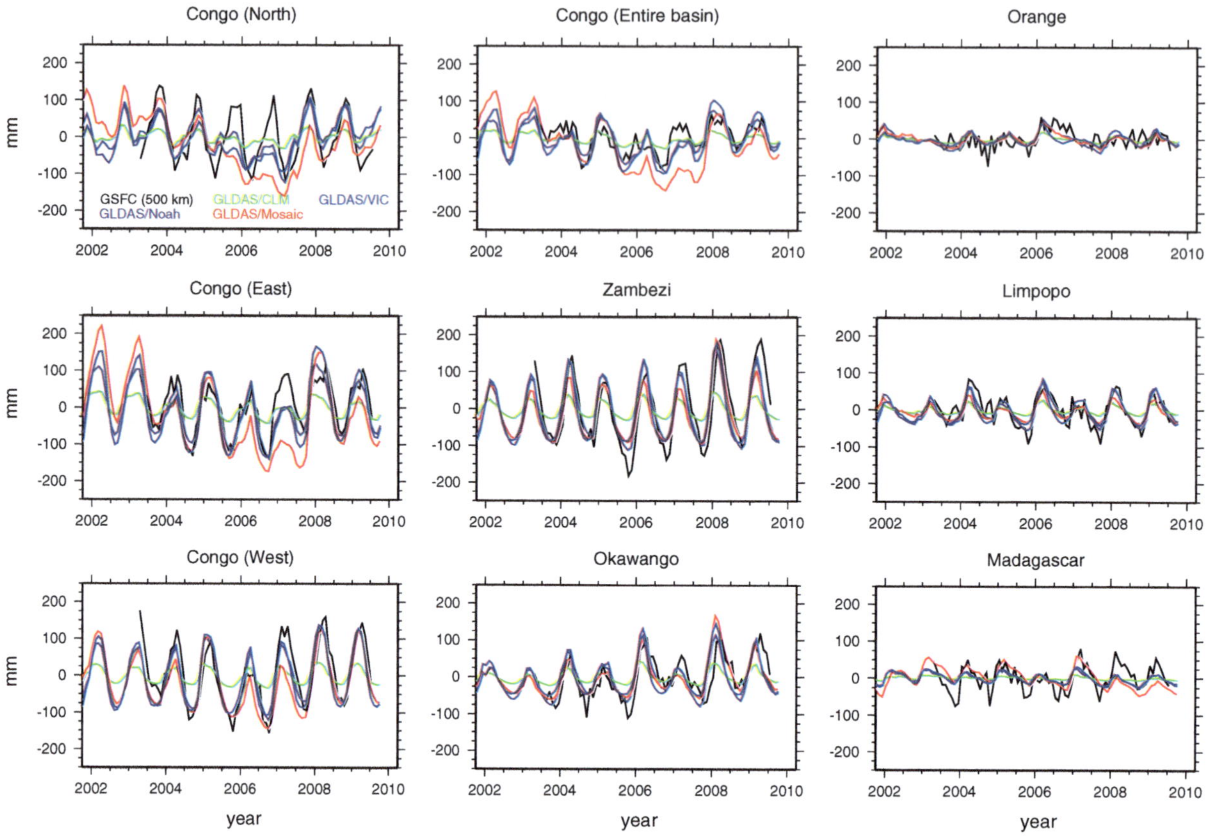

Figure 11
Same as Fig. 9, but for Congo, Zambezi, Okawango, Orange and Limpopo River basins and Madagascar

We select the three largest lakes in Africa: Malawi, Tanganika and Victoria with an area of, respectively, 29,600, 32,900 and 68,800 km^2. The water height variations are provided by the hydroweb (http://www.legos.obs-mip.fr/en/soa/hydrologie/hydroweb/) from LEGOS, Toulouse, France.

Figure 12 shows the equivalent water height changes deduced from radar altimetry and retrieved from GRACE spherical harmonic solutions. CNES/GRGS solution is shown in red. The classical GSFC-v05 solution (hydrology is not forward modelled) is shown in green, and the v06 solution (hydrology is forward modelled using GLDAS/Noah model) is shown in blue. As we are looking at much smaller wavelengths than in the previous section, a 200-km half width Gaussian filter has been applied to the unconstrained (GSFC) spherical harmonic fields as well to the altimetry datasets (we do not apply any rescaling).

Because of their azimuthal stretching, spherical harmonics, truncated at degree 50 (CNES/GRGS) or 60 (CSR, GFZ and GSFC), are not an optimal parametrization compared to, for example, localized approaches, such as mascons (Rowlands *et al.*, 2010). GRACE recovered water height variations (with or without Gaussian filtering) suffer from large leakage errors, which are not significantly reduced when hydrology is forward modelled (GSFC v06 versus GSFC v05). Because of its larger surface and its shape, Lake Victoria appears to be the only lake in Africa where GRACE can be successfully compared to altimetry observations. Because of the "striping" in unconstrained solutions, GSFC derived water height variations show higher high-frequency noise,

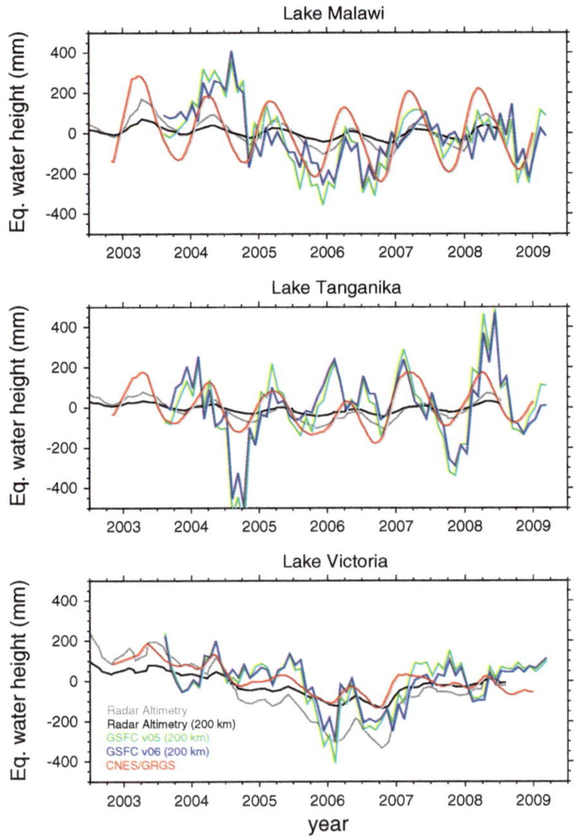

Figure 12
Comparison between water height variations deduced from radar altimetry (in *grey* without Gaussian smoothing, in *black* with 200-km half width Gaussian filtering) and derived from GRACE solutions (*green*: GSFC v05, *blue*: GSFC v06 and *red*: GRGS) for Lakes Malawi, Tanganika and Victoria. All GRACE solutions have been smoothed using a temporal 1/2 year Gaussian filter

despite the 1/2-year temporal filtering, compared to CNES/GRGS solution.

The impact of the Gaussian filtering on the water height variations can be seen on Fig. 12. The grey curves show altimetric observations when truncated at degree 60 (same as GRACE unconstrained solutions), whereas the black curves show the variations when a 200-km half width Gaussian filter has been applied. For example, for Lake Victoria the Gaussian filtering significantly reduces the long-term variations.

5. *Comparison of Precipitation Datasets*

In Sect. 3, we have investigated the sensitivity of the GLDAS models to the choice of the Land-Surface-Model (CLM2.0, MOSAIC, Noah and VIC). We have also shown that forcing GLDAS/Noah model with CMAP (Xie *et al.*, 2003) or TRMM (Huffman *et al.*, 2007) does not significantly change the seasonal continental water storage variations, although the run using TRMM as a precipitation forcing seems to be slightly in better agreement with GRACE recovered equivalent water height. However, other precipitation datasets estimated from space measurements and/or ground rain gauge stations are available. As we are not running any hydrology models, we illustrate the sensitivity of hydrology models to precipitation forcing using an over-simplified hypothesis, as done previously by Crowley *et al.* (2008).

We show in Fig. 13 the annual amplitude and phase of CMAP (pentad) and TRMM (3B42) estimated precipitation, in addition to another space derived measurement CMORPH (CPC Morphing Technique) (Joyce *et al.*, 2004) and to monthly precipitation derived from rain gauge station according to GPCC (Global Precipitation Climatology Centre) (Rudolf and Schneider, 2005).

Annual amplitude of space-derived precipitation varies from about a factor of 2 between CMAP to CMORPH estimates within the Tropics, TRMM being slightly larger than CMAP, and also closest to the ground data. The phases of the four products are generaly in good agreement.

The terrestrial water mass balance links the time-derivative of continental water storage W to precipitation P, evapo-transpiration E and run-off R:

$$\frac{\partial W}{\partial t} = P - E - R \qquad (1)$$

As precipitation is the main forcing of continental water storage variations, the integrated precipitation (after detrending) should be, at least to the first order, close to the equivalent water height variations, as retrieved by GRACE (Crowley *et al.*, 2008). It is equivalent to assume that $P + R$ is time-independent. We plot in Figs. 14 and 15 time variations of GRACE recovered water storage to the detrended integrated precipitation (same datasets as in Fig. 13) for several main river basins in Africa.

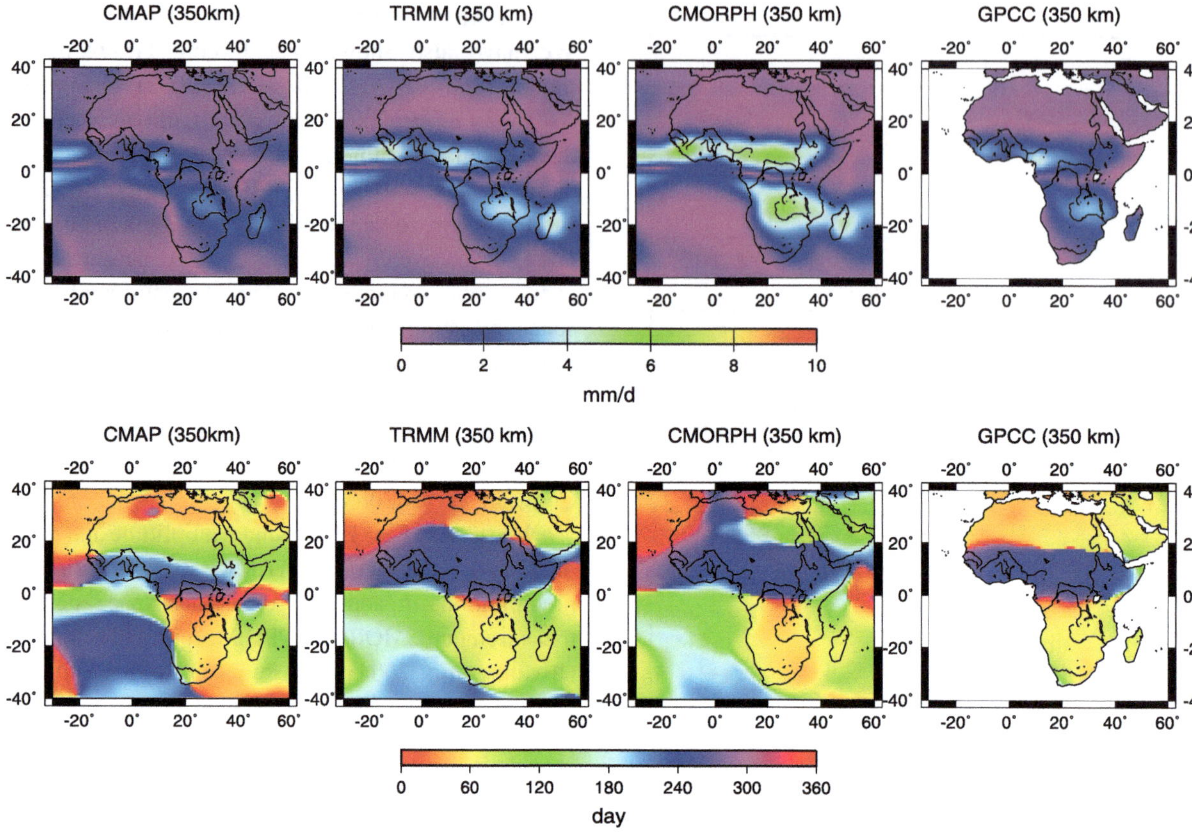

Figure 13
Amplitude (*top*) and phase (*bottom*) of annual precipitation according to CMAP (Xie *et al.*, 2003), TRMM (Huffman *et al.*, 2007), CMORPH (Joyce *et al.*, 2004) and GPCC monitoring product (Rudolf and Schneider, 2005). A Gaussian filter with a 350-km half width has been applied to all datasets

Although the influence of precipitation forcing on continental water storage is over-simplified, we show that the significant differences between the four state-of-the-art precipitation datasets at seasonal timescales lead to even larger differences in soil water content when time-integrated.

6. Discussion and Conclusion

Despite their different processing schemes and temporal resolutions, all GRACE spherical harmonic solutions from UT/CSR, GFZ, NASA/GSFC and CNES/GRGS show similar continental water storage estimates at seasonal timescales, both on continental and river basin scales. There are some differences concerning longer term variations, e.g. trends; however, all solutions show a significant increase of water storage in the Okawango River basin (southwestern Africa) and decrease in Lake Victoria drainage basin (eastern Africa). The errors of GRACE retrieved continental water storage are estimated to be about 10 to 20 mm as a result of the null test over the Sahara and Libya deserts. For other areas, GRACE recovered seasonal variations are also in agreement (both in amplitude and phase, except for GLDAS/CLM) with global hydrology models from GLDAS and ECMWF/ERA-interim. This is however not the case for the longer-term variations.

We show, for most basins in Africa, GRACE derived continental water storage variations are in better agreement with GLDAS/Noah model forced by TRMM precipitation than the classical model

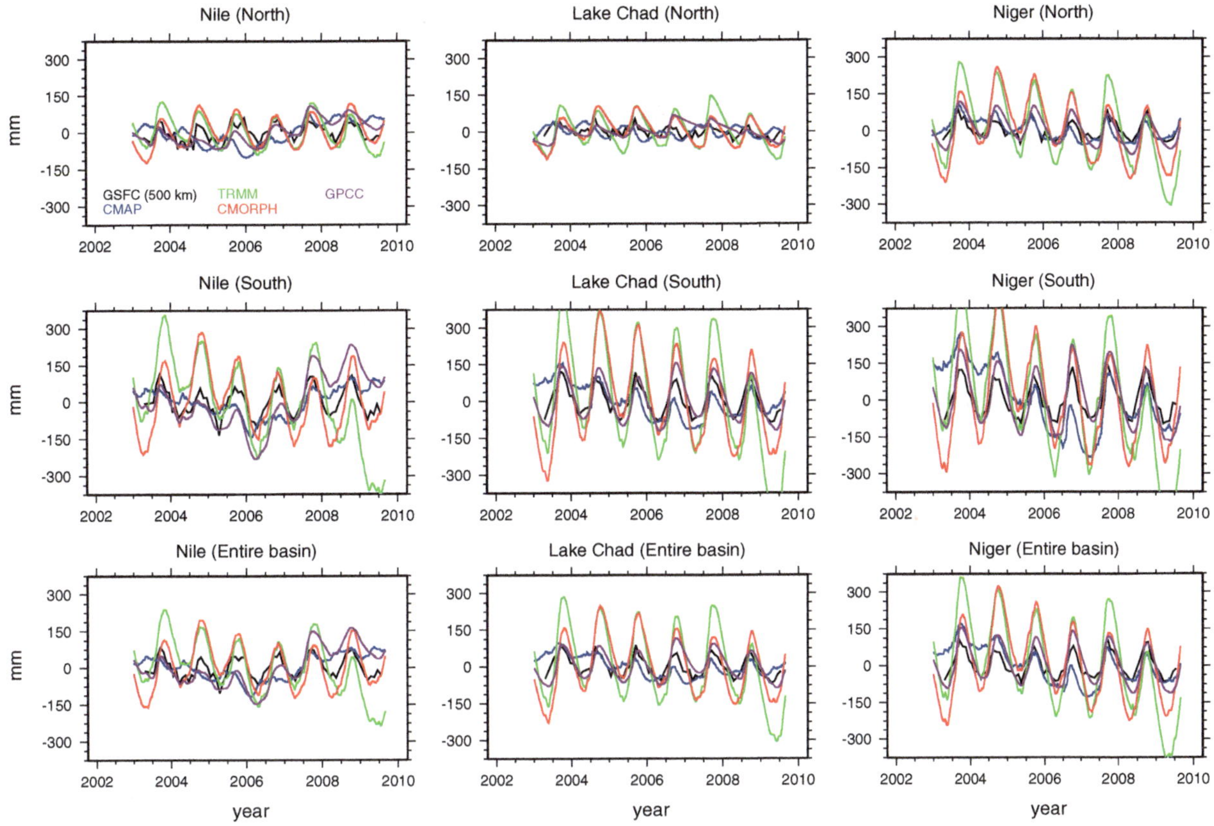

Figure 14
Comparison between GSFC solutions (500 km Gaussian smoothing applied) and detrended integrated precipitation, for Nile, Lake Chad and Niger basins

forced by CMAP datasets. However, we have shown that CMAP and TRMM precipitation do not significantly differ in Africa at least at seasonal timescales. Some other precipitation products, derived from space measurements, e.g. CMORPH, or ground rain gauges, e.g. GPCC monitoring product, show large differences with CMAP or TRMM datasets. The estimation of the sensitivity of hydrology models to the different precipitation products should be better quantified and could be validated by comparing them to GRACE observations at basin-scale.

We also compare GRACE retrieved equivalent water height of the three major lakes (Malawi, Tanganika and Victoria) in Africa to radar altimetry observations. Because of their small extent in the east-west direction, spherical harmonics are clearly not the optimal paramatrization for the retrieval of mass changes of Lakes Malawi and Tanganika. A localized approach, such as mascons (Rowlands *et al.*, 2010) should be more appropriate for monitoring lake and reservoir mass variations, although GRACE has not been specifically designed to determine surface water from space (Alsdorf *et al.*, 2007). For Lake Victoria, despite post-processing has a significant impact on water height retrieval, GRACE observations are in agreement with radar altimetry measurements. We have not been able to prove that the forward modelling of hydrology (using GLDAS/Noah model) improves the retrieval of lake mass variations. This conclusion might be different when using a mascon approach, as this technique significantly decreases leakage errors (Rowlands *et al.*, 2010).

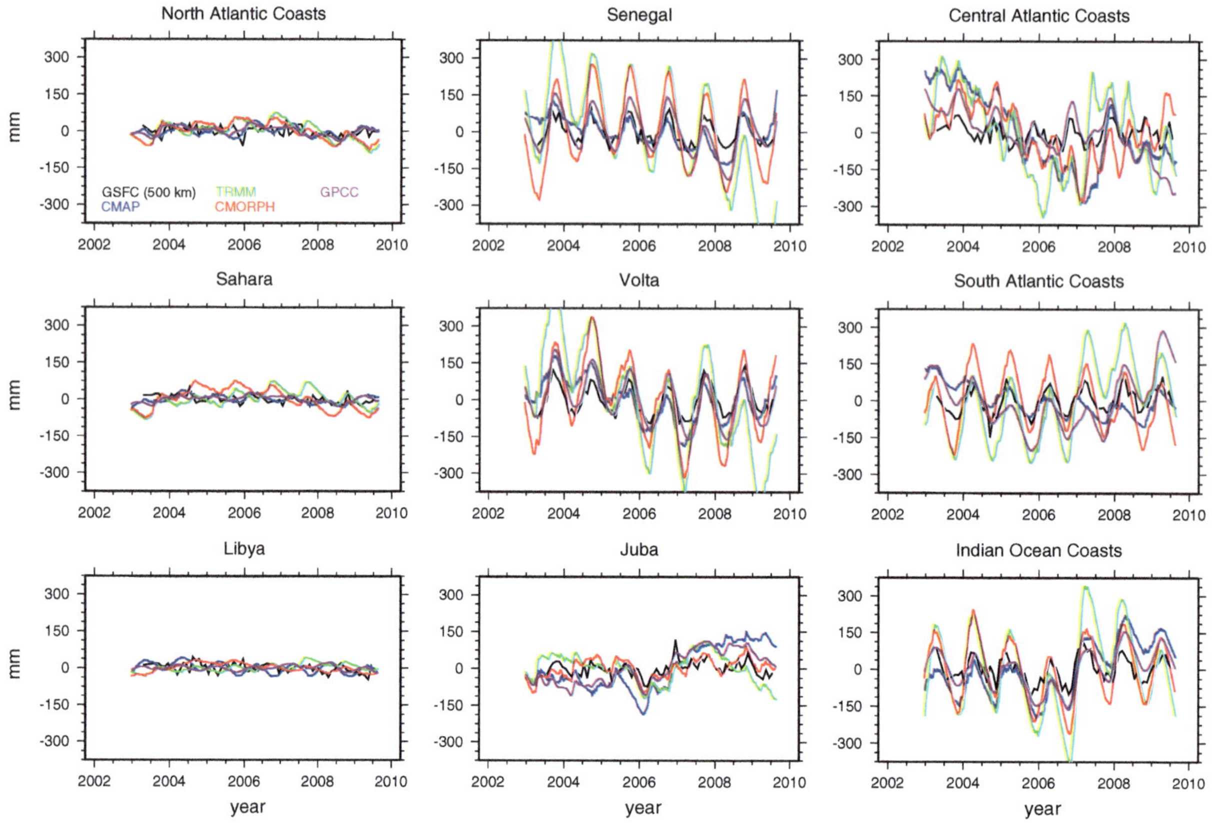

Figure 15
Same as Fig. 14, but for Congo, Zambezi, Okawango, Orange and Limpopo River basins and Madagascar island

Acknowledgments

Jean-Paul Boy was visiting NASA Goddard Space Flight Center, with a Marie Curie International Outgoing Fellowship (No. PIOF-GA-2008-221753). This project is partly funded by the French Agence Nationale de la Recherche (ANR) (GHYRAF project) and partly by a Centre National d'Etudes Spatiale (CNES) grant. The GLDAS data used in this study were acquired as part of the mission of NASA's Earth Science Division and archived and distributed by the Goddard Earth Sciences (GES) Data and Information Services Center (DISC). We thank B. F. Chao and an anonymous reviewer for their comments that helped improving the manuscript.

References

Alsdorf, D. E., Rodriguez, E. & Lettenmaier, D. P., 2007. *Measuring surface water from space*, Rev. Geophys., *45*, RG2002, doi:10.1029/2006RG000197.

Awange, J. L., Sharifi, M. A., Ogonda, G., Wickert, J., Grafarend, E. W. & Omulo, M. A., 2008. *The falling lake Victoria water level: GRACE, TRIMM and CHAMP satellite analysis of the lake basin*, Water Resour. Manage., *22*, 775–796.

Balsamo, G., Viterbo, P., Beljaars, A., van den Hurk, B., Hirschi, M., Betts, A.K. & Scipal, K., 2009. *A Revised Hydrology for the ECMWF Model: Verification from Field Site to Terrestrial Water Storage and Impact in the Integrated Forecast System*, J. Hydrometeor., *10*(3), 623–642.

Bonan, G.B., 1998. *The land surface climatology of the NCAR Land Surface Model coupled to the NCAR Community Climate Model*, J. Climate, *11*, 1307–1326.

Boy, J.-P. & Chao, B. F., 2005. *Precise evaluation of atmospheric loading effects on Earth's time-variable gravity field*, J. Geophys. Res., *110*, B08412, doi:10.1029/2002JB002333.

Bruinsma, S. Lemoine, J.-M., Biancale, R. & Valés, N., 2010. *CNES/GRGS 10-day gravity field models (release 2) and their evaluation*, Adv. Space Res., *45*, 587–601.

Carrère, C. & Lyard, F., 2003. *Modelling the barotropic response of the global ocean to atmospheric wind and pressure forcing—Comparisons with observations*, Geophys. Res. Lett., *30*(6), 1275, doi:10.1029/2002GL016473.

Chao, B. F., 2005. *On inversion for mass distribution from global (time-variable) gravity field*, J. Geodynamics, *39*, 223–230.

Chen, F., Mitchell, K., Schaake, J., Xue, Y., Pan, H., Koren, V., Duan, Y., Ek, M. & Betts, A., 1996. *Modeling of land-surface evaporation by four schemes and comparison with FIFE observations*, J. Geophys. Res., *101*(D3), 7251–7268.

Crowley, J. W., Mitrovica, J. X., Bailey, R. C., Tamisiea, M. E. & Davis, J. L., 2006. *Land water storage within the Congo Basin inferred from GRACE satellite gravity data*, Geophys. Res. Lett., *33*, L19402, doi:10.1029/2006GL027070.

Crowley, J. W., Mitrovica, J. X., Bailey, R. C., Tamisiea, M. E. & Davis, J. L., 2008, *Annual variations in water storage and precipitation in the Amazon Basin*, J. Geod., *82*, 9–13.

Flechtner, F., 2007. GFZ Level-2 Processing Standards Document For Level-2 Product Release 0004. *GRACE 327-743 (GR-GFZ-STD-001)*, Rev. 1.0.

Han, S.-C., H. Kim, I.-Y. Yeo, P. Yeh, T. Oki, K.-W. Seo, D. Alsdorf & S. B. Luthcke, 2009. *Dynamics of surface water storage in the Amazon inferred from measurements of inter-satellite distance change*, Geophys. Res. Lett., *36*, L09403, doi: 10.1029/2009GL037910.

Hinderer, J. *et al.*, 2009. *The GHYRAF (Gravity and Hydrology in Africa) experiment: description and first results*, J. Geodyn., *48*, 172–181.

Huffman, G. J., Adler, R. F., Bolvin, D. T., Gu, G., Nelkin, E. J., Bowman, K. P., Hong, Y., Stocker, E. F. & Wolff, D. B., 2007. *The TRMM Multi-satellite Precipitation Analysis: Quasi-Global, Multi-Year, Combined-Sensor Precipitation Estimates at Fine Scale*, J. Hydrometeor., *8*(1), 38–55.

Jenness, J., Dooley, J., Aguilar-Manjarrez, J. & Riva, C., 2008. African Water Resource Database. GIS-based tools for inland aquatic resource management. 2., *Technical manual and workbook*, CIFA Technical Paper, 33, Part 2. Rome, 308 p.

Joyce, R. J., J. E. Janowiak, P. A. Arkin, & P. Xie, 2004. *CMORPH: A method that produces global precipitation estimates from passive microwave and infrared data at high spatial and temporal resolution*, J. Hydromet., *5*, 487–503.

Kalnay, E., *et al.*, 1996. *The NCEP/NCAR 40-year reanalysis project*, Bull. Amer. Meteor. Soc., *77*, 437–470.

Koster, R. D. & Suarez, M. J., 1996. *Energy and Water Balance Calculations in the MOSAIC LSM*, NASA Technical Memorandum, *104606*, 9, 76 pp.

Lemoine, J.-M., Bruinsma, S., Loyer, S., Biancale, S., Marty, J.-C., Perosanz, F. & Balmino, G., 2007. *Temporal gravity field models inferred from GRACE data*, Adv. Space Res., *39*, 1620–1629.

Liang, X., Lettenmaier, D. P., Wood, E. F. & Burges, S. J. , 1994. *A Simple hydrologically Based Model of Land Surface Water and Energy Fluxes for GSMs*, J. Geophys. Res., *99*(D7), 14,415–14,428.

Luthcke, S. B., Rowlands, D. D., Lemoine, F. G.,Klosko, S. M., Chinn, D. S. & McCarthy, J. J., 2006. *Monthly spherical harmonic gravity field solutions determined from GRACE inter-satellite range-rate data alone*, Geophys. Res. Lett., *33*, L02402, doi:10.1029/2005GL024846.

Lyard, F., Lefevre, F., Letellier, T. & Francis, O., 2006. *Modelling the global ocean tides: insights from FES2004*, Ocean Dynamics, *56*, 394–415.

Rabier, F., Järvinen, H., Klinker, E., Mahfouf, J.-F. & Simmons, A., 2000. *The ECMWF operational implementation of four-dimensional variational assimilation. I: Experimental results with simplified physics*, Q. J. R. Met. Soc., *126*, 1143–1170.

Ramillien, G., Frappart, F., Cazenave, A. & Güntner, B., 2005. *Time variations of land water storage from an inversion of 2 years of GRACE geoids*, Earth Planet. Sci. Lett., *235*, 283–301.

Ramillien, G. & Boronina, A., 2008. *Application of AVHRR imagery and GRACE measurements for calculation of actual evapotranspiration over the Quaternary aquifer (Lake Chad basin) and validation of groundwater models*, J. Hydrol., *348*, 98–109.

Ray, R. D., 1999. A global ocean tide model from Topex/Poseidon altimetry: GOT99.2, *NASA Tech. Memo.* 209478, Goddard Space Flight Center, 58 pp.

Ray, R. D. & Luthcke, S. B., 2006. *Tide model errors and GRACE gravimetry: Towards a more realistic assessment*, Geophys. J. Int., *167*, 1055–1059.

Rodell, M., Houser, P. R., Jambor, J., Gottschalck, J., Mitchell, K., Meng, C.-J., Arsenault, K., Cosgrove, B., Radakovich, J., Bosilovich, M., Entin, J. K., Walker, J. P., Lohmann, D. & Toll, D., 2004. *The Global Land Data Assimilation System*, Bull. Amer. Meteor. Soc., *85*(3), 381–394.

Rowlands, D. D., Luthcke, S. B., McCarthy, J. J., Klosko, S. M., Chinn, D. S., Lemoine, F. G., Boy, J.-P. & Sabaka, T. J., 2010. *Global mass flux solutions from GRACE: A comparison of parameter estimation strategies: Mass concentrations versus Stokes coefficients*, J. Geophys. Res., *115*, B01403, doi: 10.1029/2009JB006546.

Rudolf, B., & U. Schneider, 2005. Calculation of Gridded Precipitation Data for the Global Land-Surface using in-situ Gauge Observations, *Proceedings of the 2nd Workshop of the International Precipitation Working Group IPWG*, Monterey October 2004, EUMETSAT, ISBN 92-9110-070-6, ISSN 1727-432X, 231–247.

Schmidt, M., Seitz, F. & Shum, C. K., 2008. *Regional four-dimensional hydrological mass variations from GRACE, atmospheric flux convergence, and river gauge data*, J. Geophys. Res., *113*, B10402, doi:10.1029/2008JB005575.

Simmons, A., Uppala, C., Dee, D. & Kobayashi, S., 2007. *ERA-Interim: New ECMWF reanalysis products from 1989 onwards*, ECMWF Newsletter, *110*, 25–35.

Strassberg, G., Scanlon, B. R. & Rodell, M., 2007. *Comparison of seasonal terrestrial water storage variations from GRACE with groundwater-level measurements from the High Plains Aquifer (USA)*, Geophys. Res. Lett., *34*, L14402, doi: 10.1029/2007GL030139.

Syed, T. H., Famiglietti, J. S., Chen, J. , Rodell, M., Seneviratne, S. I., Viterbo, P. & Wilson, C. R., 2005. *Total basin discharge for the Amazon and Mississippi River basins from GRACE and a land-atmosphere water balance*, Geophys. Res. Lett., *32*, L24404, doi:10.1029/2005GL024851.

Tapley, B. D., Bettadpur, S., Watkins, M. & Reigber, C., 2004. *The gravity recovery and climate experiment: Mission overview and early results*, Geophys. Res. Lett., *31*, L09607, doi:10.1029/2004GL019920.

Thomas, M., 2002. Ocean induced variations of Earth's rotation - Results from a simultaneous model of global circulation and tides, *PhD dissertation*, University of Hamburg, Germany, 129pp.

Uppala, S., Simmons, A., Dee, D. & Kobayashi, S., 2007. The third generation ECMWF reanalysis ERA-Interim, *EMS7/ECAM8 Abstracts*, 4, EMS2007-A-00167, 2007.

Winsemius, H. C., Savenije, H. H. G., van de Giesen, N. C., van den Hurk, B. J. J. M., Zapreeva, E. A. & Klees, R., 2006. *Assessment of Gravity Recovery and Climate Experiment (GRACE) temporal signature over the upper Zambezi*, Water Resour. Res., *42*, W12201, doi:10.1029/2006WR005192.

Xie, P., Janowiak, J. E., Arkin, P. A., Adler, R., Gruber, A., Ferraro, R., Huffman, G. J. & Curtis, S., 2003. *GPCP Pentad precipitation analysis: an experimental dataset based on gauge observations and satellite estimates*, J. Clim., *16*, 2197–2214.

(Published online October 13, 2011)

Pure Appl. Geophys. 169 (2012), 1391–1410

DOI 10.1007/s00024-011-0417-9

Pure and Applied Geophysics

Land Water Storage Changes from Ground and Space Geodesy: First Results from the GHYRAF (Gravity and Hydrology in Africa) Experiment

J. Hinderer,[1] J. Pfeffer,[1] M. Boucher,[2,6] S. Nahmani,[4] C. De Linage,[3] J.-P. Boy,[1,5] P. Genthon,[2] L. Seguis,[2] G. Favreau,[2] O. Bock,[4] M. Descloitres,[6] and GHYRAF team

Abstract—This paper is devoted to the first results from the GHYRAF (Gravity and Hydrology in Africa) experiment conducted since 2008 in West Africa and is aimed at investigating the changes in water storage in different regions sampling a strong rainfall gradient from the Sahara to the monsoon zone. The analysis of GPS vertical displacement in Niamey (Niger) and Djougou (Benin) shows that there is a clear annual signature of the hydrological load in agreement with global hydrology models like GLDAS. The comparison of GRACE solutions in West Africa, and more specifically in the Niger and Lake Chad basins, reveals a good agreement for the large scale annual water storage changes between global hydrology models and space gravity observations. Ground gravity observations done with an FG5 absolute gravimeter also show signals which can be well related to measured changes in soil and ground water. We present the first results for two sites in the Sahelian band (Wankama and Diffa in Niger) and one (Djougou in Benin) in the Sudanian monsoon region related to the recharge–discharge processes due to the monsoonal event in summer 2008 and the following dry season. It is confirmed that ground gravimetry is a useful tool to constrain local water storage changes when associated to hydrological and subsurface geophysical in situ measurements.

Key words: African monsoon, Sahel, water storage, gravimetry, GPS, MRS, GRACE.

Members of the GHYRAF team are given in the Appendix.

[1] Institut de Physique du Globe de Strasbourg, UMR 7516 CNRS, Université de Strasbourg, 67084 Strasbourg, France. E-mail: Jacques.Hinderer@eost.u-strasbg.fr

[2] Hydrosciences Montpellier, UMR 5569 CNRS, IRD, Université Montpellier 2, 34095 Montpellier, France.

[3] Department of Earth System Science, University of California, Irvine, Croul Hall, Irvine, CA 92697-3100, USA.

[4] LAREG/IGN, 77455 Marne la Vallée, France.

[5] Planetary Geodynamics Laboratory, code 698, NASA Goddard Space Flight Center, Greenbelt, MD 20771, USA.

[6] Laboratoire d'étude des Transferts en Hydrologie et Environnement, UMR 5564 CNRS, INPG, IRD, Université Joseph Fourier, 38 041 Grenoble, France.

1. Introduction

The GHYRAF (Gravity and Hydrology in Africa) project is a multi-disciplinary project that aims to better understand the water cycle in West Africa with the help of geodetic (GPS), gravity (surface and satellite-derived), geophysics (MRS) and hydrology experiments. These combined experiments are done on four specific sites on a north–south climatic gradient (see Fig. 1). The first one is located in the Sahara desert zone (Tamanrasset, Algeria) with almost no rain (less than 20 mm/year); two of them sample the Sahelian band (Wankama/Niamey and Bagara/Diffa in Niger) (indicated in orange colour on Fig. 1) with moderate annual rainfall (respectively, 560 and 350 mm/year) and the last one is in the sub-humid area of the monsoon zone (Nalohou/Djougou, Benin) with heavy rainfall (1,200 mm/year). Two of the sites (Wankama in the Niamey Square Degree mesoscale site and Nalohou close to Djougou in the upper Ouémé mesoscale site) belong to the AMMA-CATCH observation system (African Multidisciplinary Monsoon Analysis—Coupling the tropical atmosphere to the hydrological cycle) (http://www.amma-catch.org/) dedicated to evaluate the impact of anthropogenic and climatic changes on the surface water cycle (Lebel *et al.*, 2009). Table 1 summarizes the geographical coordinates of these four stations.

The project aims to set up new constraints to the problem of the soil and ground water storage changes during the monsoon cycle in Africa since our ground measurements as well as satellite ones are sensitive to the total variation of stored water (changes in the total water column of the ground, i.e. surface water, water of the unsaturated zone and groundwater). For a more detailed description of the GHYRAF project

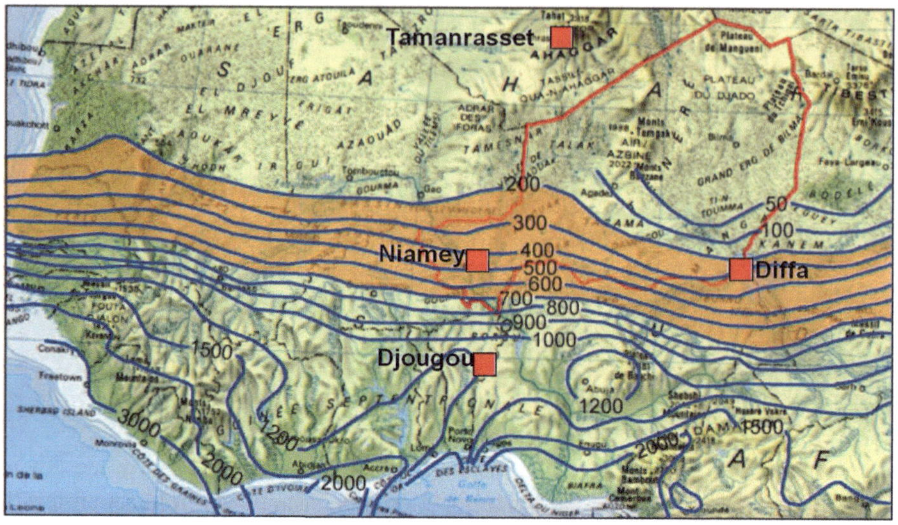

Figure 1
Location of the four principal sites investigated during the GHYRAF project. The *blue lines* are the isohyets (rainfall in mm/year, period 1950–1989 from L'Hôte and Mahé 1996) showing the strong north–south climatic gradient

characteristics we refer the reader to Hinderer *et al.* (2009).

The project started in mid-2008 and we present here the first results mainly devoted to the hydrological effects due to the 2008 monsoon period. Section 2.1 deals with the GPS observations and their ability to estimate the seasonal surface loading contribution (mainly hydrological) in West Africa. In Sect. 2.2 we show the first estimates of the water storage changes over the Niger and Lake Chad basins derived from GRACE (Gravity Recovery and Climate Experiment) satellite data and their relation to global hydrology model predictions. Section 2.3 focuses on ground gravimetry and we report on our results on each of the three sites where measurements are available. Concluding remarks are given in Sect. 3.

2. *First Results from the GHYRAF Project*

2.1. *GPS Vertical Motion*

One goal of the GHYRAF project is to measure using the GPS regional permanent network the vertical displacement due to the hydrological load caused by the monsoon. Indeed it has been shown that the precision of the current GPS processing

Table 1

Location of the four specific stations investigated in the GHYRAF project and mean annual rainfall (in mm)

Location	Country	Longitude	Latitude	Mean rainfall (mm)
Tamanrasset	Algeria	5.52 E	22.78 N	20
Diffa	Niger	12.62 E	13.32 N	350
Niamey	Niger	2.17 E	13.48 N	560
Djougou	Benin	2.62 E	9.35 N	1,200

allows the detection of the crustal deformation induced by surface loads (atmosphere, ocean and continental hydrology) (van Dam *et al.*, 2001). Figure 2 shows that the modeled vertical displacement reaches several mm according to the global hydrology model GLDAS (Rodell *et al.*, 2004).

GLDAS is a modeling platform that provides estimates of land surface fluxes and storage of water and energy. We use the storage predictions from the Noah Land Surface Model (Chen *et al.* 1996), consisting of superficial soil moisture (four layers down to 2 m depth), snow and canopy water (essential to account for evapotranspiration) at a 0.25° × 0.25° spatial resolution and a 3 h time sampling. We use two distinct simulations: the first one is forced by CMAP (CPC Merged Analysis of Precipitation) (Xie *et al.* 2003) and is available at

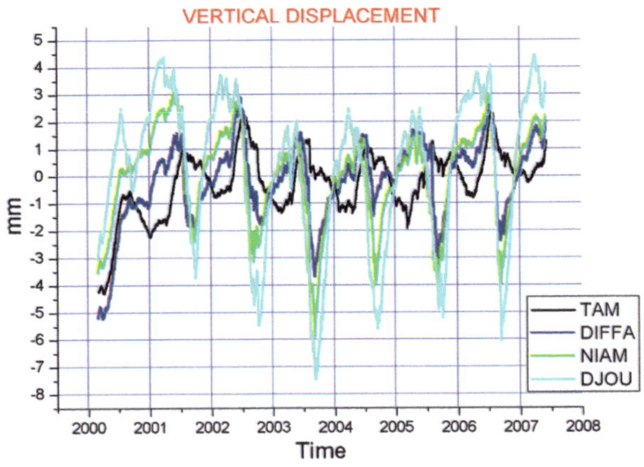

Figure 2
Predicted vertical displacement (in mm) at the four GHYRAF sites during the 2000–2008 period from the GLDAS global hydrology model

http://disc.sci.gsfc.nasa.goc/hydrology/data-holdings, and the second one is forced by TRMM (Tropical Rainfall Measuring Mission) (HUFMANN *et al.* 2007) precipitation data sets.

Therefore, accurate GPS station height solutions are required and it is mandatory to process the data using a global network to obtain coordinates within a well defined reference frame. Thus, data from the AMMA-GHYRAF GPS stations (BOCK *et al.*, 2008) were processed over period 2005.5 to 2009.0 using the GAMIT software (HERRING *et al.*, 2008) as part of the TIGA GPS analysis at the University of La Rochelle (ULR) TIGA Analysis Center (TAC) (Santamaría-Gómez *et al.*, in press). In this analysis, station coordinates, orbital parameters, Earth orientation parameters, 2-hourly zenith total delays, and atmospheric gradients (1/day for N/S and E/W) were estimated. The VMF1 mapping function and a priori hydrostatic delays derived from the ECMWF numerical weather model are used (BOEHM *et al.*, 2006). GPS height time series (shown in Fig. 3) were obtained in the ITRF2005 (ALTAMIMI *et al.*, 2007) by stacking the weekly position estimates (Santamaría-Gómez *et al.*, 2009).

Figure 3 shows the GPS vertical displacements for two stations (Djougou and Niamey) where we superimposed the sum of hydrologic and atmospheric load effects computed from GLDAS water content and NCEP surface pressure data (PETROV and BOY, 2004). Despite the stronger variability in the GPS solutions, it is very obvious that there is an annual modulation of about 15–20 mm peak-to-peak. This modulation is in close agreement with the models and is well anti-correlated with rainfall at both stations. In fact, in a monsoon context, small and large scales are well correlated, so local rainfall is likely to be

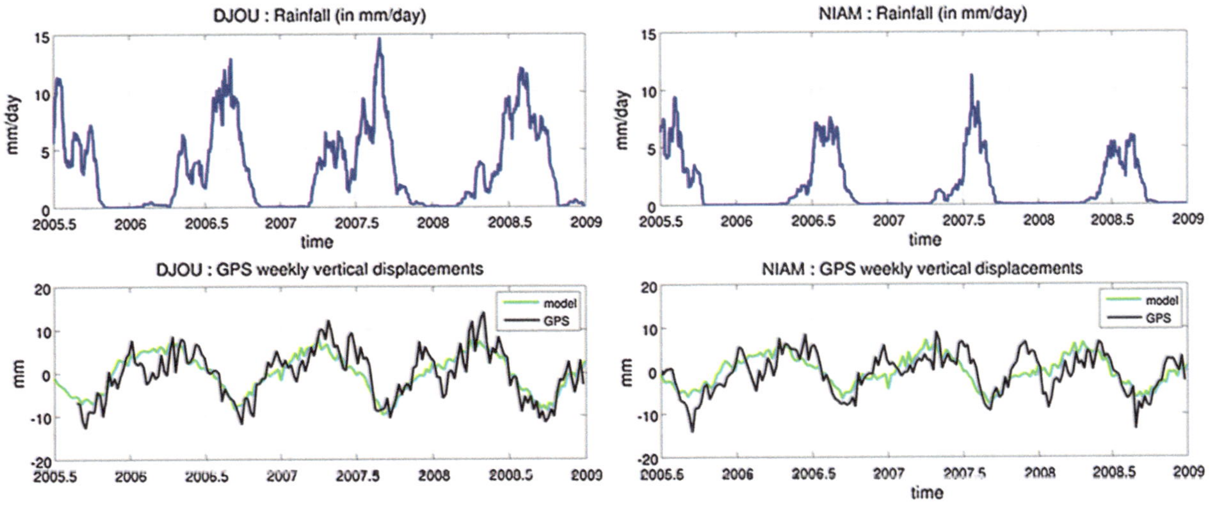

Figure 3
The rainfall (*top*) (in mm/day) and vertical displacements (*bottom*) (in mm, smoothed by 15-day running average filter) for the stations of Djougou (*left*) and Niamey (*right*) during the 2005.5–2009 period. Vertical displacements are from the predicted load (atmosphere + hydrology) (in *green*) and the observed load (in *black*) from the AMMA-GHYRAF permanent GPS network

representative of what happens at larger scales and, therefore, GPS can detect such a hydrological loading signal at very large-scale (thousands of km) generated by water infiltration and storage after rain.

The correlation coefficients between GPS and GLDAS are as large as 65 and 80% for Niamey and Djougou stations respectively. The differences could be either due to limitations in the models and errors in the GPS solutions. Among the latter, some have been evidenced in studies for other parts of the world: unmodeled sub-daily signals (King *et al.*, 2008), and systematic errors due to the combination procedure (Collilieux *et al.*, 2007), non-tidal ocean mass loading, bedrock thermal expansion, mis-modeling of receiver antenna phase center variations (see Dong *et al.*, 2002 for more details). Further work is ongoing on the analysis of these error sources.

Another interest of observing by GPS the vertical displacement at our specific stations is that we will be able to correct our ground gravity observations for this effect (free air component). Such a correction is indeed needed in any ground/satellite comparison since space gravity observations are not sensitive to this geometrical effect (Hinderer *et al.*, 2006a, b; de Linage *et al.*, 2007; Neumeyer *et al.*, 2006). An alternative of using GPS estimates to correct the gravity data would be to use a deformation model applied to the GRACE solutions of total water storage and convert the recovered vertical deformations into gravity via the free-air gradient of -0.3086 μGal/mm. Such a method would require a precision level of at least 3 mm (1/0.3086) for the GRACE-recovered deformation in order to reach the precision level of the superconducting gravimeter (SG) gravity measurements (better than 1 μGal).

2.2. *Space Gravimetry (GRACE)*

Since its launch in April 2002, the GRACE satellites allow the recovery of Earth's time variable gravity field with temporal and spatial resolutions of, respectively, 10–30 days and larger than 400 km (Tapley *et al.*, 2004). Mass variations at the Earth's surface can be uniquely retrieved from Stokes coefficients (Chao, 2005).

Figure 4 shows the annual amplitude and phase of equivalent water height variations in Africa, according to the CNES/GRGS Release 2 10-day constrained solution (Bruinsma *et al.*, 2010), monthly unconstrained GSFC (Luthcke *et al.*, 2006), CSR and GFZ (Flechtner, 2007) solutions. A 500 km half-width Gaussian filter has been applied to the monthly unconstrained solutions; such a post-processing is not required for the CNES/GRGS constrained solutions.

All solutions show similar features of seasonal variations in the equatorial band, with amplitudes reaching 200 mm of equivalent water height; the northern and southern hemispheres are out of phase. The N–S phase dichotomy in Africa is due to the seasonal oscillation of the Intertropical Convergence Zone across the equator. Because of the filtering, the unconstrained solutions exhibit slightly smaller amplitudes than CNES/GRGS.

The noise level in GRACE recovered continental water storage variations is estimated to a few centimeters by looking at the seasonal variations in Sahara and in the oceans, where seasonal amplitudes should be small.

In addition, GRACE solutions can be used to estimate water storage at river basin scale. Figure 5 shows the location of major river basins in Africa (derived from Jenness *et al.*, 2008), and water storage variations in Lake Chad and Niger basins from GRACE and global hydrology models. The same treatment (spherical harmonic decomposition till degree $n = 60$ and 500 km half-width Gaussian filter) is applied to the different GRACE solutions and hydrology models.

Despite the different processing schemes, both solutions (CNES/GRGS and CSR) show similar temporal variations in Lake Chad and Niger river basins, with seasonal amplitude reaching about up to 150 mm of equivalent water height. GRACE observations are in agreement with GLDAS/Noah (Rodell *et al.*, 2004) global hydrology model. ECMWF hydrology model (Viterbo and Beljaars, 1995) shows smaller variations than GRACE data and GLDAS model.

The Lake Chad (2.5 Mkm^2) and Niger (2.3 Mkm^2) basins are among the largest African basins as shown in Fig. 5. Because of the large scale of these basins they cannot be covered by conventional hydrological observations. Thus, satellite gravity data provide a unique opportunity to constrain the

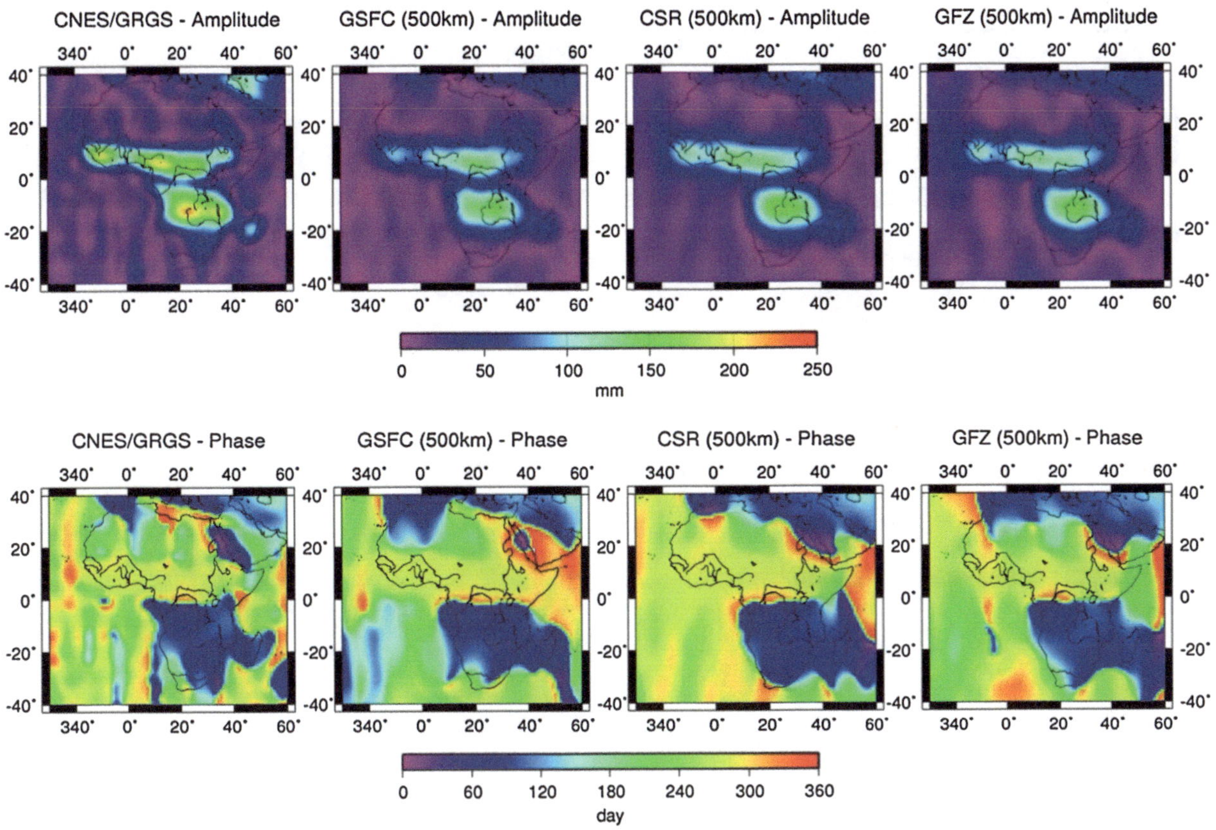

Figure 4
Amplitude (in mm) and phase (in days from 1 January) of the annual component in Africa inferred from GRACE observations according to four different processing centers (CNES/GRGS, GSFC, CSR, GFZ)

large scale water storage changes of these basins. But this is only true when averaging over large basins. By adding surface gravity measurements like in this study and relating them to water storage changes, we aim at having a better insight into hydrology models at finer scales.

As precipitation is the main forcing of soil-moisture variations, we can directly compare precipitation datasets to GRACE observations (see, for example, Crowley *et al.*, 2006). The conservation of mass provides the relation between continental water storage variations W, precipitation P, evapotranspiration E and runoff R (or more precisely the outflow in the case of Sahelian endorheic basins):

$$\frac{\partial W}{\partial t} = P - E - R \qquad (1)$$

An endorheic basin is a closed basin that does not allow any outflow to other bodies of water such as rivers or oceans. This applies to the Diffa site (outflow ends up in nearby Lake Chad) and the Wankama site (outflow/runoff end up in ponds with no outflow to the Niger River).

P is the main driver of soil moisture variations. If we assume that the sum $E + R$ is small or constant in time (e.g. E and R are periodical but out-of-phase), the time-integration of P can be directly compared to W (as observed by GRACE) after detrending. The validity of this hypothesis was shown, for instance, by Crowley *et al.* (2008) for the Amazon river basin and is demonstrated here at large wavelength scales by Fig. 5, for the Lake Chad and Niger river basins (it might not be the case at smaller spatial scales). The seasonal water storage changes deduced from CMAP dataset (also used to force the GLDAS/Noah model) are indeed the largest in amplitude but are in fair agreement with GRACE observations.

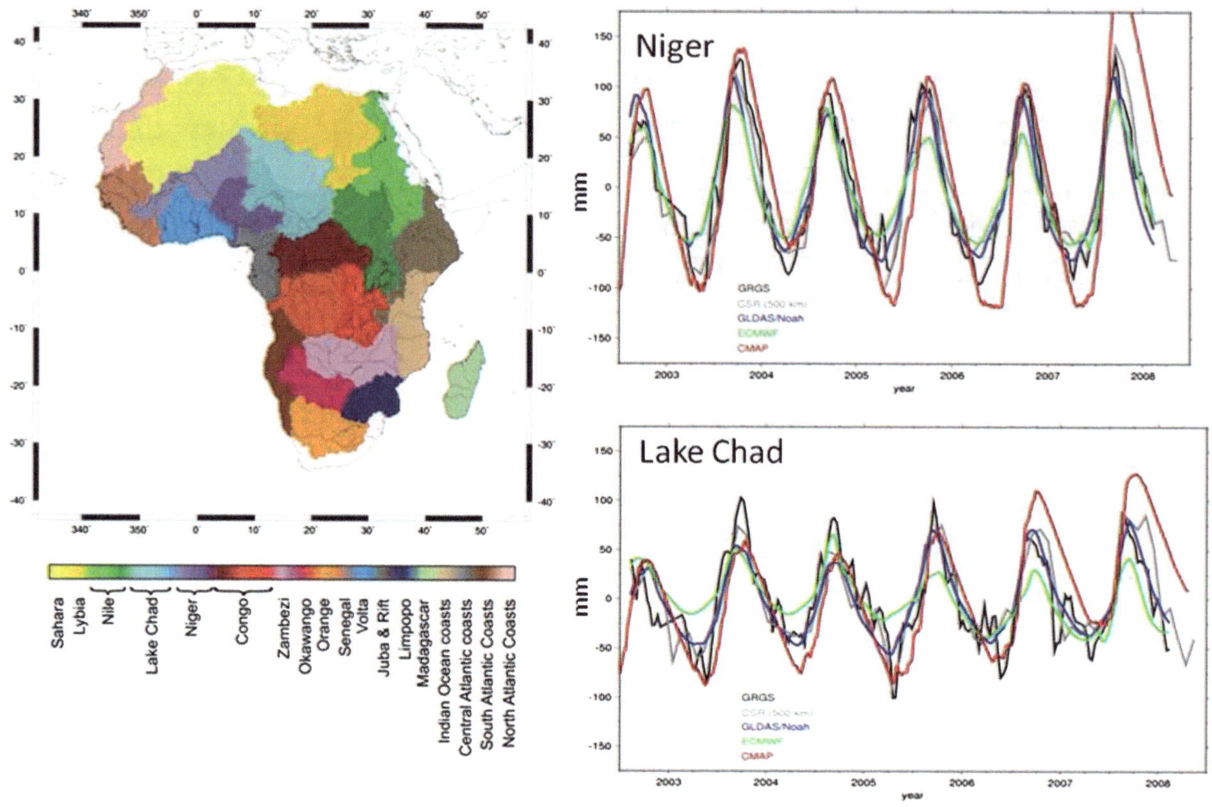

Figure 5
Location of the major hydrological basins in Africa (*left*) and estimates of the water storage changes (in mm equivalent water height) for the Niger (*right top*) and Lake Chad basins (*right bottom*) from GRACE data (CNES/GRGS and CSR solutions are respectively shown in *black and gray*), hydrology model predictions according to GLDAS/Noah (*blue*) (Rodell *et al.*, 2004) and ECMWF operational (*green*) (Viterbo and Beljaars, 1995). In *red* is shown the detrended time-integration of CMAP dataset (Xie *et al.*, 2003)

In general, E is not small compared to P depending on the integration period; at the annual time scale, for instance, E represents 80% of P in the Sudanian zone and 90% in the Sahelian zone. In contrast to equatorial areas, the runoff contribution R is likely to be negligible in arid (Sahelian) regions.

In the equatorial monsoon zone the rain is larger than the potential evapotranspiration all year long. The real evapotranspiration is, hence, equal to the potential one and is independent on rain. Hence, the cumulative rain changes are indeed directly comparable to water storage changes.

In arid (Sahelian) regions the rain is smaller than the potential evapotranspiration meaning that the real evapotranspiration is a function of the rain; as a consequence, the comparison between rain and water storage is more complicated.

A detailed comparison of GRACE spherical harmonic solutions, global hydrology models and precipitation datasets for the major African basins is given in Boy *et al.* (2011, this issue).

2.3. *Ground Gravimetry*

Surface gravimetric measurements can be used to investigate the underground water storage dynamics (Kroner *et al.*, 2007). This can be done by repeating absolute gravimeter measurements at specific points (Jacob *et al.*, 2008) or by performing relative gravimetry campaigns (Naujoks *et al.*, 2008). There are also a number of studies investigating various hydrological influences on fixed gravimeters such as the superconducting gravimeters of the GGP network (see e.g. Kroner and Jahr,

2006; Hasan *et al.*, 2006; Boy and Hinderer, 2006; Hinderer *et al.*, 2007).

2.3.1 Length Scales in Hydrology

The gravity changes (in μGal) due to the soil moisture content (GLDAS model) for the time span 2000–2008 are computed for the stations of Table 1 and shown on Fig. 6. One notices that, as expected from the climatic gradient present in West Africa (see Fig. 1), there is almost no gravity change in Tamanrasset (<1 μGal) and that, as expected, the seasonal gravity changes increase as a function of the input rainfall reaching >15 μGal peak to peak in the monsoon region (Djougou).

Since the first survey in Sahara near Tamanrasset (Algeria) was postponed due to security reasons, we have no result to report here on the so-called 'null test' where observations are done in a region where almost no gravity changes are predicted due to the lack of water storage changes. Hence, we report here only preliminary results coming from the repetition of absolute gravity measurements on the station Wankama (Niamey), Bagara (Diffa) and Nalohou (Djougou). The observed gravity changes related to the 2008 monsoonal recharge and discharge are depicted in Fig. 7 where we have also plotted the large scale predictions of the GLDAS/Noah model

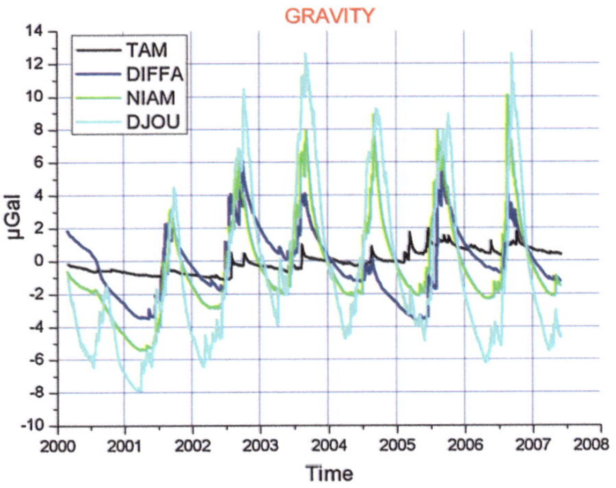

Figure 6
Predicted surface gravity changes (in μGal) at the four GHYRAF sites during the 2000–2008 period from the GLDAS global hydrology model

forced by two different meteorological forcing fields: CMAP and TRMM.

The question of the length scales is always important in hydrology. In Fig. 8 (top) we have computed for the GLDAS model the predictions of ground gravity changes for Niamey (13°N, 2°E). We have separated the total contribution of the GLDAS model into a local one (which is only the Newtonian attraction effect of the soil moisture present in the local pixel (simply by multiplying the local value of the soil moisture by the coefficient 42 μGal/m of equivalent water height) and a non-local one caused by the combination of the Newtonian attraction of remote masses outside the local pixel and elasto-gravitational loading effect (effect of vertical displacement + mass redistribution inside the Earth) (see de Linage *et al.* 2007, 2009). In fact, the local pixel is defined by the integration step in the loading computation (here 50 m), but the value of the local soil water content is identical to the model value for the pixel including the station (0.25° × 0.25°). Therefore, we neglect the contribution of the masses located at distances between 50 m and 25 km.

One can easily see from Fig. 8 (top) that the local contribution is by far the largest, the non-local term being only about 20% in amplitude (this latter term being dominated by the effect of vertical displacement). However, if we neglect this term we would overestimate the local gravity changes in terms of local soil moisture.

Figure 8 (bottom) shows that the non-local gravity changes can be derived either from the hydrology model (GLDAS) or from the GRACE observations, even if these seem to be slightly larger in Niamey.

We applied this separation scheme between local and non-local contributions to GRACE-GRGS total water content estimates (Bruinsma *et al.*, 2010) converted into ground gravity changes. In this case, the local effect corresponds to a 400 km × 400 km cell (GRACE effective resolution). The difference in the size of the so-called local cell between GRACE (say 400 km) and the GLDAS hydrology model (50 m for the integration cell and 25 km for the uniform value of the soil water content) may explain the different amplitudes shown in Fig. 8 but this needs to be further investigated. However, it is well known that regional masses (a few tens of km) have a

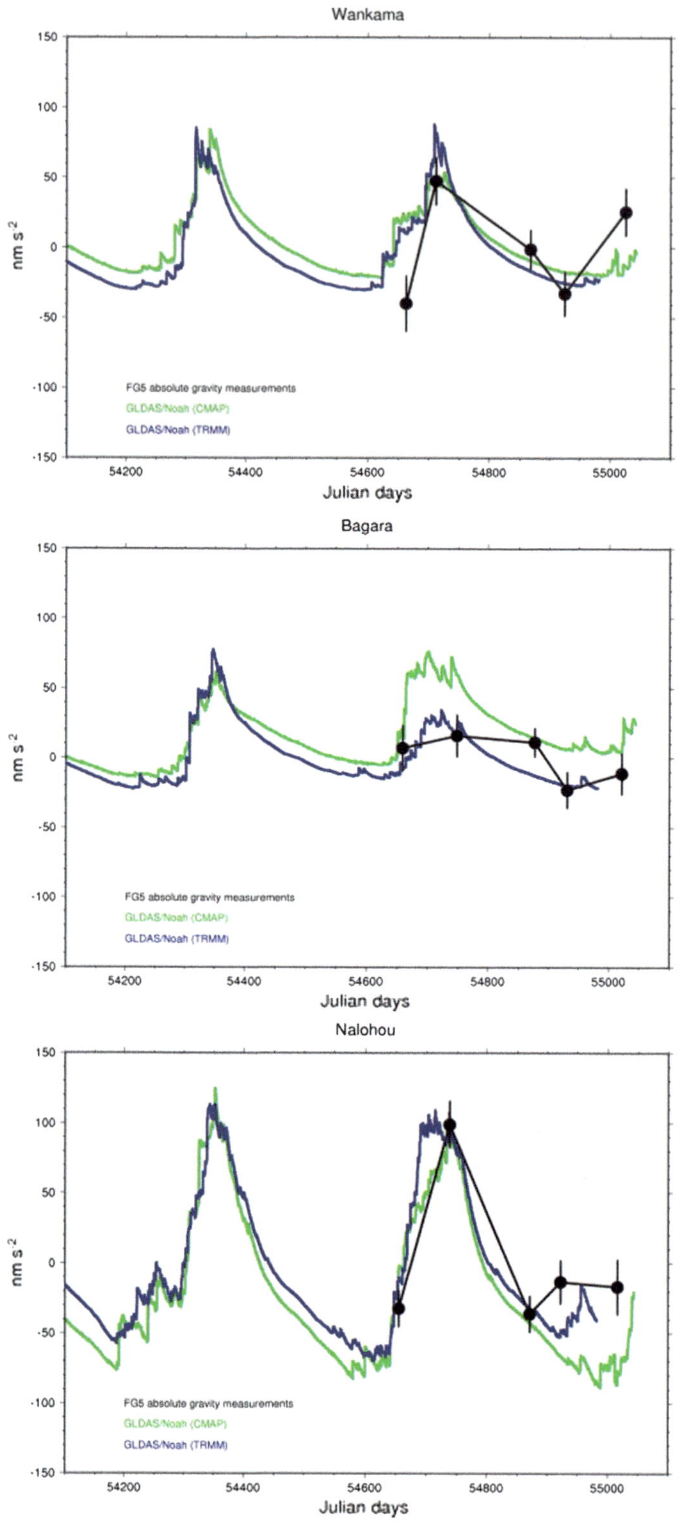

Figure 7

Predicted (GLDAS/Noah hydrology model forced by CMAP in *green* and GLDAS/Noah forced by TRMM in *blue*) and observed (FG5 measurements in *black*) gravity changes at Wankama (*top*), Bagara (*center*) and Nalohou (*bottom*) sites related to the 2008 monsoonal rainfall

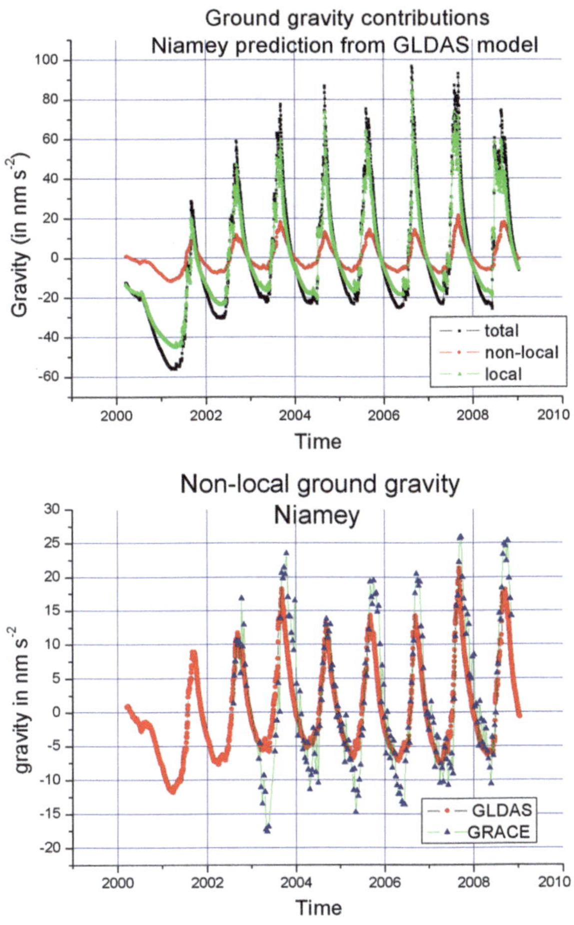

Figure 8
Top ground gravity prediction (in nm s^{-2}) from GLDAS model for Niamey (the local contribution is in *green*, the non-local one in *red* and the total in *black*). *Bottom* comparison of the non-local contribution predicted by GLDAS (*red circles*) and inferred from GRACE data (*blue triangles*)

small effect on local gravity measurements (Llubes *et al.* 2004) because the associated loading is small and because the attraction is merely horizontal and hence does not change gravity.

2.3.2 *Water Storage Changes in Wankama (Southwest Niger)*

The Wankama site is 70 km from Niamey (Niger) and belongs to the 1° × 1° Sahelian mesoscale site of AMMA-CATCH observatory (Cappelaere *et al.*, 2009). This site benefits from long-term observations that have started in the early 1990s with HAPEX–Sahel experiments (Goutorbe *et al.*, 1997). Various types of hydrologic variables are monitored: automatic rain gauges record rainfall every 5 min, the pond water level is measured every 20 min with a pressure sensor and every 5 min with a float, soil moisture capacity probes buried from 0.1 to 2.5 m below the surface measure every min the soil volumetric water content; the piezometric level is measured every 20 min with pressure sensors, and the runoff on small gullies every 30 s; in addition there are also regular measurements of the vegetation cycle, heat and water fluxes between atmosphere and soil.

These data are used for ecohydrologic modeling (Boulain *et al.*, 2009) and for estimating the water balance components (Ramier *et al.*, 2009).

The local variations of surface water (pond), shallow soil moisture and deeper aquifer storage generate a change of gravity due to the direct attraction effect. Time variations of gravity associated with water storage from May 2008 to May 2009 were modeled at FG5 point using observations of pond level and water table level (between 10 and 20 m below the soil surface as a function of the season/time period and the distance from the Wankama pond), the changes of soil moisture being assumed weak at distance from the pond due to soil crusting and Hortonian runoff (overland flow occurring when the rainfall rate is larger than the infiltration rate of the non-saturated surface layer).

This assumption is reasonable because no rainfall events occur before FG5 measurements dates and the FG5 point is located in a bare soil zone.

The water table shape is constrained by water level observations at four piezometers (Fig. 9). The water level is assumed linear between piezometers, symmetric with respect to the pond axis, and constant in the pond axis direction. The volume of the pond is calculated using its water level and a stage-area relationship derived from topography. More information on seasonal changes in the water table near the pond can be found in Favreau *et al.* (2009). In the model, the geometry of the pond was simplified by a rectangular shape which results in negligible effect on gravity. As the topography of the site is rather flat, the contribution of pond level variations on gravity change is under 3 nm s^{-2} at FG5 point, which is one order of magnitude beneath the instrumental detection limit. The specific yield of the aquifer defined as

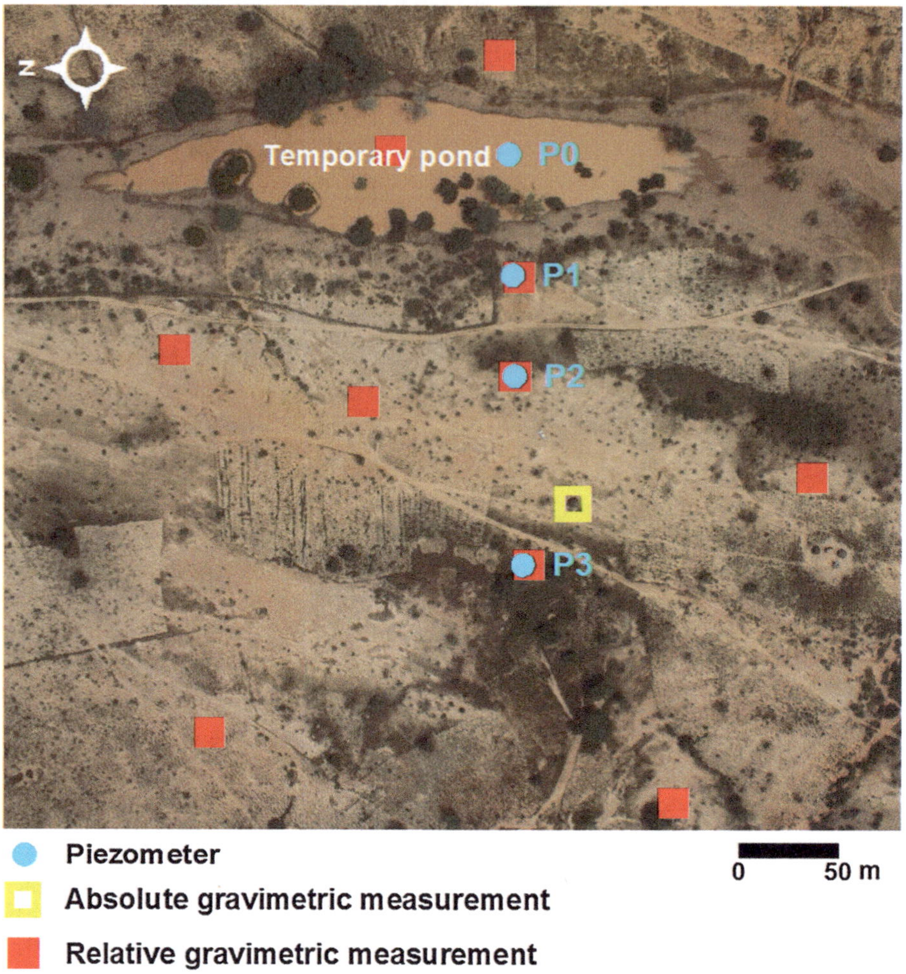

Figure 9
Aerial view (October 2008, J.-L. Rajot IRD) of the Wankama site near Niamey (Niger). The *yellow square* indicates the location of the FG5 absolute gravimetric measurement, the *blue disks* the piezometer location along a profile starting in the temporary pond (P0) and the *red squares* the location of the relative gravimetric measurements in the vicinity of the FG5 point

the ratio between the volume of water that can be extracted by gravity and the total volume of drained rock is assumed constant and varies from 1 to 15% (dashed lines from blue to red in Fig. 10a). As expected, the amplitude of the gravimetric signal is proportional to the specific yield.

We remind here that we only consider the direct attraction of the pond and water table variations in this model. Large scale effects associated with the elastic deformation of the earth and Newtonian attraction of distant loads (>500 m) are not taken into account in this calculation. To compare gravimetric measurements with the model of local water storage, we have to correct from large scale hydrological effects, as discussed in the beginning of Sect. 2.3.1. This correction is performed using either large scale hydrological models (e.g. GLDAS), or GRGS solutions of GRACE satellite data (see Fig. 8). The total gravimetric change predicted by GLDAS is computed as the convolution of Green functions with their estimate of water mass distribution. To obtain solely the large scale contribution, local effects are removed and estimated as the Bouguer anomaly due to the equivalent layer width contained in a pixel (about 25 km × 25 km for GLDAS model and about 400 km × 400 km for GRACE data).

The estimates of large scale hydrological effects on gravity only differ by 5 nm s^{-2} as can be seen

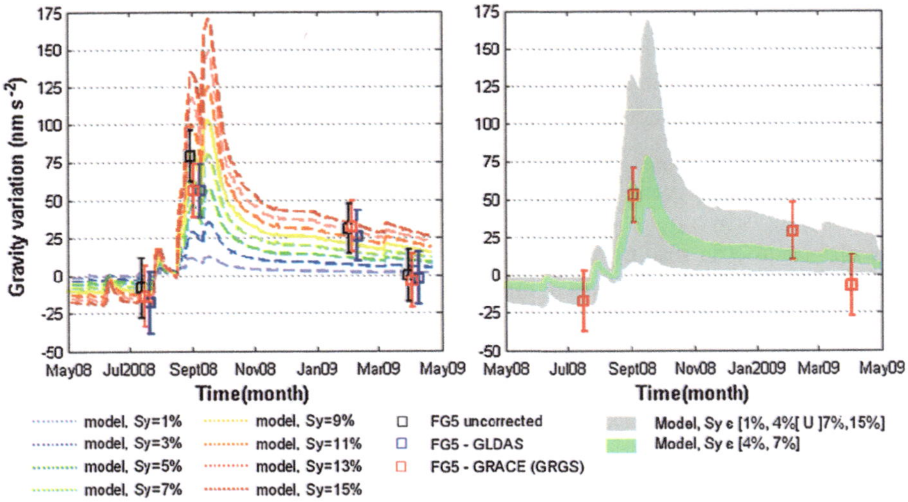

Figure 10

Synthetic gravity variations (in nm s^{-2}) at the Wankama/Niamey site caused by the changes in the water table level after the seasonal monsoon recharge in 2008. The specific yield is denoted Sy. The *squares* are the FG5 observations uncorrected for the non-local hydrology contribution (in *black*), corrected using the GLDAS model (in *blue*) and corrected using the GRACE data from CNES/GRGS (in *red*). For visibility purpose, the different values for a specific measurement considering different corrections are slightly shifted in time on the *left part* from Fig. 10 left (differences between the blue and red squares).

The model of local water storage can now be compared to FG5 measurements (Fig. 10 left), corrected or not for large scale hydrological effects considering GLDAS and GRACE (GRGS) data. The model is within the error bars of corrected FG5 measurements for specific yield values of the aquifer ranging from 4 to 7% (Fig. 10 right). The minimum standard deviation between model and measurements reaches 13 nm s^{-2} and is obtained for a specific yield of 7% considering GRACE correction. A standard deviation of 16 nm s^{-2} is obtained for the same specific yield considering GLDAS correction. The standard deviation between model and measurements is higher when considering the corrections of large scale effects, and reaches 24 nm s^{-2} for a specific yield of 10%.

The first result derived from our study at Wankama is an estimate of the specific yield of the aquifer that ranges from 4 to 7% according to gravimetry when using a model only dependent on water table fluctuations. If one also considers a soil moisture contribution, the specific yield values may be extended to the 2 to 7% range (Pfeffer *et al.*, 2011). In addition to gravity measurements, magnetic resonance soundings (MRS) were performed in the vicinity of the FG5 site. The MRS method (Legchenko *et al.* 2002; Vouillamoz *et al.* 2008) allows measuring a MRS water content which is expected to be a good estimate of effective porosity (Lubczynski and Roy 2005; Boucher *et al.* 2009a, b) itself being generally slightly higher than specific yield. In this paper, we first consider that the aquifers are unconfined. Second, we assume that the values of MRS water content derived from field measurements are representative of the top part of the aquifer, where water table fluctuations occur, thus influencing gravity data. Third, we assume that the MRS water content value maximizes the value of the specific yield of the aquifer. Sometimes called drainage porosity (for unconfined aquifer), the specific yield quantifies the amount of water released from the aquifer by gravity forces, and thus is a key parameter for gravimetric measurements understanding. This third assumption is supported by several studies that compare MRS water content parameters with pumping tests. In a sedimentary context, the MRS water content is found to be generally higher than the specific yield estimated by pumping tests (Boucher *et al.*, 2009a, b). In a crystalline context, the study of Vouillamoz *et al.* (2005) has shown that the MRS water content can be used to give an estimate of the specific yield (in unconfined situation). In the present study, the MRS water content parameter derived from the field measurements is also called "MRS porosity" and is expressed in % of the total volume. This MRS water content is used as an input factor for gravimetric numerical modeling.

On the Wankama site, MRS results display a MRS water content ranging between 4 and 11%. This good agreement with gravimetric results is evidence for the consistency of both methods. The gravimetric method can thus be used for constraining parameterization of local hydrological modeling. To enhance our knowledge of water storage variations at this site, additional measurements with relative spring gravimeters have been performed in the rainy season in 2009 (Fig. 9) and will be analyzed to better characterize spatial heterogeneity in water storage variability.

2.3.3 Water Storage Changes in Diffa (East Niger)

The Bagara site near Diffa, in the eastern part of Niger, is located in the representative region of the Sahara–Sahel transition zone (350 mm/year of rain). Specific hydrology observations are done immediately near the location of the absolute gravimeter in Bagara and subsurface geophysics (MRS) has been applied to infer the local MRS porosity. One objective is to better understand the water recharge/discharge processes from the Komadugu-Yobe river of the sedimentary layers in Bagara and how these processes relate to the more general hydrology models like GLDAS in the region.

The Bagara site is close to the Komadugu-Yobe River, in eastern Niger (Fig. 11). The mean annual rainfall is 350 mm, but here rainfall presents a large interannual variability, for example 305, 530, 344, 279 mm, respectively, for the 2006–2009 period. Groundwater recharge is governed mainly by the Komadugu-Yobe temporary river (Gaultier, 2004), which flows between July and November, through direct infiltration at its sandy bed and from its floodzone, which is nearly 1 km wide (Fig. 11a). Rain infiltration occurs directly or via temporary ponds, and the vegetation consists of a sparse thorny savannah during most of the year. A thick (0.5–1 m) herbaceous cover grows during the rainy season,

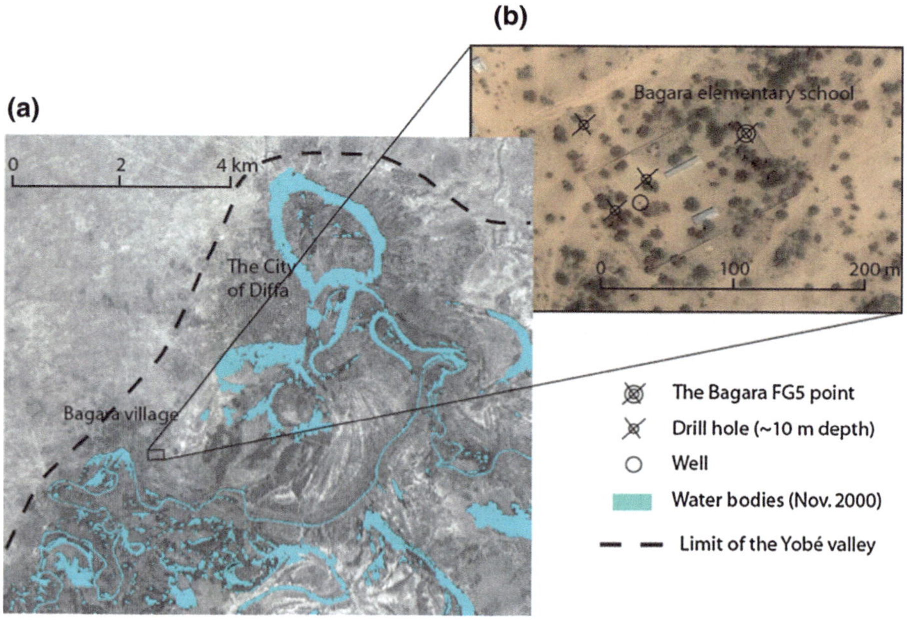

Figure 11
a Location of the Bagara site (near Diffa) in the eastern part of Niger close to Lake Chad. Bagara is close to the Komadugu-Yobe river which flows into Lake Chad at a distance of about 100 km. Flooded zone in high waters of the Komadugu-Yobe (LANDSAT data, Nov. 2000) are in blue. **b** Near surroundings of the FG5 absolute gravity measurement point. Note the scarce vegetal cover (in March) and the high density of logging data

lasting from July to September, and dries out in October.

The underground water level record is clearly not continuous during the GHYRAF observation period (Fig. 12). However, previous data, collected in 1995–1997, and this record allow to define the local hydrological cycle which is characterized by a nearly 0.4 m annual amplitude and extrema in January (maximum level) and July (minimum level), which are offset relatively to the rainfall and, therefore, also to the computed GLDAS hydrological signal.

Magnetic resonance sounding (MRS) results indicate a MRS water content of nearly 25% for the whole aquifer. However, MRS content is poorly constrained in the uppermost aquifer, which is only concerned by water level fluctuations. MRS water content estimates are in the range 14–23% there, depending on the parameters adopted for the inversion, and the manual estimate is 24%. Adopting a mean MRS water content of 20% for the uppermost part of the aquifer, and making the hypothesis that it corresponds to the specific yield of the layer where the water table fluctuations are located, leads then to

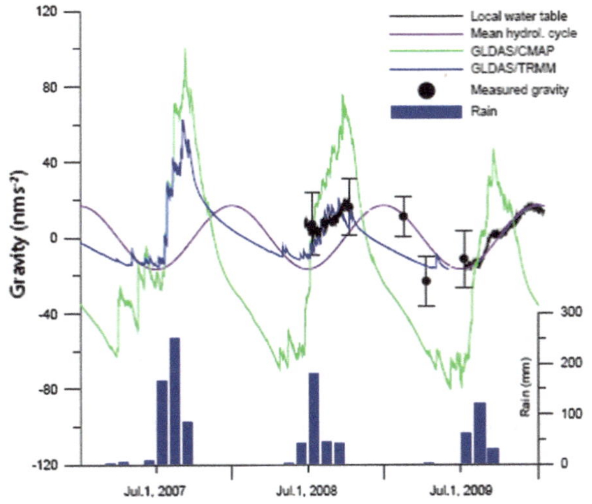

Figure 12
Seasonal gravity changes (in nm s^{-2}) for the Bagara site (near Diffa, Niger) where the black circles are the FG5 AG measurements with their uncertainties, the global hydrological model GLDAS/Noah forced by CMAP in *green*, GLDAS/Noah forced by TRMM in *blue*. The gravity changes due to the available local piezometric changes assuming a porosity of 20% are the two continuous black lines and the changes due to the mean hydrological cycle are in *pink*. We have also added the rainfall (in mm) for the period 2007–2009

a 30 nm s^{-2} amplitude gravity signal, which is displayed in Fig. 12 together with the observed gravity from FG5 AG measurements and the GLDAS model. These signals are of similar amplitude, but both the gravity and the mean hydrological signal are offset in time, which cannot be accounted for by uncertainties in gravity or in water level measurements. This discrepancy persists during the dry season, when the superficial soil layer contains almost no water.

The GLDAS models produce a sharp rise in the gravity signal, which is triggered by the onset of the rainy season and also a rather sharp recession curve, while both the water level and the gravity signal seem to present a smoother signal.

However, clayey layers have been commonly observed in drillholes in this area, one of these layers being found at the top of groundwater at the Bagara site. Statistical analyses of Le Coz *et al.* (2011) show that these clayey layers are of metric thickness and have a mean 300 m lateral extension. These layers correspond to the bottom of a temporary pond fed either by rainwater or by the spades of the Komadugu-Yobe river. During the dry season the pond presents a highly cracked bottom, which indicates volume change of these clayey layers with their water contents.

Therefore, it can be suspected that poroelasticity effects may occur during underground water level changes. These effects as well as diffusion effects arising from infiltration from the Komadugu-Yobe are still to be assessed with 3D models, when a record of at least one year duration will be available for this site.

2.3.4 *Water Storage Changes in Nalohou (Benin)*

The Nalohou site near Djougou (Benin) is a Soudanian site belonging to the upper Ouémé catchment (14,600 km^2) with average yearly rainfall of 1,200 mm. The upper Ouémé catchment is the wet mesoscale site of the long term hydrometeorological AMMA-Catch Observatory. At the mesoscale, rainfall and streamflow are controlled over a set of embedded catchments. To document the water budget at the local scale, several catena selected according to the land-use/cover are instrumented with piezometers

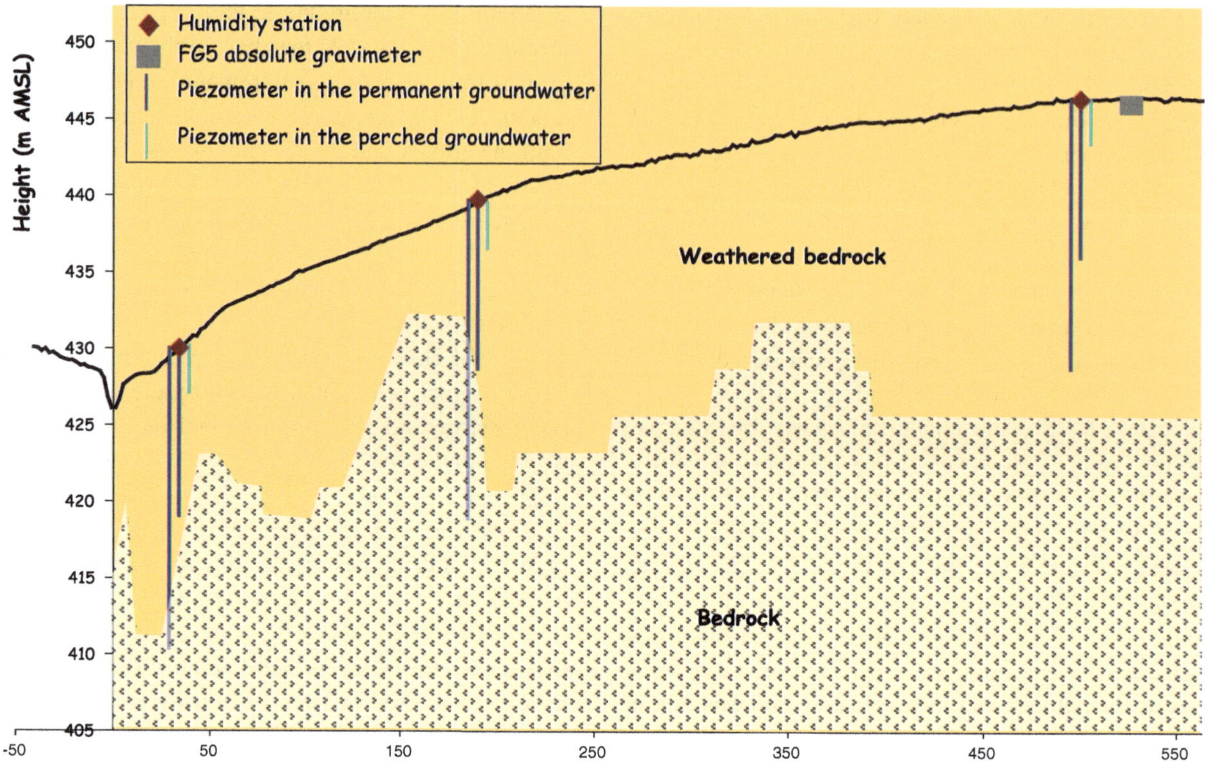

Figure 13
A sketch of the Nalohou site (near Djougou, Benin). The piezometers measuring the water table in deep permanent groundwater are in *dark blue* while those monitoring the perched and seasonal shallow groundwater are in *light blue*. The *brown diamond-shaped* indicate the humidity stations where soil moisture is observed. Finally, the location of the FG5 absolute gravimeter measurement is shown by a *grey rectangle*

screened at different depths, humidity probe stations (to monitor soil moisture of the first meter of the vadose zone) and flux stations (Guyot *et al.*, 2009). The Nalohou catena (Fig. 13), where absolute gravimetric measurements are carried out, exhibits a fallow-culture land-cover.

Combining geophysical with hydrological and geochemical data contributed to drawing up a hydrological functioning scheme (Kamagaté *et al.*, 2007). Recharged from July to September, permanent groundwater with low transmissivity and specific yield is located on the saprolite layer of the weathering profile, i.e. just above the crystalline bedrock. From September to the following June, the groundwater depletion is regular and slow (1.2 cm day^{-1}).

The general drying up of rivers in October–November follows the end of the rains and does not coincide with the lowest level of the water table (in June), reflecting the weakness or even the lack of permanent groundwater input to the base flow. Distinct geochemical signatures between base flow and the permanent groundwater provide other evidence of this feature. The origin of the base flow water has been found in a non permanent perched shallow aquifer located in the seasonally waterlogged headwaters of the streams (Séguis *et al.*, 2004, 2011). The shallow groundwater exfiltration is the dominant component of the annual discharge in the Upper Ouémé (Kamagaté *et al.*, 2007). The hypothesis proposed to explain the depletion of the permanent ground water during the dry season is tree transpiration. In a detailed analysis of evapotranspiration after an isolated rainfall event in the dry season, Guyot *et al.* (2009) demonstrate that, 1 month after the rainfall event, observed evapotranspiration rate could only be explained by a contribution of water uptake from deep soil layers (vadose or saturated zone).

Subsurface geophysics has been used to constrain better changes in water storage in the Nalohou catena. MRS survey conducted over several geological units and their respective weathered regolith has shown that significant lateral variations of MRS water content can exist over the measurement area. We found values ranging between 1.5 and 3% for most of the regolith in the area, within a noticeable exception, 10%, over quartz dykes (Descloitres *et al.*, 2011). Another attempt to estimate the specific yield has been made using mercury porosimetry method. Five samples of weathered regolith have been analysed. The results show a 20 (±5) % specific yield value. Between two dates, the water storage variations of each part (deep and shallow ground water) are the product of the difference of the piezometric levels by the specific yield. The water storage in the first meter of the vadose zone is directly estimated by the humidity probes (see Fig. 13).

Absolute gravimetric measurements have been carried out at key dates of the hydrologic cycle. In early July 2008, the permanent groundwater exhibits a low piezometric level, the perched groundwater is not yet present and the firstmeter of the vadose zone is partly dampened (see Fig. 14). Close to the end of the

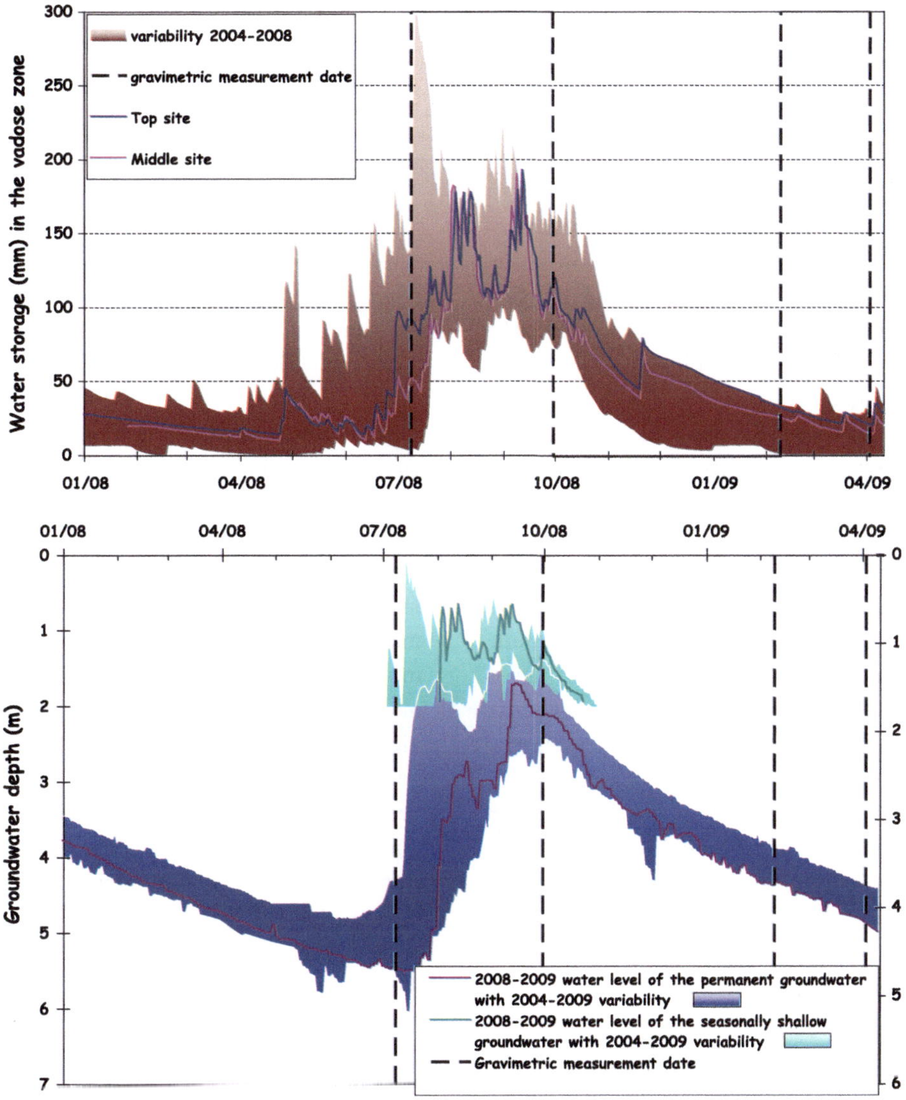

Figure 14
Dynamics of the three hydrological compartments observed close to the gravimeter site: piezometric levels of the seasonally shallow groundwater and permanent groundwater in 2008–2009 (*bottom*), water storage in the first meter of the vadose zone in top and middle sites of the catena (*top*)

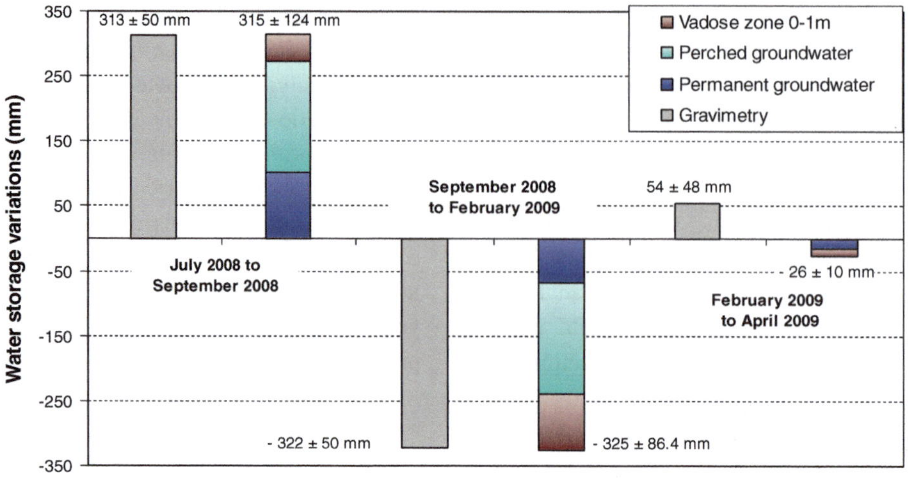

Figure 15
Water storage variations (in mm) as inferred from hydrology measurements (humidity stations in the 0–1 m vadose zone, piezometers in the perched groundwater and permanent groundwater) and from gravimetric measurements

rainy season, in mid-September 2008, the recharge of the deep groundwater has occurred, the shallow layers of the weathered mantle are saturated and the first meter of the vadose zone is now dampened. The underground storages are then the highest. Conversely, from this date to January 2009, the shallow groundwater disappears, the vadose zone is the driest and the permanent groundwater starts its depletion. In April 2009, the only change is the continuation of the permanent groundwater depletion. However, on this site, the geological units are organized in a complicated manner, i.e. the metamorphic rock and their respective regolith are organized in stripes elongated north–south, and a small dip angle is suspected (20° east) (Descloitres *et al.* 2011). This geometry could also influence the gravimetric measurements at the local scale and will be further investigated.

In Fig. 15, a good agreement between water storage variations is inferred from hydrology measurements and from gravimetric measurements. The similarity between the two estimations is high whatever the type of variation: recharge (July to September 2008) or drying-up (September 2008 to February 2009). A small discrepancy in the range of the uncertainty is observed on the last period (February to April 2009) characterised by a small depletion of the groundwater.

In the context of the GHYRAF project and the AMMA-CATCH observatory a new superconducting gravimeter (SG) has been installed in 2010 at Nalohou. This meter will be part of the GGP project (Crossley *et al.* 1999; Hinderer and Crossley 2004) and its location is very interesting in terms of network coverage (there are very few stations around the equator and only presently one in Africa). The high sampling rate of the gravimeter (1 s) will clearly enhance our domain of investigation in the spectral band and enable us to follow quite rapid hydrological events related to the monsoonal rainfalls. Further hydrogeological investigations will also be undertaken in the future (as pumping tests and vadose zone moisture monitoring using neutron probes) to obtain more independent data for comparison purpose. These investigations will lead to estimates of the specific yield independent from the ones inferred from MRS and to a better knowledge of the soil moisture in the deep vadose zone (below the first meter); this information will then allow a better estimate of the water storage changes to be compared with the surface gravity changes.

3. *Conclusion*

We have presented in this paper the first results from the GHYRAF (Gravity and Hydrology in Africa) experiment which started in 2008 in Niger and Benin. This project uses geodetic tools (GPS, surface and satellite gravimetry) in close connection to hydrology observations to investigate water storage changes along a climatic rainfall gradient from the Sahara to the equatorial monsoon region. The measurements from the AMMA-GHYRAF permanent GPS network in West Africa were analyzed and the comparison during the 2005–2008 period between hydrological loading predictions and observations clearly shows a nice agreement for the stations in Niamey (Niger) and Djougou (Benin). This demonstrates that GPS has the ability to retrieve annual signals of the order of 20 mm peak to peak due to continental hydrology loading. Different GRACE solutions were also used to investigate the annual signal in gravity over West Africa. The water storage changes were estimated for the Niger and Lake Chad basins showing again a nice agreement between the GLDAS/Noah global hydrology predictions, GRACE solutions and CMAP cumulative precipitation field. We have then focused on the results coming from our ground gravimetric measurements (using a FG5 absolute gravimeter) done during the 2008 monsoonal event. After showing the respective importance of local and non-local hydrology contributions to surface gravity and correcting the surface measurements for the latter contribution using GRACE satellite observations, we investigate in more details the local water storage changes in Wankama (Niger), Diffa (Niger) and Nalohou (Benin) specific sites.

In Wankama, a major result from the observed gravity changes mainly caused by changes in the piezometric level after indirect recharge processes is that the combination of both gravimetry and MRS techniques converge to infer a common specific yield value (of the order of 4–7%); this is evidence for the consistency of the two independent methods to better constrain important underground hydrodynamical parameters.

In Diffa, which is located in the Lake Chad basin and close to Komadugu-Yobe river, the observed gravity changes are smaller than the one predicted by global hydrology models but in close agreement (both in amplitude and phase) with the local piezometric level changes. It is clearly a local recharge–discharge process linked to the Komadugu-Yobe river which accounts for this rather than the mean rainfall field over the basin which explains that the soil moisture changes (and consequently the hydrology model) is shifted by several months with respect to the piezometric level.

Finally, in Nalohou, the observed gravity changes from July 2008 to April 2009 are well explained in terms of water storage changes in three specific compartments: the 0–1 m vadose zone, water level changes in temporary perched aquifers and finally water level changes in a deep permanent aquifer. The water storage budget is very consistent between the values inferred from gravimetry and the one computed from the hydrological in situ measurements.

Gravimetry appears to be a useful tool to solve some hydrological questions compared to borehole tests (being both local and invasive) or MRS (again somehow local with specific limitations due to the method). Gravimetry leads to a rapid and integrative estimate of the water storage changes provided that some simple hypotheses are made for the underground geometry. On the contrary, estimating the water storage by classical hydrology techniques requires the knowledge of piezometric changes (using boreholes), specific yield (e.g. using pumping tests or calibrated MRS) and soil moisture in the entire vadose zone (e.g. using calibrated neutron probes).

Gravimetry is also one of the few integrative methods present in hydrology (except for the surface flow measurement at the catchment outlet) and needs further investigation for a better integrated estimate of underground water storages.

Acknowledgments

This project is funded by the French Agence Nationale de la Recherche (ANR) for 4 years (2008–2011). The strong logistic and manpower support found in Niger and Benin is available thanks to the Institut de Recherche pour le Développement (IRD). Assistance of local water authorities in Niger and Benin is also warmly acknowledged. We also acknowledge the support of INSU-CNRS to run the AMMA-CATCH observatory in West Africa (http://www.amma-catch.org).

Appendix: Members of the GHYRAF team

F. Masson, Institut de Physique du Globe de Strasbourg (UMR 7516 CNRS, Université de Strasbourg) 67084 Strasbourg, France; Y. Rogister, Institut de Physique du Globe de Strasbourg (UMR 7516 CNRS, Université de Strasbourg) 67084 Strasbourg, France; F. Littel, Institut de Physique du Globe de Strasbourg (UMR 7516 CNRS, Université de Strasbourg) 67084 Strasbourg, France; B. Luck, Institut de Physique du Globe de Strasbourg (UMR 7516 CNRS, Université de Strasbourg) 67084 Strasbourg, France; M. Calvo Institut de Physique du Globe de Strasbourg (UMR 7516 CNRS, Université de Strasbourg) 67084 Strasbourg, France; B. Cappelaere, Hydrosciences Montpellier (UMR 5569 CNRS, IRD, Université Montpellier 2) 34095 Montpellier, France; C. Peugeot, Hydrosciences Montpellier (UMR 5569 CNRS, IRD, Université Montpellier 2) 34095 Montpellier, France; F. Delclaux, Hydrosciences Montpellier (UMR 5569 CNRS, IRD, Université Montpellier 2) 34095 Montpellier, France; M. Oi Hydrosciences Montpellier (UMR 5569 CNRS, IRD, Université Montpellier 2) 34095 Montpellier, France; A. Santamaria-Gomez, LAREG/IGN 77455 Marne la Vallée, France; R. Bayer, Géosciences Montpellier (UMR 5243 CNRS, Université Montpellier 2) 34095 Montpellier, France; C. Champollion, Géosciences Montpellier (UMR 5243 CNRS, Université Montpellier 2) 34095 Montpellier, France; P. Collard, Géosciences Montpellier (UMR 5243 CNRS, Université Montpellier 2) 34095 Montpellier, France; N. Le Moigne, Géosciences Montpellier (UMR 5243 CNRS, Université Montpellier 2) 34095 Montpellier, France; M. Diament, Institut de Physique du Globe de Paris (UMR 7154 CNRS, IPGP, Université Paris 7) 75252 Paris, France; S. Deroussi, 7 Institut de Physique du Globe de Paris (UMR 7154 CNRS, IPGP, Université Paris 7) 75252 Paris, France; O. de Viron, Institut de Physique du Globe de Paris (UMR 7154 CNRS, IPGP, Université Paris 7) 75252 Paris, France; R. Biancale, Dynamique terrestre et planétaire (UMR 5562 CNRS, Université Paul Sabatier) 31400 Toulouse, France; J.-M. Lemoine, Dynamique terrestre et planétaire (UMR 5562 CNRS, Université Paul Sabatier) 31400 Toulouse, France; P. Gegout, Dynamique terrestre et planétaire (UMR 5562 CNRS, Université Paul Sabatier) 31400 Toulouse, France; S. Galle, Laboratoire d'étude des Transferts en Hydrologie et Environnement (UMR 5564 CNRS, INPG, IRD, Université Joseph Fourier) 38 041 Grenoble, France; J.-P. Laurent, Laboratoire d'étude des Transferts en Hydrologie et Environnement (UMR 5564 CNRS, INPG, IRD, Université Joseph Fourier) 38 041 Grenoble, France; M.-N. Bouin, Météo France, Centre de Météorologie Marine, 29604 Brest, France; M. Calvo, Instituto Geografico Nacional, Madrid, Spain; A. Santamaria-Gomez, Instituto Geografico Nacional, Madrid, Spain; G. Woppelmann, Université de La Rochelle, CNRS, UMR 6250 LIENSS, La Rochelle, France; A. Zannou, Direction Générale de l'Eau, Cotonou, Bénin; Y. Nazoumou, Département de Géologie, Université Abdou Moumouni, Niamey, Niger.

References

Altamimi, Z., X. Collilieux, J. Legrand, B. Garayt, and C. Boucher (2007), *ITRF2005: A new release of the International Terrestrial Reference Frame based on time series of station positions and Earth Orientation Parameters*, J. Geophys. Res., *112*, B09401, doi:10.1029/2007JB004949.

Bock, O., M.N. Bouin, E. Doerflinger, P. Collard, F. Masson, R. Meynadier, S. Nahmani, M. Koité, K. Gaptia Lawan Balawan, F. Didé, D. Ouedraogo, S. Pokperlaar, J.-B. Ngamini, J.P. Lafore, S. Janicot, F. Guichard, M. Nuret, 2008. *The West African Monsoon observed with ground-based GPS receivers during AMMA*, J. Geophys. Res., *113*, D21105, doi:10.1029/2008JD010327.

Boehm, J., B. Werl, and H. Schuh (2006), *Troposphere mapping functions for GPS and very long baseline interferometry from European Centre for Medium-Range Weather Forecasts operational analysis data*, J. Geophys. Res., *111*, B02406, doi:10.1029/2005JB003629.

Boulain N., Cappelaere B., Séguis L., Favreau G., Gignoux J. (2009). *Water balance and vegetation change in the Sahel: a case study at the watershed scale with an ecohydrological model.* J. of Arid Environments, *73*, 1125–1135.

Boucher, M., Favreau, G., Descloitres, M., Vouillamoz, J.-M., Massuel, S., Nazoumou, Y., Cappelaere, B., & Legchenko A., 2009a. *Contribution of geophysical surveys to groundwater modelling of a porous aquifer in semiarid Niger: an overview.* C.R.Geoscience, *341*, 800–809.

Boucher, M., Favreau, G., Vouillamoz, J.-M., Nazoumou, Y., & Legchenko A., 2009b. *Estimating specific yield and transmissivity with magnetic resonance sounding in an unconfined sandstone aquifer (Niger).* Hydrogeology Journal, *17*, 1805–1815, doi:10.1007/s10040-009-0047-x

Boy, J.-P., & Hinderer, J. (2006) *Study of the seasonal gravity signal in superconducting gravimeter data.* J Geodyn *41*(1–3): 227–233. doi:10.1016/j.jog.2005.08.035

Boy, J.-P., de Linage, C. and Hinderer, J., 2010. *Retrieval of large-scale hydrological signal in Africa from GRACE time-variable gravity fields*, Pure appl. Geophys., this issue.

Bruinsma, S. Lemoine, J.-M., Biancale, R. and Valès, N., 2010. *CNES/GRGS 10-day gravity field models (release 2) and their evaluation*, Adv. Space Res., *45*, 587–601.

Cappelaere B., Descroix L., Lebel T., Boulain N., Ramier D., Laurent J.P., Favreau, G., Boubkraoui S., Boucher M., Bouzou Moussa I., Chaffard V., Hiernaux P., Issoufou H.B.A., Le Breton E., Mamadou I., Nazoumou Y., Oi M., Ottlé C., Quantin G. (2009) *The AMMA-CATCH experiment in the cultivated Sahelian area of south-west Niger – Investigating water cycle response to a fluctuating climate and changing environment*, Journal of Hydrology *375*(1–2), 34–51.

Chao, B. F., 2005. *On inversion for mass distribution from global (time-variable) gravity field*, J. Geodyn., *39*(3), 223–230.

Chen, F., Mitchell, K., Schaake, J., Xue, Y., Pan, H., Koren, V., Duan, Y., Ek, M. & Betts, A., 1996. *Modeling of land-surface evaporation by four schemes and comparison with FIFE observations*, J. Geophys. Res., *101*(D3), 7251–7268.

Coe, M. T., and C. M. Birkett (2004), *Calculation of river discharge and prediction of lake height from satellite radar altimetry: Example for the Lake Chad basin*, Water Resour. Res., *40*, W10205, doi:10.1029/2003WR002543.

Collilieux, X., Z. Altamimi, D. Coulot, J. Ray, and P. Sillard (2007), *Comparison of very long baseline interferometry, GPS, and satellite laser ranging height residuals from ITRF2005 using spectral and correlation methods*, J. Geophys. Res., *112*, B12403, doi:10.1029/2007JB004933.

Crossley, D., Hinderer, J., Casula, G., Francis, O., Hsu, H.-T., Imanishi, Y., Jentzsch, G., Kaarianen, J., Merriam, J., Meurers, B., Neumeyer, J., Richter, B., Shibuya, K., Sato, T., and T. van Dam, 1999. *Network of superconducting gravimeters benefits a number of disciplines*, EOS, Transactions, AGU, *80*, no11, 121, 125–126.

Crowley, J. W., Mitrovica, J. X., Bailey, R. C., Tamisiea, M. E. and Davis, J. L., 2006. *Land water storage within the Congo Basin inferred from GRACE satellite gravity data*, Geophys. Res. Lett., *33*, L19402, doi:10.1029/2006GL027070.

Crowley JW, Mitrovica JX, Bailey RC, Tamisiea ME, Davis JL (2008) *Annual variations in water storage and precipitation in the Amazon basin*. Journal of Geodesy *82*:9–13

Descloitres, M., Séguis, L Legchenko, A., Wubda, M, Guyot, A., & Cohard, J.M., 2011. *The contribution of MRS and resistivity methods to the interpretation of Actual Evapo-Transpiration measurements: a case study in metamorphic context in north Benin*, Near Surface Geophysics, special issue on Magnetic Resonance Soundings, Vol. *9*, N°2, pp. 187–200, doi: 10.3997/1873-0604.2011003

de Linage, C., Hinderer, J., & Rogister, Y., 2007. *A search for the ratio between gravity variation and vertical displacement due to a surface load*, Geophys. J. Int., *171*, 986–994.

de Linage, C., J. Hinderer and J.-P. Boy, 2009. *Variability of the gravity-to-height ratio due to surface loads*, Pure and Applied Geophysics, doi:10.1007/s00024-004-0506-0

Dong, D., P. Fang, Y. Bock, M. K. Cheng, and S. Miyazaki (2002), *Anatomy of apparent seasonal variations from GPS-derived site position time series*, J. Geophys. Res., *107*(B4), 2075, doi: 10.1029/2001JB000573.

Favreau, G., Cappelaere B., Massuel S., Leblanc M., Boucher M., Boulain N., Leduc C. (2009). *Land clearing, climate variability and water resources increase in semiarid southwest Niger: a review*. Water Resources Research, *45*, W00A16, doi: 10.1029/2007WR006785.

Flechtner, F., 2007. *GFZ Level-2 Processing Standards Document For Level-2 Product Release ~ 0004*, GRACE 327-743 (GR-GFZ-STD-001), Rev. 1.0.

Gaultier G. (2004), *Recharge et paléorecharge d'une nappe libre en milieu Sahélien (Niger oriental): approche géochimique et hydrodynamique*, PhD thesis, Université de Paris-XI, Orsay, France.

Goutorbe J. P., Lebel T., Dolman A. J., Gash J. H. C., Kabat P., Kerr Y. H., Monteny B., Prince S. D., Stricker J. N. M., Tinga A., Wallace J. S. (1997) *An overview of HAPEX-Sahel: a study in climate and desertification*. Journal of Hydrology, *188–189*, 4–17.

Guyot A., Cohard J-M., Anquetin S. *et al.*, 2009. *Combined analysis of energy and water budgets to consolidate latent heat flux estimation using an infrared scintillometer*. Journal of Hydrology, *375*:227–240, doi:10.1016/j.jhydrol.2008.12.027.

Hasan, S., Troch, P., Boll, J., & Kroner, C. (2006). *Modeling of the hydrological effect on local gravity at Moxa, Germany*. J Hydrometeor *7*(3):346–354. doi:10.1175/JHM488.1

Herring, T. A., King R. W., and McClusky S. C. (2008), *Introduction to GAMIT/GLOBK*, Mass Inst. of Technol., Cambridge.

Hinderer, J., & Crossley, D., 2004. *Scientific achievements from the first phase (1997–2003) of the Global Geodynamics Project using a worldwide network of superconducting gravimeters*, J. Geodynamics, *38*, 237–262.

Hinderer, J., de Linage, C., & Boy, J.-P., 2006a. *How to validate satellite-derived gravity observations with gravimeters at the ground?*, Bull. Inf. Marées Terr., *142*, 11433–11441.

Hinderer, J., Andersen, O., Lemoine, F., Crossley, D., & Boy, J.-P., 2006b. *Seasonal changes of the Earth's gravity field from GRACE: a comparison with ground measurements from superconducting gravimeters and with hydrology model predictions*, J. Geodynamics, *41*, 59–68.

Hinderer, J., Crossley, D., & Warburton, R., 2007. *Superconducting gravimetry, in Treatise on Geophysics*, vol. 3 Geodesy, vol. ed. T. Herring, ed. in chief G. Schubert, Elsevier, 65–122.

Hinderer, J. *et al.*, 2009. *The GHYRAF (Gravity and Hydrology in Africa) experiment: description and first results*, J. of Geodynamics, Volume *48*, Issues 3–5, 172–181.

Huffman, G. J., Adler, R. F., Bolvin, D. T., Gu, G., Nelkin, E. J., Bowman, K. P., Hong, Y., Stocker, E. F. & Wolff, D. B., 2007. *The TRMM Multi-satellite Precipitation Analysis: Quasi-Global, Multi-Year, Combined-Sensor Precipitation Estimates at Fine Scale*. Hydrometeor., *8* (1), 38–55.

Jacob, T., Bayer, R., Chery, J., Jourde, H., Le Moigne, N., Boy, J.P., Hinderer, J., Luck, B. and Brunet, P., 2008. *Absolute gravity monitoring of water storage variation in a karst aquifer on the Larzac plateau (Southern France)*. Journal of Hydrology, *359*(1–2): 105–117, doi:10.1016/j.jhydrol.2008.06.020.

Jenness, J., Dooley, J., Aguilar-Manjarrez, J. and Riva, C., 2008. African Water Resource Database. GIS-based tools for inland aquatic resource management. 2., *Technical manual and workbook*, CIFA Technical Paper, **33**, Part 2. Rome, 308 ~ p.

Kamagaté B., Séguis L., Favreau G., Seidel J.L., Descloitres M., Affaton P., 2007. *Hydrological processes and water balance of a tropical crystalline bedrock catchment in Benin (Donga, upper Ouémé River)*. C. R. Geoscience, *339*, 418–429.

King, M. A., C. S. Watson, N. T. Penna, and P. J. Clarke (2008), *Subdaily signals in GPS observations and their effect at semi-annual and annual periods*, Geophys. Res. Lett., *35*, L03302, doi:10.1029/2007GL032252.

Kroner, C, & Jahr, T. (2006) *Hydrological experiments around the superconducting gravimeter at Moxa Observatory*. J Geodyn *41*(1–3):268– 275. doi:10.1016/j.jog.2005.08.012

KRONER, C, JAHR, T, NAUJOKS, M, & WEISE, A (2007) *Hydrological signals in gravity—foe or friend?* In: RIZOS C, TREGONING P (eds) Dynamic planet—monitoring and understanding a dynamic planet with geodetic and oceanographic tools, IAG Symposia Series, vol 130. Springer, Heidelberg, pp 504–510.

LEBEL, T., CAPPELAERE, B., GALLE, S., HANAN, N., KERGOAT, L., LEVIS, S., VIEUX, B., DESCROIX, L., GOSSET M., MOUGIN, E., PEUGEOT, C., SÉGUIS, L. (2009). *The AMMA-CATCH studies in the Sahelian region of West-Africa: an overview*. J. of Hydrology, *375*, 3–13.

LE COZ, M., GENTHON, P., & ADLER, P., 2011, *Multiple-point statistics for modeling facies heterogeneities in a porous media: the Komadougou-Yobe alluvium, Lake Chad bassin*, Math. Geosci., doi:10.1007/s11004-9353-6.

LEGCHENKO A, BALTASSAT JM, BEAUCE A, BERNARD J (2002) *Nuclear resonance as a geophysical tool for hydrogeologists*. Journal of Applied Geophysics, *50* (1–2): 21–46.

L'HÔTE Y.,& MAHÉ G. (1996). *Afrique de l'Ouest et Centrale, Précipitations moyennes annuelles (période 1951–1989)*. Echelle 1/6 000 000 ème. *Collection des cartes ORSTOM*, ORSTOM Ed.

LLUBES, M., FLORSCH, N., HINDERER, J., LONGUEVERGNE, L., & AMALVICT, M., 2004. *Local hydrology, the Global Geodynamics Project and CHAMP/GRACE perspective: some case studies*, J. Geodynamics, *38*, 355–354.

LUBCZYNSKI M., ROY J. (2005) *MRS contribution to hydrogeological system parametrization*. Near Surface Geophysics *3* (3): 131–139.

LUTHCKE, S. B., ROWLANDS, D. D., LEMOINE, F. G.,KLOSKO, S. M., CHINN, D. S. and MCCARTHY, J. J., 2006. *Monthly spherical harmonic gravity field solutions determined from GRACE inter-satellite range-rate data alone*, Geophys. Res. Lett., *33*, L02402, doi:10.1029/2005GL024846.

NAUJOKS, M., WEISE, A., KRONER, C., & JAHR, T., 2008. *Detection of small hydrological variations in gravity by repeated observations with relative gravimeters*, J. Geod, *82*, 543–553.

NEUMEYER, J., F. BARTHELMES, O. DIERKS, F. FLECHTNER and M. *HARNISCH,Combination of temporal gravity variations resulting from superconducting gravimeter (SG) recordings, GRACE satellite observations and global hydrology models*, Journal of Geodesy, 2006, Volume *79*, Numbers 10–11, 573–585.

PFEFFER, J., BOUCHER, M., HINDERER, J., FAVREAU, G., BOY, J.-P., DE LINAGE, C., CAPPELAERE, B., LUCK, B., OI, M. and LE MOIGNE, N. (2011), *Local and global hydrological contributions to time-variable gravity in Southwest Niger*. Geophysical Journal International, *184*: 661–672. doi:10.1111/j.1365-246X.2010.04894.x

PETROV L., J.-P. BOY (2004), *Study of the atmospheric pressure loading signal in VLBI observations, Journ*al of Geophysical Research, doi:10.1029/2003JB002500, Vol. 109, No. B03405.

RAMIER D., BOULAIN N., CAPPELAERE B., TIMOUK F., RABANIT M., LLOYD C.R., BOUBKRAOUI S., MÉTAYER F., DESCROIX L., WAWRZYNIAK V. (2009). *Towards an understanding of coupled physical and biological processes in the cultivated Sahel - 1. energy and water*. J. of Hydrology, *375*, 204–216.

RODELL, M., P. R. HOUSER, U. JAMBOR, J. GOTTSCHALCK, K. MITCHELL, C.-J. MENG, K. ARSENAULT, B.COSGROVE, J. RADAKOVICH, M. BOSILOVICH, J. K. ENTIN, J. P. WALKER, D. LOHMANN, & D. TOLL, 2004.*The Global Land Data Assimilation System*, Bull. Amer. Meteor. Soc., *85* (3), 381–394.

SANTAMARÍA-GÓMEZ A., M.-N. BOUIN, & G. WÖPPELMANN, 2011. *Improved GPS data analysis strategy for tide gauge*, Proceedings of the 2009 IAG General Assembly, Buenos Aires, Argentina, August 31–September 4 2009, Springer, in press.

SÉGUIS, L., CAPPELAERE, B., MILÉSI, G., PEUGEOT, C., MASSUEL, S., & FAVREAU, G., 2004. *Simulated impacts of climate change and land-clearing on runoff from a small Sahelian catchment*, Hydrol. Process. *18*, 3401–3413, doi:10.1002/hyp.1503.

SÉGUIS, L., KAMAGATÉ, B., FAVREAU, G., DESCLOITRES, M., SEIDEL, J.-L., GALLE S., PEUGEOT, C., GOSSET, M., LE BARBÉ, L., MALINUR, F., VAN EXTER, S. ARJOUNIN, M., & WUBDA, M., 2011. *Origins of streamflow in a crystalline basement catchment in a sub-humid Sudanian zone: the Donga basin (Benin, West Africa). Inter-annual variability of water budget*. Journal of Hydrology, 402, 1–13.

TAPLEY, B., BETTADPUR, S., RIES, J., THOMPSON, P. & WATKINS, M., 2004. *Grace measurements of mass variability in the Earth system*, Science, *305*, 503–505, doi:10.1126/science.1099192.

VAN DAM, T., J. WAHR, P. C. D. MILLY, A. B. SHMAKIN, G. BLEWITT, D. LAVALLÉE, and K. M. LARSON (2001), *Crustal displacements due to continental water loading*, Geophys. Res. Lett., *28*(4), 651–654.

VITERBO, P. and BELJAARS, A. C. M., 1995. *An improved land surface parametrization scheme in the ECMWF model and its validation*, J. Clim., *8*, 2716–2748.

VOUILLAMOZ, J. M., DESCLOITRES, M., TOE, G., LEGCHENKO, A., 2005. *Characterization of crystalline basement aquifers with MRS: comparison with boreholes and pumping tests data in Burkina Faso*. Near Surface Geophysics, *3*: 107–111.

VOUILLAMOZ, J.M., G. FAVREAU, S. MASSUEL, M. BOUCHER, Y. NAZOUMOU, and A. LEGCHENKO (2008), *Contribution of magnetic resonance sounding to aquifer characterization and recharge estimate in semiarid Niger*, J. Appl. Geophys., *64*, 99–108.

XIE, P., JANOWIAK, J. E., ARKIN, P. A., ADLER, R., GRUBER, A., FERRARO, R., HUFFMAN, G. J. and CURTIS, S. 2003. *GPCP Pentad precipitation analysis: an experimental dataset based on gauge observations and satellite estimates*, J. Clim., *16*, 2197–2214.

(Published online October 14, 2011)

Pure Appl. Geophys. 169 (2012), 1411–1423

DOI 10.1007/s00024-011-0400-5

Pure and Applied Geophysics

Combination of Multisatellite Altimetry and Tide Gauge Data for Determining Vertical Crustal Movements along Northern Mediterranean Coast

F. García,[1] M. I. Vigo,[1] D. García-García,[1] and J. M. Sánchez-Reales[1]

Abstract—Sea level variations (SLV) can be measured by tide gauges (TG) at the coast and by altimeters onboard satellites. The former measures the SLV relative to the coast, whereas altimetry provides the SLV with respect to a geocentric reference frame. The differences between SLV measurements from these two techniques can be used as an indirect assessment of vertical crustal motions at the TG sites. In this study, we exploit this idea, analyzing differences between sea level signals as measured by altimetric missions (TOPEX/Poseidon and Jason-1) and by 47 TG stations along northern Mediterranean coasts for the period 1993–2007. This allows us to estimate the vertical land motion along these coasts at the TG sites in this time window. For those sites where the TG is co-located or has a nearby global positioning system (GPS) station, these estimates are compared with the vertical rates derived from GPS measurements. Our results on vertical ground motion along the Mediterranean coast provide a useful source of data for studying, contrasting, and constraining tectonic models for the region.

Key words: Mediterranean Sea, sea level, altimetry, tide gauge, vertical crustal motion.

1. Introduction

Sea level variations (SLV) are primarily determined by two methods. For almost two centuries, the long-term SLV has been typically estimated from tide gauge (TG) measurements (e.g., Barnett, 1984; Douglas, 1991). Alternatively, during the last two decades, altimeters onboard satellites, whose orbits are precisely determined, have provided SLV measurements with high accuracy. Satellite altimeters measure SLV with respect to an ellipsoid of reference, whereas TG measurements are relative to the land on which the TG rests. By itself, a tide gauge cannot tell the difference between local crustal motion and sea level changes. However, the combination of the two techniques can provide information for estimating vertical crustal motions at the TG sites, which is complementary to and independent of global positioning system (GPS) measurements at these sites.

Cazenave *et al.* (1999) combined altimetry and TG data, for the period 1993–1997, to estimate the vertical ground motion at 53 TG sites, mainly located in the Pacific Ocean. The basic idea is to form the difference time series for altimetry minus TG, and examine its long-term behavior. This approach can provide information about the long-term vertical ground motion in the absolute sense. Six of these estimates were verified by Doppler orbitography and radio-position integrated by satellite (DORIS) data. A recent study by Ray *et al.* (2010), following the same idea but extending the study to a 16-year period (1993–2009), analyzed the vertical ground motion of 28 TG and compared the results with recent DORIS-derived vertical rates. The first estimations for the northern Mediterranean and Black Sea coasts, following a similar technique, are presented in Garcia *et al.* (2007) with results corresponding to a 9-year period (1993–2001). Results for most areas are in accordance with the tectonic setting of the Mediterranean, which is dominated by subduction in the Hellenic and Calabrian Arc and by the collision between the African and Arabian Plates with Eurasia, but with several regions of extension, such as the Alboran Sea, the Algero-Provencal Basin, and the Tyrrhenian and Aegean Seas (Jimenez-Munt *et al.*, 2003). The tectonic-seismic activity is higher in the eastern Mediterranean and becomes quieter towards the north.

[1] Departamento de Matemática Aplicada, Escuela Politécnica Superior, Universidad de Alicante, San Vicente del Raspeig, 03080 Alicante, Spain. E-mail: Fernando.GC@ua.es

Furthermore, the results also provide significant information for other areas, such as the distinct land subsidence on the west coast of Greece and the east coast of the Adriatic Sea, which can constitute an indicator that may be related to the Adriatic lithosphere subduction beneath the Eurasian Plate along the Dinarides Fault (Bennett *et al.* 2008). However, though the main contribution to the vertical land motion here reported should be tectonic, there can be other anthropogenic origins that might be dominant at specific locations.

In this study, we revisit the problem presented in Garcia *et al.* (2007), but the approach here presented has been significantly improved in several aspects: (1) the time period has been extended in most cases up to 15 years (1993–2007); (2) the altimetric measurements used have been improved near the coast; and (3) in order to make both data types as consistent as possible, TG measurements are corrected for ocean, atmospheric, and hydrological loading, and for the ocean pole tide. Besides, the glacial isostatic adjustment (GIA) has been corrected following Peltier (2004); although the vertical displacement produced at the Mediterranean coast by the GIA is minimal, it can represent a significant portion of the observed vertical rate at some specific sites (Stocchi and Spada, 2009).

2. Data

2.1. Altimetry Data

The altimetry data used in this study are along track measurements from TOPEX/Poseidon (T/P) and Jason-1. Altimetry was originally designed for the open ocean, and the quality of altimetric measurements decreases in proximity to the coast, where our interest lies. However, in recent years, great improvement of satellite altimetry quality in the coastal zones has been achieved by the development of data products specially processed for coastal applications. In this work we use altimetry sea level anomalies from a regional solution for the Mediterranean Sea which is part of the coastal products developed, validated, and distributed by the Centre de Topographie des Océans et de l'Hydrosphère (CTOH) in France. The altimetric measurements are based on the round trip of a radar pulse between the satellite and the sea surface. The atmosphere influences the radar pulse, and then the signal is corrected for the ionosphere, and dry/wet troposphere effects. Besides, the altimetric measurements are corrected for known geophysical processes as solid, ocean, and pole tides, loading effect of the ocean tides, sea state bias, and the atmospheric inverse barometer (IB) response of the ocean. Detailed information of the corrections can be found at http://ctoh.legos.obs-mip.fr/.

2.2. Tide Gauge Data

In this study, revised local reference (RLR) TG data from 47 stations of the Permanent Service for Mean Sea Level (PSMSL; Spencer and Woodworth, 1993) were used. Figure 1 shows the TG locations numbered from west to east along the coast, and Table 1 gives information for each TG (PSMSL code, name, latitude and longitude coordinates, and time span). In particular, we use the data from 5 TG along the Atlantic southern coast of Spain (numbers 1–5 in Table 1; Fig. 1), 41 from the northern coast of the Mediterranean (numbers 6–46), and 1 from the east coast (number 47). Data are monthly mean time series, spanning mostly from 10 to 15 years in the period 1993–2007. The southern Mediterranean coast is not studied because of the unavailability of TG measurements for the studied period.

As our primary aim is to isolate vertical crustal motion by comparing altimetric and TG measurements, the signal from both techniques should correspond to the same SLV. So, it is important to minimize possible inconsistencies due to the different corrections applied to each dataset. Thus, TG time series were corrected for several geophysical processes. Although the influence of some of the corrections in the studied trends is almost negligible, they have been included for consistency with the altimetry data. For example, the vertical displacements at the TG benchmark due to oceanic, atmospheric, and hydrological loading are corrected. The oceanic tide loading has been corrected for each TG according to the GOT4.7 tidal model (courtesy of M.S. Bos and H.-G. Scherneck; http://froste.oso.chalmers.se/loading/). The effects of the atmospheric and hydrological loading have been computed using

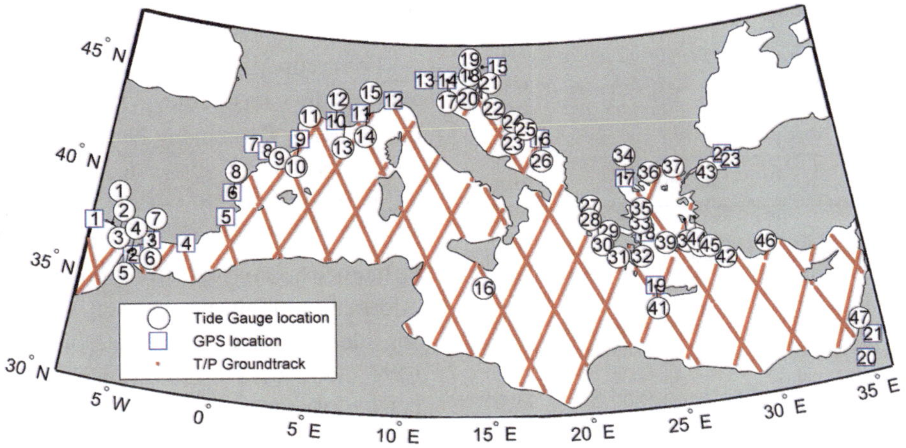

Figure 1
Data locations used in this study: (1) TG sites are represented by *numbered circles*, according to Table 1, following the coast from west to east; (2) altimeter tracks are depicted by *solid red lines*; (3) GPS sites are indicated by *blue squares* numbered according to Table 2

ECMWF reanalysis (ERA-interim; Berrisford *et al.*, 2009) data at the different TG locations (Petrov and Boy, 2004). In particular, the oceanic, atmospheric, and hydrological loading represent around 0.1, 2, and 2 mm, respectively, of the average standard deviation of all the TG, which is 70 mm. The addition of these corrections only accounts for around 4.3 mm, which is 6% of the signal.

The ocean and solid Earth tides are corrected with models in altimetry, which measures SLV every 10 days at the same point. The TG measures the sea level at a unique point every hour, although more recently the sampling frequency of some TG has been increased to 15 min, 6 min, and even higher rates (Joint Technical Commission for Oceanography and Marine Meteorology, 2006). Then, the TG time series were averaged monthly, which acts as a low-pass filter that eliminates the six largest constituents of the (ocean and solid Earth) tidal signal (Chelton and Enfield, 1986). So, both datasets are corrected for these tides, and although the corrections are different, only inconsistencies at the correction error level may arise. The pole tide is corrected in altimetry following Wahr (1985), and then we apply the same correction to the TG time series. This correction accounts for 1.5 mm of the average standard deviation of all the TG, which is around 2% of the signal (Desai, 2002). Note that this correction ignores, in both altimetry and TG, the effects of loading and self-gravitation of the ocean pole tide. The sea state bias must be corrected in altimetry because there is only a measurement every 10 days at each point of the track. This effect is also corrected in the TG data because of the monthly averaging process mentioned above. The IB response of the ocean to atmospheric pressure changes is originally corrected in altimetry, but not in TG. As far as long-term variations are concerned, the IB effect is not important (Le Traon and Gauzelin, 1997). However, we decide to eliminate the IB correction in altimetry for consistency with the TG data. In any case, the trends of the altimetry minus TG time series have been computed for datasets both with and without IB correction, and they are very similar, although the error estimate is smaller when the IB correction is not applied. Therefore, after all the corrections, the TG data show the SLV, including the IB response of the ocean due to changes in the atmospheric pressure, and the signature of vertical crustal motion at the TG benchmark, such as those produced by the GIA. Although the latter is almost negligible for most sites (Table 3), it is corrected for each TG according to the ICE-5Gv1.2 (VM4) model (Peltier, 2004).

This study is based on the fact that two independent techniques measure the SLV at a same location. However, the measurements from altimetry are made along the track of the satellite, which rarely coincides with the location of a TG (Fig. 1). So, the altimetry measurements are interpolated to each TG location. To do this, the altimetry data were convolved with a two-dimensional Gaussian function centered at the

Table 1

Information for each tide gauge. Columns from left to right correspond to: numbering as shown in Fig. 1, PSMSL station code, city and country, latitude and longitude coordinates, and span of time used in this study

No.	Code	Site	Location	Span
01	220005	Huelva (Spain)	37 08 N 06 50 W	97–06
02	220008	Bonanza (Spain)	36 48 N 06 20 W	93–07
03	220003	Cádiz III (Spain)	36 32 N 06 17 W	93–07
04	220011	Algeciras (Spain)	36 07 N 05 26 W	93–02
05	220021	Tarifa (Spain)	36 00 N 05 36 W	93–07
06	340001	Ceuta (Spain)	35 54 N 05 19 W	93–07
07	220032	Málaga II (Spain)	36 43 N 04 25 W	93–07
08	220056	Valencia (Spain)	39 28 N 00 20 W	93–05
09	220061	Barcelona (Spain)	41 21 N 02 10 E	93–07
10	220081	L'Estartit (Spain)	42 03 N 03 12 E	93–01
11	230021	Sète (France)	43 24 N 03 42 E	96–05
12	230051	Marseille (France)	43 18 N 05 21 E	93–07
13	230061	Toulon (France)	43 07 N 05 55 E	93–07
14	230081	Nice (France)	43 42 N 07 16 E	98–07
15	230011	Monaco C.	43 44 N 07 25 E	01–07
16	265001	Valletta (Malta)	35 54 N 14 31 E	00–04
17	270054	Venice P.S. (Italy)	45 26 N 12 20 E	93–00
18	270061	Trieste (Italy)	45 39 N 13 45 E	93–07
19	279003	Luka Koper (Slovenia)	45 34 N 13 45 E	93–03
20	280006	Rovinj (Croatia)	45 05 N 13 38 E	93–06
21	280011	Bakar (Croatia)	45 18 N 14 32 E	93–06
22	280013	Zadar (Croatia)	44 07 N 15 14 E	93–06
23	280021	Split RT Marjana (Croatia)	43 30 N 16 23 E	93–06
24	280031	Split Harbor (Croatia)	43 30 N 16 26 E	93–06
25	280046	Sucuraj (Croatia)	43 08 N 17 12 E	93–05
26	280081	Dubrovnik (Croatia)	42 40 N 18 04 E	93–06
27	290001	Preveza (Greece)	38 57 N 20 46 E	93–06
28	290004	Levkas (Greece)	38 50 N 20 42 E	93–06
29	290014	Patrai (Greece)	38 14 N 21 44 E	93–06
30	290017	Katakolon (Greece)	37 38 N 21 19 E	93–07
31	290021	Kalamai (Greece)	37 01 N 22 08 E	93–01
32	290030	North Salaminos (Greece)	37 57 N 23 30 E	93–00
33	290034	Khalkis North (Greece)	38 28 N 23 36 E	93–07
34	290051	Thessaloniki (Greece)	40 37 N 23 02 E	93–07
35	290037	Skopelos (Greece)	39 07 N 23 44 E	00–07
36	290061	Kavalla (Greece)	40 44 N 24 25 E	93–07
37	290065	Alexandroupoli (Greece)	40 51 N 25 53 E	93–07
38	290071	Chios (Greece)	38 23 N 26 09 E	93–07
39	290081	Siros (Greece)	37 26 N 24 55 E	98–07
40	290091	Leros (Greece)	37 05 N 26 53 E	93–07
41	290097	Soudhas (Greece)	35 30 N 24 03 E	93–01
42	290110	Rodhos (Greece)	36 26 N 28 14 E	93–07
43	310038	Erdek (Turkey)	40 23 N 27 51 E	99–04
44	310042	Mentes/İzmir (Turkey)	38 26 N 26 43 E	99–04
45	310046	Bodrum II (Turkey)	37 02 N 27 25 E	99–04
46	310052	Antalaya II (Turkey)	36 50 N 30 37 E	99–04
47	320016	Hadera (Israel)	32 28 N 34 53 E	93–07

TG location. The function takes the value 1/2 at a distance of 50 km from the TG and vanishes beyond 500 km. Altimetry data are further averaged to monthly values.

2.3. GPS Data

For comparison purposes we have included in the analysis the measurements from 23 GPS stations from the EUREF Permanent Network (EPN), which are co-located at, or close to, a TG station (Fig. 1; Table 2). Most of these GPS time series did not start until the late 1990s or early 2000s. Since the differences between altimetry and TG only give information about the vertical land motion, we only consider the GPS vertical rate of change. We use the EPN cumulative solution expressed in the global ITRF2005 reference frame (see Kenyeres and Bruyninx 2004) as provided by EUREF. Details about the estimation procedure are available through the "Time Series Analysis" web page at the EPN Central Bureau (http://epncb.oma.be/_dataproducts/products/timeseriesanalysis/index.php)

3. Analysis and Results

SLV values in the Mediterranean present a strong seasonal signal with amplitudes of the annual cycle that range from 4 to 16 cm, explaining most of the SLV variability (Vigo *et al.*, 2011). This is produced by both changes in the density of the water column and in the water mass budget (Garcia-Garcia *et al.*, 2010). For the studied period, a model accounting for the annual, semiannual, and a linear trend was adjusted to the TG and altimetry time series using a least-squares fit to the following model:

$$\begin{aligned}\text{signal} &= a + bt + A_{\text{a}}\sin(\omega_{\text{a}}t - \varphi_{\text{a}})\\ &\quad + A_{\text{sa}}\sin(\omega_{\text{sa}}t - \varphi_{\text{sa}}),\end{aligned} \tag{1}$$

where t is time, A is amplitude, ω is frequency, φ is phase, and the subscripts "a" and "sa" refer to annual and semiannual cycles, respectively.

Then, we obtain the nonseasonal time series by subtracting all terms in this model (except for the trend one) from the original time series. After that, for each TG location, we get the difference time series for altimetry minus TG by subtracting the TG values from the geographically interpolated altimetry time series.

If all measurements were perfect and no errors were introduced by the altimetry geographical

Table 2

Vertical velocities of the EUREF GPS stations

No.	Code	Site	Location	Span	Trend	Sigma	
01	HUEL	Huelva (Spain)	37 08 N 06 50 W	07–09	−0.13	0.57	≈
02	CEU1	Ceuta (Spain)	35 54 N 05 19 W	08–09	−3.10	0.50	↓
03	MALA	Málaga (Spain)	36 43 N 04 25 W	06–09	0.43	0.34	↑
04	ALME	Almería (Spain)	36 51 N 02 28 W	01–09	−0.32	0.15	↓
05	ALAC	Alicante (Spain)	38 20 N 00 29 W	99–09	−1.47	0.16	↓
06	VALE	Valencia (Spain)	39 28 N 00 20 W	01–09	−1.26	0.23	↓
07	EBRE	Ebre (Spain)	41 49 N 00 30 E	97–09	−0.72	0.11	↓
08	BELL	Bellmunt (Spain)	41 36 N 01 24 E	99–09	−0.41	0.16	↓
09	CREU	Cap de Creus (Spain)	42 19 N 03 18 E	99–09	−0.36	0.24	↓
10	MARS	Marseille (France)	43 18 N 05 21 E	98–09	−0.99	0.25	↓
11	GRASS	Caussels (France)	43 45 N 06 55 E	96–09	0.50	0.14	↑
12	GENO	Genova (Italy)	44 25 N 08 55 E	99–09	−0.58	0.16	↓
13	PADO	Padova (Italy)	45 24 N 11 52 E	01–09	−0.67	0.19	↓
14	VENE	Venice P.S. (Italy)	45 26 N 12 20 E	02–07	−1.51	0.29	↓
15	GSR1	Geoservis GR. 1 (Slovenia)	46 03 N 14 33 E	02–09	0.04	0.24	≈
16	DUBR	Dubrovnik (Croatia)	42 40 N 18 04 E	00–09	−1.42	0.20	↓
17	AUT1	Thessaloniki (Greece)	40 37 N 23 02 E	05–09	−2.53	0.23	↓
18	NOA1	Athens (Greece)	38 00 N 23 48 E	06–09	−1.24	0.37	↓
19	TUC2	Chania (Greece)	35 32 N 24 04 E	04–09	−0.82	0.30	↓
20	RAMO	Mitzpe Ramon (Israel)	30 36 N 34 46 E	98–09	0.63	0.20	↑
21	DRAG	Metzoke Dragot (Israel)	31 36 N 35 24 E	00–09	5.08	0.38	↑
22	ISTA	Istanbul (Turkey)	41 06 N 29 01 E	00–09	1.03	0.14	↑
23	TUBI	Gebze (Turkey)	40 47 N 29 27 E	99–09	−1.89	0.14	↓

Columns from left to right correspond to: numbering shown in Fig. 1, EUREF station code, city and country, latitude and longitude coordinates, span of time, vertical velocities (mm/year), and error estimate. The last column classifies the trends as negative values (meaning land subsidence, indicated by ↓), positive values (meaning land uplift, indicated by ↑), and values indistinguishable from zero within 1 sigma (meaning no significant motion, indicated by ≈)

interpolation, SLV as measured by both TG and altimetry should be identical and thereby cancel in the altimetry minus TG time series. In this case, the remaining signal in the residuals should be produced by vertical motion at the TG location. Unfortunately, neither measurements nor procedure are perfect, and there are several source of errors, including loss of accuracy close to the coasts due to altimetry interpolation to the TG site.

Nevertheless, both datasets are remarkably consistent over their common time interval, and have close similarity in their variations of sea level at low and high frequencies. The agreement of the records provides confidence in the quality of their data. However, when examining the long-term behavior of SLV, discrepancies between the two of them arise, and these are our sought signal. As an illustration of this comparison, the time series corresponding to Marseille, Trieste, and Rovinj are depicted in the first, second, and third rows of Fig. 2, respectively. In the left column, we present the original observation data from the two sources; in the middle one, the same signals are presented, but removing their seasonal component as described above; in the right column, we can see the respective nonseasonal altimetry minus TG time series.

Then, their linear trends are estimated by a robust multilinear regression analysis, which iteratively reweights the least squares via a bi-square weighting function to minimize the effect of outliers in the time series. The weights are given by

$$\omega = \begin{cases} (1-r^2)^2, & \text{if } r<1; \\ 0, & \text{otherwise;} \end{cases} \quad \text{and} \qquad r = \frac{e}{4{,}685 s\sqrt{1-h}}. \tag{2}$$

In the last formula, e is the residual vector of the first iteration, h is the vector of leverage values from a least-squares fit, and s is computed as MAR/0.6745, where MAR is the median absolute residual.

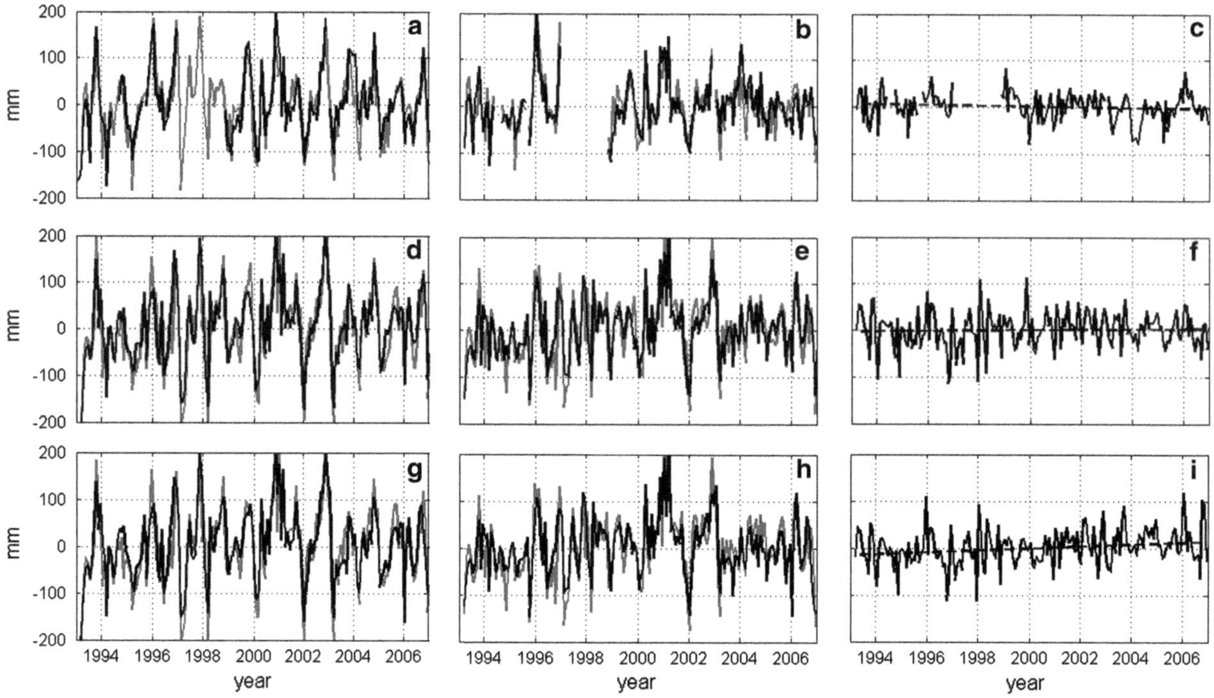

Figure 2
The *first* and *second columns* show seasonal and nonseasonal altimetry (in *black*) and TG (*grey*) time series at Marseille (*first row*), Trieste (*second row*), and Rovinj (*third row*). The *third column* shows the nonseasonal altimetry minus TG time series and its linear term. Units are mm

The resultant linear trends of the nonseasonal altimetry minus TG time series can be seen in Table 3 (column 3) and in Fig. 3. Positive linear trends represent land uplift while negative ones represent land subsidence, so Fig. 3 presents positive trends using triangles pointing up and negative ones by triangles pointing down. Circles in the figure represent trends with values indistinguishable from zero within 1 standard deviation, which account for 32% of the total TG. Table 3 also presents the GIA correction applied at each TG site, which is negligible for most of the locations.

For comparison purposes we compare our estimates of vertical crustal motion with estimates obtained from independent GPS measurements nearby the TG sites (Fig. 1). In general, their time span is shorter than that of the altimetry minus TG time series, since most of these GPS time series started in the late 1990s or early 2000s. For this reason, the period of study for both datasets does not exactly match, which can lead to differences in the estimated vertical motions. On the other hand, the EPN cumulative solution, even though it is in the global frame, is slightly biased as it cannot perfectly reproduce the global frame (A. Kenyeres and J. Legrand, 2009, personal communication). Moreover, the geophysical corrections applied to the GPS data are different from those used for the TG and altimetry data. Table 3 shows that the vertical motions estimated from altimetry minus TG and from GPS only agree in 4 of 13 cases. These discrepancies are in part due to the time span of the GPS time series, which is not coincident with that of the altimetry minus TG time series, or is very short. In fact, the qualitative agreement between both techniques increases to four of five cases when the linear trends are estimated for a common time span of at least 5 years (Table 4).

4. Discussion and Conclusions

The main objective of this work is to provide an estimation of the vertical land motion along the Mediterranean coastline from altimetry and TG sea

Table 3

Vertical velocities

TG		Altimetry minus TG (GIA corrected)	GIA correction		GPS		
No.	Site	Trend (mm/year)	ICE-5Gv1.2 (VM4) Trend (mm/year)		Station	Trend (mm/year)	
01	Huelva	−3.47 ± 1.15	−0.09	↓	HUEL	−0.13 ± 0.57	≈
02	Bonanza	−4.21 ± 0.74	−0.09	↓			
03	Cádiz III	0.25 ± 1.05	−0.07	≈			
04	Algeciras	−2.55 ± 1.69	−0.08	↓			
05	Tarifa	−4.91 ± 0.69	−0.07	↓			
06	Ceuta	0.82 ± 0.78	−0.08	↑	CEU1	−3.10 ± 0.50	↓
07	Málaga II	−3.58 ± 0.81	−0.14	↓	MALA	0.43 ± 0.34	↑
08	Valencia	−14.91 ± 1.16	−0.08	↓	VALE	−1.26 ± 0.23	↓
09	Barcelona	−5.25 ± 0.71	0.01	↓	BELL	−0.41 ± 0.15	↓
10	L'Estartit	−0.54 ± 1.18	0.01	≈	CREU	−0.36 ± 0.24	↓
11	Sète	−3.51 ± 1.18	−0.16	↓			
12	Marseille	−0.45 ± 0.62	−0.07	≈	MARS	−0.99 ± 0.25	↓
13	Toulon	−0.13 ± 0.66	−0.02	≈			
14	Nice	−3.09 ± 1.32	−0.12	↓	GRASS	0.50 ± 0.14	↑
15	Monaco C.	−5.08 ± 2.33	−0.13	↓			
16	Valletta	−6.98 ± 2.18	0.09	↓			
17	Venice P. S.	0.86 ± 2.03	−0.34	≈	VENE	−1.51 ± 0.29	↓
18	Trieste	1.10 ± 0.73	−0.34	↑			
19	Luka Koper	9.74 ± 1.64	−0.34	↑			
20	Rovinj	2.62 ± 0.77	−0.34	↑			
21	Bakar	0.89 ± 0.81	−0.34	↑			
22	Zadar	0.81 ± 1.04	−0.31	≈			
23	Split RT M	−0.49 ± 0.68	−0.29	↓			
24	Split H	−0.86 ± 0.73	−0.29	↓			
25	Sucuraj	0.59 ± 0.85	−0.29	≈			
26	Dubrovnik	−0.82 ± 0.67	−0.30	↓	DUBR	−1.42 ± 0.20	↓
27	Preveza	0.67 ± 0.88	−0.27	≈			
28	Levkas	0.64 ± 0.86	−0.25	≈			
29	Patrai	−9.17 ± 0.95	−0.25	↓			
30	Katakolon	0.09 ± 0.89	−0.17	≈			
31	Kalamai	−0.06 ± 1.79	−0.14	≈			
32	N Salaminos	9.68 ± 2.14	−0.25	↑	NOA1	−1.24 ± 0.37	↓
33	Khalkis N	−2.93 ± 0.76	−0.30	↓			
34	Thessaloniki	−0.67 ± 0.87	−0.42	↓	AUT1	−2.53 ± 0.23	↓
35	Skopelos	0.16 ± 2.43	−0.35	≈			
36	Kavalla	2.30 ± 1.93	−0.41	↑			
37	Alexandroupoli	0.12 ± 1.01	−0.40	≈			
38	Chios	0.89 ± 0.84	−0.32	↑			
39	Siros	2.48 ± 1.34	−0.19	↑			
40	Leros	4.91 ± 0.82	−0.20	↑			
41	Soudhas	5.17 ± 1.52	0.01	↑	TUC2	−0.82 ± 0.20	↓
42	Rodhos	−3.26 ± 0.72	−0.16	↓			
43	Erdek	2.45 ± 4.42	−0.42	≈			
44	Mentes/İzmir	0.16 ± 3.33	−0.36	≈			
45	Bodrum II	−12.23 ± 2.20	−0.23	↓			
46	Antalaya II	−11.90 ± 2.72	−0.33	↓			
47	Hadera	−4.72 ± 0.85	−0.22	↓			

The first column gives the TG number, and the second one the TG station name. The third column gives the linear rate of change and the formal error estimate (1 sigma) of the altimetry minus TG time series. The forth column gives the applied GIA correction. The fifth column gives the qualitative estimated motion (↑: uplift; ↓: subsidence; ≈: nonsignificant motion). The sixth, seventh, and eighth columns give, when possible, the closest GPS station from EUREF, its vertical rate, and the qualitative motion, respectively. All units are mm/year

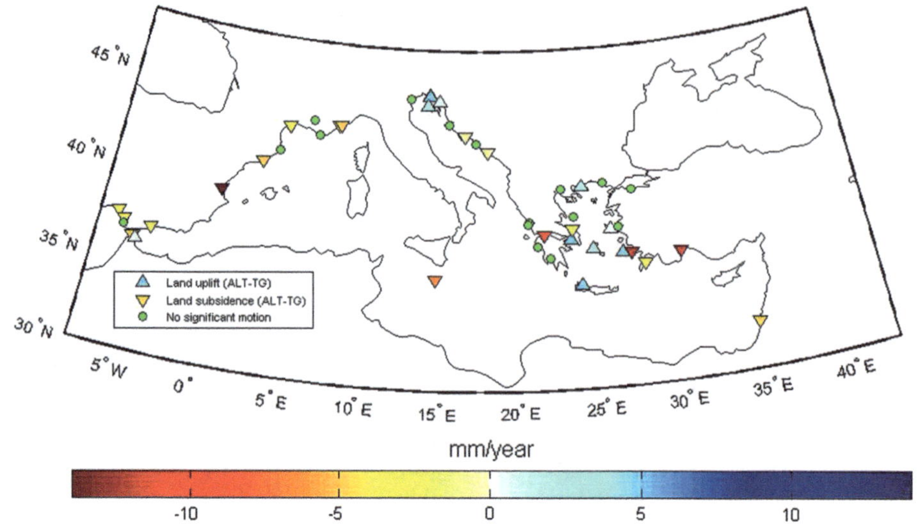

Figure 3

Colored triangles indicate the vertical land movement along the northern Mediterranean coast derived from the altimetry minus TG time series according to the third column of Table 3. *Triangles* pointing up and down represent ground uplift and subsidence, respectively. *Circles* indicate values indistinguishable from zero within 1σ

Table 4

As Table 3, but the vertical velocities are estimated in the common time span of the altimetry minus TG and the closest GPS time series

Site TG	Span	Altimetry minus TG (GIA corrected, mm/year)	GPS (mm/year)	GPS station
Valencia	01–05	−26.12 ± 3.47	−4.44 ± 0.37	VALE
Barcelona	99–07	−7.03 ± 1.41	−2.12 ± 0.14	BELL
Marseille	99–07	−0.57 ± 1.24	−3.04 ± 0.09	MARS
Nice	98–07	−3.09 ± 1.34	−0.50 ± 0.08	GRASS
Dubrovnik	01–05	−7.55 ± 2.64	−3.69 ± 0.23	DUBR

level measurements. The estimated vertical motions are potentially representative of tectonic uplift or subsidence at most of the sites. Due to the proximity to the former Alpine and Fennoscandian ice sheets, the northern Mediterranean coasts are potentially affected by the GIA (Stocchi *et al.*, 2005). Nevertheless, GIA-related deformation of the whole basin is mainly driven by the response of the solid Earth and the geoid to loading effects of melt water since the deglaciation, which contributes to significant and widespread subsidence (Stocchi and Spada, 2007, 2009). In a recent work, Stocchi and Spada (2009) performed a comprehensive analysis with different GIA model predictions, providing upper and lower bounds on the current rate of SLV associated with GIA in the Mediterranean. The GIA-induced rate of SLV in the eastern Mediterranean is close to zero, but at the western margin produces a falling sea level with maximum values close to −0.3 mm/year. On the other hand, the central Mediterranean is the most sensitive to GIA, reaching maximum values around 0.8 mm/year (eastern coast of Calabria and Sicily). This study accounts for the GIA correction according to Peltier (2004), although this does not constitute a significant portion of the observed motion (Table 3). However, at quite stable sites such as Trieste (northern Italy), the GIA accounts for around 30% of the reported vertical land motion.

The results obtained in this work are in accordance with the present geodynamic framework of the Mediterranean region, whose geology has been shaped by the interplay between two plates, the

African and Eurasian ones, and smaller intervening microplates. As a result, this region is characterized by three main subsidence zones, i.e., the Alps–Betic, the Apennines–Maghrebides, and the Dinarides–Hellenides–Taurides, closely related to which are the Carpathian subsidence and the Pyrenees (see Fig. 4, modified from CARMINATI and DOGLIONI, 2004 with permission from the authors). These subsidence zones present variable rates of activity, being higher in the eastern Mediterranean and becoming quieter towards the north. However, many fundamental issues of the very complex tectonic regime of microplates remain unresolved, especially with respect to vertical motions.

Some of the present plate boundaries, especially in the eastern Mediterranean, appear to be so diffuse and so anomalous that they cannot be categorized into the three classical types of plate (boundaries-subduction, spreading, and transform). According to measurements from space-geodetic techniques, the Eurasian and African Plates converge along NW–SE direction, both rotating anticlockwise (JIMENEZ-MUNT *et al.*, 2003; see also NASA database on present global plate motion http://sideshow.jpl.nasa.gov/mbh/series.html). As one of the most seismotectonic active areas in Europe, the Southern Aegean and the Mediterranean ridge should be taken into special consideration. In this area, including western Turkey and Greece, many destructive earthquakes have occurred throughout history. The Island of Crete is situated on the north side of the Hellenic Arc, which is the result of the ongoing collision between the African and Eurasian Plates, with the consequent escape of the Anatolian Microplate (Turkey) sideways from the Arabia–Eurasia collision (JIMENEZ-MUNT and SABADINI, 2002; KREEMER and CHAMOT-ROOKE, 2004).

During the last 15 years, continuous and campaign-type GPS measurements have been used to determine the crustal motion and deformation of the eastern Mediterranean (e.g., COCARD *et al.*, 1999; BRIOLE *et al.*, 2000; KAHLE *et al.*, 2000). In a recent study by HOLLENSTEIN *et al.* (2008) the crustal dynamics of the area of Greece and the Aegean Sea were estimated from a detailed high-quality solution

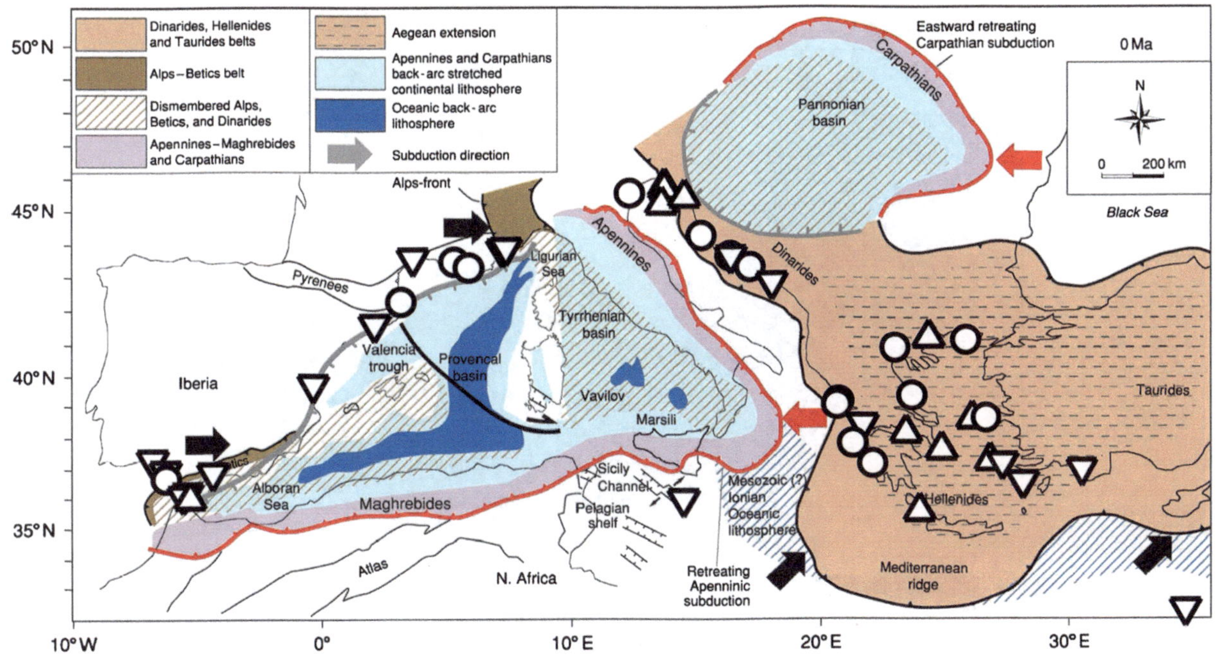

Figure 4

Geodynamic framework for the Mediterranean region as provided by CARMINATI and DOGLIONI, (2004), on which we have indicated the TG sites considered in our work as *triangles* and *circles* representing the estimated vertical motions from the altimetry minus TG time series. *Triangles* pointing up and down represent ground uplift and subsidence, respectively. *Circles* indicate values indistinguishable from zero within 1σ

based on combined processing of existing and new continuous and campaign-type GPS measurements carried out between 1993 and 2003. Hollenstein *et al.* (2008) have also estimated the two-dimensional dilatation rates from the strain rate fields, providing maps of crustal extension and compression of the region (see Fig. 5, modified from Hollenstein *et al.* 2008 with permission from the authors). In general, crustal compression has an effect of land uplift, whereas crustal extension produces land subsidence. Bearing this fact in mind, our results in the region are in very good agreement with those provided by Hollenstein *et al.* (2008). Of 19 TG in the region, 9 present significant vertical motion. Of the latter, three TG presenting subsidence (numbers 29, 42, and 45) are in extension zones, three TG presenting uplift (numbers 32, 39, and 41) are in compression zones, and only three TG (numbers 36, 38, and 40) do not agree with the extension/compression zones. Among the other ten TG with values indistinguishable from zero (within 1 standard deviation), five of them (numbers 27, 28, 30, 31, and 34) are close to

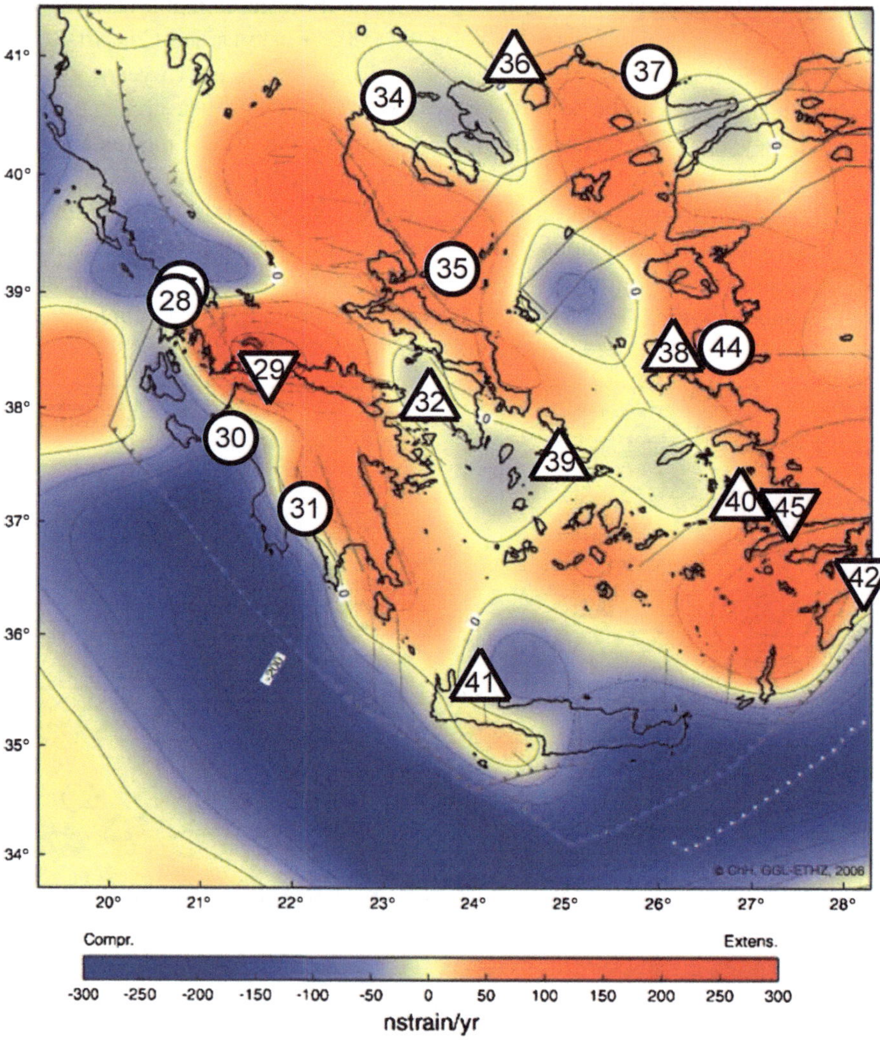

Figure 5

Crustal extension/compression map from Hollenstein *et al.* (2008) with qualitative estimates of the vertical crustal motion from the altimetry minus TG time series superimposed. Extension areas (representing subsidence) are shown in *red*, and compression areas (representing uplift) in *blue*. Units are number strain per year. *Numbers* correspond to TGs (Table 1). *Triangles* pointing up and down represent ground uplift and subsidence, respectively. *Circles* indicate values indistinguishable from zero (within 1σ)

transition zones from extension-compression. This remarkable consistency between both estimates for the vertical crustal displacement, in such an active region, based on completely independent techniques, provides confidence in terms of the reliability of the results. Note that the African block is subsided northward, producing uplift of the Island of Crete (Rahl *et al.*, 2004), which is in agreement with our results found at Soudhas (TG number 41).

Land movement at the coast of the Adriatic Sea is caused by the oblique contacts between the Adriatic Microplate and the Dinarides. There exist three sections of the Adriatic Microplate, differing in size and rate of movement. This results in different land movement behavior along the Dinarides zone in the northern, central, and the southern part (Kuk *et al.*, 2000). Our results for the coast of the Adriatic Sea show a different land movement behavior in the northern, central, and southern part of the Dinarides zone.

It should be noted that not all the vertical crustal motion reported in this study has a geological origin; for example, Venice must be considered as a special case due to the large influence of anthropogenic effects (e.g., Carminati and Di Donato, 1999). In fact, in spite of being located in a subsidence area (Rutigliano *et al.*, 2000), neither this study nor Becker *et al.* (2002) showed such subsidence. Another case is Valencia, where the altimetry minus TG time series shows an uplift of 3.2 mm/year for the period 1995–1999, which is reversed and increased by a factor of 10, reaching −29.8 mm/year, for the period 2000–2005 (Fig. 6). The origin of this abrupt change is not geology but the construction of a bridge weighing more than 2,000 tons in the Port of Valencia, close to the TG site.

In summary, in spite of several possible error sources in the altimetry minus TG time series, with a sufficiently long time span reaching 15 years, we are able to obtain estimates for the vertical ground motion at the TG sites along the northern Mediterranean Sea coast. These estimates are in agreement with independent techniques such as GPS. When no continuous GPS measurements are available, we provide new and useful information for those sites. These results provide useful source data for studying, contrasting, and constraining tectonic models of the region.

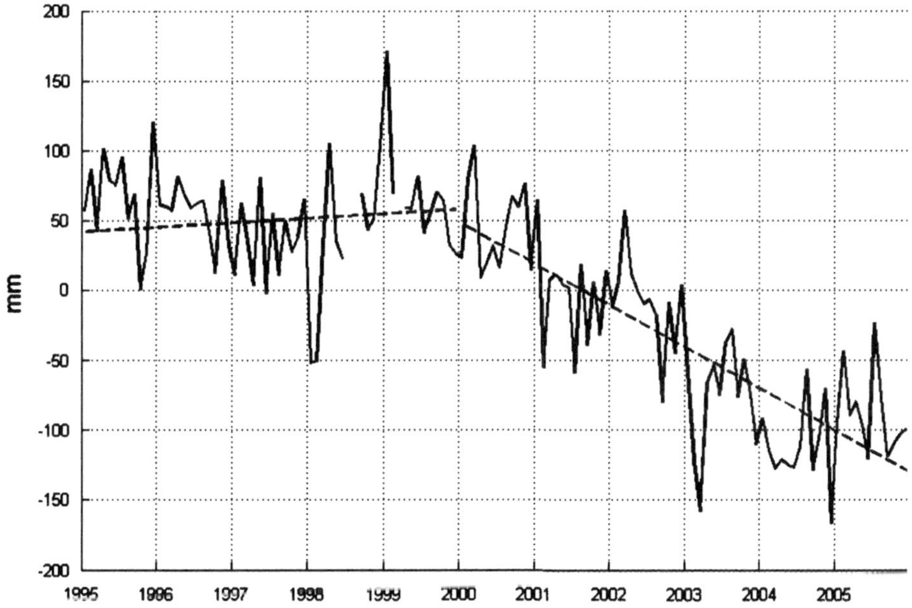

Figure 6
The *solid line* represents the altimetry minus TG time series of Valencia (east coast of Spain). *Dashed lines* represent the linear rates estimated for the periods 1995–2000 and 2000–2006, whose values are 3.2 and −29.8 mm/year, respectively

Acknowledgments

We thank Jean-Paul Boy for atmospheric and hydrological loading analysis data support and M. Carmen Martinez-Belda for constructive discussion about the GPS data. Daniel García-Castellanos is acknowledged for his valuable comments on Mediterranean tectonics. The authors would also like to acknowledge all the institutions that have participated in the development and distribution of the data products used in this analysis. We obtained tide gauge data from the PMSLS. Altimetry data used in this study were developed, validated, and distributed by CTOH/LEGOS, France. We used GPS vertical crustal velocities provided by EPN. This work was partly funded by the Spanish Ministry of Education and Science under research projects CGL2010-12153-E and AYA2009-07981.

References

Barnett, T.P. (1984), *The Estimation of "Global" Sea Level Change: A Problem of Uniqueness*, J. Geophys. Res. *89* (C5), 7980-7988.

Becker, M., Zerbini, S., Baker, T., Brki, B., Galanis, J., Garate, J., Georgiev, I., Kahle, H.-G., kotzev, V., Lobazov, V., Marson, I., Negusini, N., Richter, B., Veis, G., and Yuzefovich., P. (2002), *Assessment of height variations by GPS at Mediterranean and Black Sea coast tide gauges from the SELF projects*, Global Planet. Change, *34*, 5–35.

Bennett, R.A., Hreinsdóttir, S., Buble, B., Bai, T., Bai, E., Marjanovi, M., Casale, G., Gendaszek, G., and Cowan, D. (2008), *Eocene to present subduction of southern Adria mantle lithosphere beneath the Dinarides*, Geology v. *36* no. 1 p. 3-6 doi:10.1130/G24136A.1.

Berrisford, P., Dee, D., Fielding K., Fuentes, M., Kaallerg, P., Kobayashi, S., and Uppala, S.M. (2009), *The ERA-Interim Archive*, Tech rep., ERA Report Series No 1.

Briole P., Rigo A., Lyon-Caen H., Ruegg J.C., Papazissi K., Mitsakaki C., Balodimou A., Veis G., Hatzfeld D., and Deschamps A. (2000), *Active deformation of the Corinth rift, Greece: Results from repeated Global Positioning System surveys between 1990 and 1995*: J. Geophys. Res. *105* (25), 605-625.

Carminati, E., and Di Donato, G. (1999), *Separating natural and anthropogenic vertical movements in fast subsiding areas: The Po Plain (N. Italy) Case*, Geophys. Res. Lett., *26* (15), 2291–2294.

Carminati, E., and Doglioni, C. Mediterranean tectonics. Encyclopedia of Geology (Elsevier, 2004).

Cazenave, A., Dominh, K., Ponchaut, F., Soudarin, L., Cretaux, J.F., and Le Provost, C. (1999), *Sea level changes from Topex-Poseidon altimetry and tide gauges, and vertical crustal motions from DORIS*. Geophys. Res. Lett., *26*, 2077-2080.

Chelton, D.B., and Enfield, D.B. (1986), *Ocean Signals in Tide Gauge Records*, J. Geophys. Res. *91* (B9), 9081-9098.

Cocard, M., Kahle, H.G., Peter, Y. Geiger, A., Veis, G., Felekis, S., Paradisis, D. and Billiris, H. (1999), *New constraints on the rapid crustal motion of the Aegean region: recent results inferred from GPS measurements (1993-1998) across the West Hellenic Arc, Greece*, Earth Planet. Sci. Lett., *172*, 39-47.

Desai, S.D. (2002), *Observing the pole tide with satellite altimetry*, J. Geophys. Res., *107*(C11), 3186, doi:10.1029/2001JC001224.

Douglas, B.C. (1991), *Global Sea Level Rise*, J. Geophys. Res., *96* (C4), 6981–6992.

Garcia, D., Vigo, I., Chao, B.F. and Martínez, M.C. (2007), *Vertical Crustal Motion along the Mediterranean and Black Sea Coast Derived from Ocean Altimetry and Tide Gauge Data*, Pure Appl. Geophys. 164, 851–863.

Garcia-Garcia, D., Chao, B.F., and Boy, J.-P. (2010), *Steric and Mass-Induced Sea Level Variations in the Mediterranean Sea, Revisited*, J. Geophys. Res., in press, doi:10.1029/2009JC005928.

Hollenstein, C.H., Müller, M.D., Geiger, A., and Kahle, H.G. (2008), *Crustal motion and deformation in Greece from a decade of GPS measurements, 1993–2003*, Tectonophysics 449, 17–40.

Joint Technical Commission for Oceanography and Marine Meteorology (2006), Manual on Sea Level Measurement and Interpretation Volume IV: An Update to 2006. Technical Report No. 31WMO/TD. No. 1339. http://www.jcomm.info/.

Jimenez-Munt, I., and Sabadini, R. (2002), *The block-like behavior of Anatolia envisaged in the modeled and geodetic strain rates*, Geophys. Res. Lett. *29* (20), 1978, doi:10.1029/2002GL015995.

Jimenez-Munt, I., Sabadini, R., Gardi, A., and Bianco, G. (2003), *Active deformation in the Mediterranean from Gibraltar to Anatolia inferred from numerical modeling and geodetic and seismological data*, J. Geophys. Res., *108*(B1), 2006, doi: 10.1029/2001JB001544.

Kahle, H.-G., Cocard, M., Peter, Y., Geiger, A., Reilinger, R., Barka, A., and Veis, G. (2000), *GPS-derived strain rate field within the boundary zones of Eurasian, African, and Arabian plates*, J. Geophys. Res., 105 (23), 23353-23370.

Kenyeres, A., and Bruyninx, C. (2004), Monitoring of the EPN Coordinate Time Series for Improved Reference Frame Maintenance GPS solutions, Vol 8, No 4, 200-209.

Kreemer, C., and Chamot-Rooke, N. (2004), *Contemporary kinematics of the southern Aegean and the Mediterranean Ridge*. Geophysical Journal International, v. *157*, no. 3, pp 1377–1392.

Kuk, V., Prelogovic, E., and Dragicevic, I. (2000), *Seismotectonically Active Zones in the Dinarides*. Geol. Croat, *53/2*, 295-303.

Le Traon, P., and Gauzelin, P. (1997), *Response of the Mediterranean mean sea level to atmospheric pressure forcing*, J. Geophys. Res., *102*(C1), 973–984.

Peltier W.R. (2004), *Global Glacial Isostasy and the Surface of the Ice-Age Earth: The ICE-5G(VM2) model and GRACE*, Ann. Rev. Earth Planet. Sci., *32*, 111-149.

Petrov L., and Boy, J.-P. (2004) *Study of the atmospheric pressure loading signal in VLBI observations*, J. Geophys. Res., Vol. *109*, No. B03405. doi:10.1029/2003JB002500.

Rahl, J. M., Fassoulas, C., and Brandon, M.T. (2004), Exhumation of high-pressure metamorphic rocks within an active convergent margin, Crete, Greece: A field guide, 32th International Geological Congress, Vol. 2 – from B16 to B33, art. No B32.

Ray, R.D., Beckley, B.D., and Lemoine, F.G. (2010), *Vertical crustal motion derived from satellite altimetry and tide gauges, and comparisons with DORIS measurements*, Adv. Space Res., *45*, 1510–1522.

RUTIGLIANO, P., FERRARO, C., DEVOTI, R., LANOTTE, R., LUCERI, V., NARDI, A., PACIONE, R., and SCIARRETTA, C. (2000), Vertical motions in the Western Mediterranean area from geodetic and geological data, in the proceedings of The Tenth General Assembly of the Wegener Project.

SPENCER, N.E., and WOODWORTH, P.L. (1993), Data Holdings of the Permanent Service for Mean Sea Level, Bidston, Birkenhead: Permanent Service for Mean Sea Level. 81.

STOCCHI, P., SPADA, G. and CIANETTI, S. (2005), *Isostatic rebound following the Alpine deglaciation: impact on the sea level variations and vertical movements in the Mediterranean region*, Geophys. J. Int., *162*, doi:10.1111/j.1365-246X.2005.02653.x.

STOCCHI, P., and SPADA, G.(2007), *Glacio and hydro-isostasy in the Mediterranean Sea: Clark's zones and role of remote ice sheets*, Ann. Geophys., *50* (6), 741–761.

STOCCHI P., and SPADA, G. (2009), *Influence of glacial isostatic adjustment upon current sea level variations in the Mediterranean*. Tectonophysics, *474*, 56–68.

VIGO, M.I., SÁNCHEZ-REALES, J.M., TROTTINI, M., and CHAO B.F., *Mediterranean Sea level variations: Analysis of the satellite altimetric data, 1992–2008*. J. GEODYN. (2011), doi:10.1016/j.jog.2011.02.00).

WAHR, J.W. (1985), *Deformation Induced by Polar Motion*, J. Geophys. Res., *90* (B11), 9363–9368.

(Received January 3, 2010, revised July 14, 2010, accepted February 7, 2011, Published online September 15, 2011)

Pure Appl. Geophys. 169 (2012), 1425–1441

DOI 10.1007/s00024-011-0401-4

Pure and Applied Geophysics

Using a Mesoscale Meteorological Model to Reduce the Effect of Tropospheric Water Vapour from DInSAR Data: A Case Study for the Island of Tenerife, Canary Islands

ANTONIO EFF-DARWICH,[1,4] JUAN C. PÉREZ,[2] JOSÉ FERNÁNDEZ,[3] BEGOÑA GARCÍA-LORENZO,[4,5] ALBANO GONZÁLEZ,[2] and PABLO J. GONZÁLEZ[3,6]

Abstract—Measurements of ground displacement through classical Differential Interferometric SAR (DInSAR) and advanced DInSAR techniques have been carried out over the entire actively volcanic island of Tenerife, Canary Islands. However, a detailed analysis of the effect of tropospheric water vapour on DInSAR at Tenerife should be carried out to evaluate its influence, including correction models that might improve the accuracy of DInSAR derived deformation signals. Unlike water vapour correction models that are based on space platforms (e.g. MODIS and MERIS), we present an alternative approach that is based on precise water vapour estimations derived from mesoscale numerical meteorological models, in particular the Weather Research and Forecasting (WRF) model. The application of this approach to a set of DInSAR observations of the island of Tenerife shows encouraging results.

Key words: ground deformation, Tenerife, volcanic activity, water vapour.

1. Introduction

The potential of spaceborne differential interferometric synthetic aperture radar (DInSAR) for measuring ground deformation related to volcanic activity at spatial resolutions of a few tens of metres (DZURISIN, 2007) is well established. This is a powerful tool to characterize deformation styles on a large variety of volcanic edifices in conjunction with existing monitoring networks; moreover SAR imagery is also a unique tool for measuring ground deformations in remote areas (MASSONNET *et al.*, 1994; HANSSEN, 2001).

In the case of the volcanic island of Tenerife, Canary Islands, theoretical analysis carried out by YU *et al.* (2000) and EFF-DARWICH *et al.* (2008a) demonstrated the need to extend the existing geodetic network in the central area of the island (the Las Cañadas Caldera) to cover the full island for volcano monitoring purposes. Therefore, ground displacement analyses have been carried out on the entire island, by means of DInSAR techniques and complemented by GPS from 1992 to 2008 (FERNÁNDEZ *et al.*, 2003, 2004, 2005, 2009; SAMSONOV *et al.*, 2008; GONZÁLEZ *et al.*, 2010a). FERNÁNDEZ *et al.* (2009) used the Small Baseline Subset (SBAS) DInSAR algorithm (BERARDINO *et al.*, 2002) to process 55 radar images acquired from descending orbits by the ERS-1/2 satellites during 1992–2005. The SBAS technique removes artifacts due to atmospheric heterogeneities between acquisition pairs, observing that the atmospheric phase signal component is highly correlated in space but poorly in time. This filtering step also includes the compensation for the topography correlated atmospheric phase artefacts. They found four funnel-shaped areas with high displacement rates. The first one (located in the area labelled a in Fig. 1), which has the largest magnitude, affects an area of 15 km^2 with a deformation rate of about 15 mm/year. The second area, labelled b in Fig. 1, extends over

[1] Departamento de Edafología y Geología, Univ. de La Laguna, 38206 La Laguna, Tenerife, Spain. E-mail: adarwich@ull.es

[2] Departamento de Física Fundamental y Experimental, Electrónica y Sistemas, Universidad de La Laguna, 38205 La Laguna, Tenerife, Spain.

[3] Facultad de Ciencias Matemáticas, Instituto de Astronomía y Geodesia (CSIC-UCM), Ciudad Universitaria, Plaza de Ciencias, 3, 28040 Madrid, Spain. E-mail: jose_fernandez@mat.ucm.es; pjgonzal@mat.ucm.es

[4] Instituto de Astrofísica de Canarias, 38200 La Laguna, Tenerife, Spain.

[5] Departamento de Astrofísica, Universidad de La Laguna, 38205 La Laguna, Tenerife, Spain.

[6] *Present Address*: Department of Earth Sciences, University of Western Ontario, London, ON, Canada.

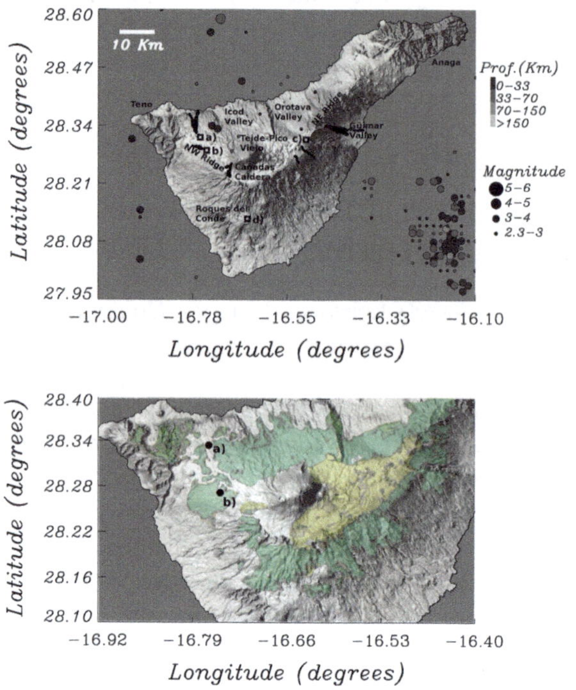

Figure 1
Upper panel shaded-relief map of the island of Tenerife. *Filled circles of different colours* and sizes represent the location and magnitudes of earthquakes registered in and around Tenerife from 1993 to 2008, whereas the *black-filled* areas represent the extension of the historical eruptions (last 500 years). Four major detected localized subsidences are represented by *black squares* labelled as *a, b, c* and *d*. See text for more details. *Lower panel* detailed shaded-relief map of the northwestern portion of Tenerife, where the spatial distribution of forests (*green areas*) and high altitude bushes (*yellow areas*) have been represented. The subsidences *a* and *b* of the upper panel are also represented in the lower panel

8 km^2 and it has a deformation rate of about 5–6 mm/year. The third zone (c in Fig. 1) has a deformation rate of 3 mm/year, whereas the last analyzed deformation (d in Fig. 1) shows a deformation rate of about 3 mm/year. These subsidences were found closer to the locations of the most recent eruptions in the island (Arenas Negras, Chahorra and Chinyero). In any case, the subsidences (in particular a and b) might spread over larger areas that could not be detected by interferometric observations due to the lack of coherence induced by the presence of densely vegetated areas, e.g. forests of pine trees and high altitude bushes (see lower panel of Fig. 1). It has also been found that the summit area of the volcanic edifice is characterized by a continuous subsidence extending well beyond Las Cañadas caldera rim and corresponding to the dense core of the island. These results, coupled with the observations of a 22-stations GPS network from 2000 to 2006 (Fernández *et al.* 2003, 2004, 2005), structural and geological information and deformation modelling, suggest that an intrusive complex beneath the central part of Tenerife is subsiding into a weak lithosphere and that the volcanic edifice is in a state of compression (Fernández *et al.* 2009). Another intriguing result obtained by Fernández *et al.* (2004) is a change in the deformation pattern in the period 2000–2002 with respect to the observed by DInSAR for the period 1992–2000. Similar changes were also detected by Fernández *et al.* (2009) for several areas (see their Fig. 2). Those changes could be related to the change in seismicity detected in the same period and could be a geodetic precursor of the 2004 volcano-tectonic crisis (Fernández *et al.* 2009). Examination of the geophysical observations on Tenerife, human activities and the results of the theoretical modelling seem to indicate that at least part of the observed deformation may be caused by changes in the groundwater level and therefore that part of the deformation might not be linked to volcanic and/or tectonic reactivation.

Ground deformation associated with volcanic activity might span from a few days, as in the case of the 2000 Miyakejima volcanic reactivation (Irwan *et al.* 2003), to months, as it was observed during the 2007 Tanzania rifting episode (Biggs *et al.* 2009). However, in time scales of days to months, or even years, DInSAR observations might be affected by physical processes that are not related to geological activity. Among them, atmospheric effects significantly limit the accuracy of DInSAR techniques, which largely result from changes in water vapour content in the troposphere (Hanssen, 2001). Because of the unpredictable character of atmospheric phase delays, it has, until recently, been difficult to separate atmospheric delays from the effects of deformation and topography without external data on the atmospheric water vapour content such as measurements provided by radio-soundings or permanent GPS arrays. However, satellite estimates of the atmospheric water vapour content, using the Moderate Resolution Imaging Spectro-radiometer (MODIS) or the Medium Resolution Imaging Spectrometer

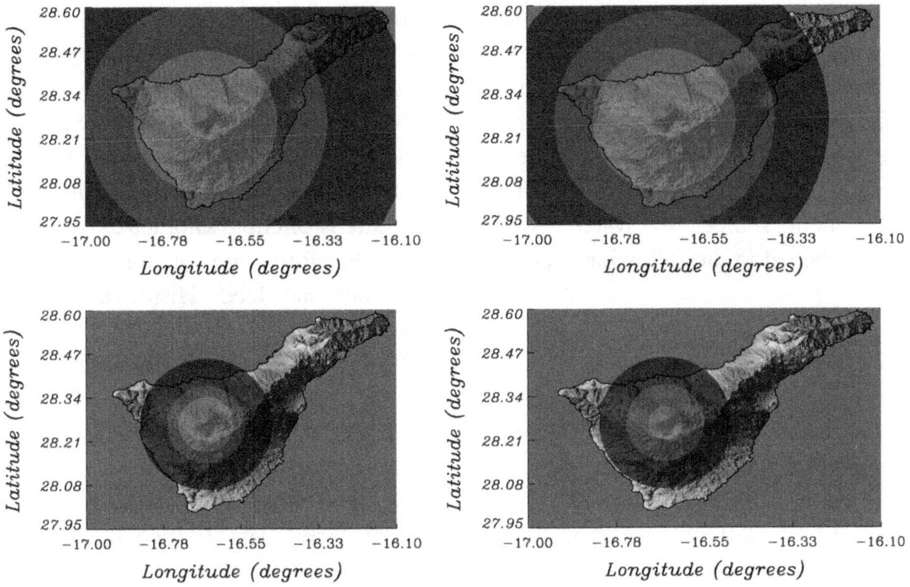

Figure 2
Areas affected by vertical displacement as the result of the activity of a magma chamber. The three grey areas (*dark to light grey*) represent crustal deformations of at least 1, 3 and 10 cm, respectively. *Upper* and *lower panels* present results for intrusions with $Pa^3 = 10^4$ MPa km^3 and $Pa^3 = 10^3$ MPa km^3, respectively (see text for details). *Left and right panels* correspond to the results obtained for magma intrusions located at a depth of 2 and 6 km below sea level, respectively

(MERIS) are now available on a global scale. These estimates of water vapour content in the troposphere have been used to successfully remove the atmospheric effects from DInSAR data (Li *et al.* 2009). However, satellite data have serious limitations, in particular water vapour content can be only retrieved over clear land masses and above the clouds, hence underestimating the amount of water vapour in cloudy days. Moreover, the date and time of acquisition of both satellite-derived water vapour and ground deformation might not coincide, hence introducing a new source of uncertainties.

Instead of satellite observations, a different approach would use a mesoscale meteorological model to perform a meteorological analysis of the atmosphere for the exact acquisition time for each scene of an SAR observational pair. In this sense, the mesoscale meteorological models are used to calculate the predicted atmospheric delay that might be used to generate a synthetic interferogram. This interferogram is then compared with the observed interferogram and subtracted from it, reducing the atmospheric noise and improving our ability to identify and resolve the geodetic signals. Foster *et al.* (2006) used the meteorological model MM5 (Grell *et al.*, 1995) to study the effect of tropospheric water vapour on SAR observations for Mauna Loa volcano, Hawaii. The same model was used by Puysségur *et al.* (2007) to analyse ground deformation on Lebanon. Moreover, Wadge *et al.* (2004) calculated atmospheric path delays over Etna volcano, Italy, through the NH3D model (Miranda and James, 1992). In these cases, the atmospheric models gave an overall picture similar to the observed SAR delays, in particular at long spatial wavelengths (tens of kilometres), hence accounting for the path delays associated with height differences and the trade winds. However, these models were not accurate enough to predict path delays with short spatial wavelengths (few kilometres).

In this work, we propose to use a state-of-the-art mesoscale numerical meteorological model, the Weather Research and Forecasting (WRF) model, to estimate the water vapour content of the troposphere. Our purpose is to assess the potential of WRF to properly calculate path delays at spatial scales of kilometres and hence, we will apply this methodology to DInSAR observations on the volcanic island of Tenerife, where abrupt changes in the topography and the effect of wet trade winds takes place within a few kilometres. In the following sections we will summarize the geology of Tenerife and the theoretical crustal deformation patterns expected from different types of volcanic reactivations. The results from the mesoscale numerical meteorological model will be compared to those obtained from MODIS observations, thus both sources of water vapour estimates will be explained in the following sections.

2. *Geological Setting*

Tenerife is the largest island of the Canarian Archipelago and one of the largest volcanic islands in the world. It is located between latitudes 28–29°N and longitudes 16–17°W, 280 km distant from the African coast (Fig. 1). It conforms an active volcanic region, its age varying from Middle Miocene to present, with no evidence of important gaps in its volcanic activity history, at least in the last 3–4 Ma (Ancochea *et al.* 1999). This activity is still evident in stationary low temperature fumarolic activity at Teide crater (<85°C), diffusive gaseous emissions, groundwater temperatures reaching up to 50°C and volcanic contamination of groundwater in the subsurface of the central region (Eff-Darwich *et al.* 2008b).

The morphology of Tenerife (see Fig. 1) is the result of a complex geological evolution: the subaerial part of the island was originally constructed by fissural eruptions of ankaramite, basanite and alkali basalts that occurred between 12 and 3.3 My (Guillou *et al.* 2004). These formations made up shield volcanoes that remain at present as three eroded massifs occupying the three corners of the island (Teno, Anaga and Roques del Conde massifs). In the central part of the island, from 3.5 My to present, the emission of basalts and differentiated volcanics gave rise to a large central volcanic complex, the Las Cañadas Edifice (Martí *et al.* 1994). After a period of mafic volcanism, several periods of phonolitic activity took place, culminating in the formation of a large elliptical depression measuring 16×9 km^2, known as Las Cañadas Caldera. In the northern sector of the caldera, the Teide-Pico Viejo complex was constructed as the product of the most recent phase of central volcanism. Teide-Pico Viejo is a large stratovolcano that has grown during the last 175 Ky. The post-shield basaltic activity, which overlaps the Las Cañadas Edifice, is mainly found on two ridges (NE and NW), which converge on the central part of the island (Ablay and Hurlimann, 2000). Large scale lateral collapses, involving rapid mass movements of hundreds of cubic kilometres of rock, are responsible for the formation of three valleys: La Orotava, Güímar and Icod. Historical eruptive activity has consisted of six strombolian eruptions (Cabrera and Hernández-Pacheco, 1987), namely Siete Fuentes (1704), Fasnia (1705), Arafo (1705), Arenas Negras (1706), Chahorra (1798) and Chinyero (1909). The last three eruptions occurred at the NW rift system, the most active area of the island together with El Teide-Pico Viejo Edifice for the last 50,000 years (Carracedo *et al.* 2003).

It is important to mention, in the context of volcanic geodetic monitoring, the possible existence of a shallow magma chamber underneath Teide-Pico Viejo. It is estimated from petrologic analyses (Araña *et al.* 1989) that the top of the magma chamber is located at sea level, having a volume of approximately 30 km^3 and a radius of 2 km (under the supposition of spherical shape). Thermodynamical (Díez and Albert, 1989) and chemical (Albert-Beltrán *et al.* 1990) modelling of the fumaroles at El Teide summit revealed that the present temperature at the top of the magma chamber would be approximately 350°C, whereas the top of the chamber coincides with that calculated from petrologic analyses (Araña *et al.* 1989). However, Araña *et al.* (2000) found long-wavelength magnetic anomalies in the central part of Tenerife that could be interpreted as the top of deep intrusive bodies or magma chambers zone ($\approx$5.7 km b.s.l.). In this sense, the possible location of the top of the magma chamber ranges from sea level to nearly 6 km below sea level.

It is also important for geodetic studies that most recent eruptions (<3 My) have been fed by dikes (Fernández *et al.* 2003). These dikes are usually associated with the two main rift zones of Tenerife, the NE and NW ridges (see Fig. 1) that converge in the central region of the island. In other cases, the dikes are located in shallow radial or circular fractures in large volcanic structures, e.g. the eruptive systems of the Teide-Pico Viejo volcano, in the central area of Tenerife. Most of the visible dikes are less than 1 m thick in the shallowest sections. However, when erosion exposes deeper sections, they are seen to be much thicker, especially those of a saline composition.

We carried out a theoretical analysis in order to study the crustal deformation patterns induced by the activity associated to a magma chamber and a dike injection, both being processes very common along the geological history of the island. For the case of the magma chamber, we followed the methodology devised by EFF-DARWICH *et al.* (2008a), that considered the elastic-gravitational deformation model described FERNÁNDEZ and RUNDLE (1994); FERNÁNDEZ *et al.* (1997, 2006). As it was already mentioned, the possible location of the magma chamber underneath El Teide-Pico Viejo is uncertain, ranging the position of the top of the chamber from approximately sea level to 6 km below sea level. In this sense, several calculations were carried out for different cases of spherical intrusion, varying its depth and the parameter Pa^3, being P and a the pressure and radius of the magma intrusion, respectively. Figure 2 presents the simulated ground displacement for intrusions located underneath El Teide-Pico Viejo stratovolcano at 2 and 6 km depth, that are characterized by two different values of Pa^3, namely 10^3 and 10^4 MPa km^3.

The second case of the theoretical analysis corresponds to the other possible kind of magmatic intrusion in Tenerife, namely dikes (YU *et al.*, 2000). To explain the observed ground deformation, we use the conventional assumption of dislocations buried in an elastic half space composed of a Poisson solid (OKADA, 1985). We specified eight parameters describing the rectangular fault patch: three centroid coordinates X, Y, h, strike angle α and dip angle δ along-strike length L and down-dip width W, and the slip vector of the tensile component (thickness of the dike) U_3, as illustrated by FEIGL and DUPRÉ (1999). These authors developed the numerical code RNGCHN that was used to calculate the ground deformation due to dike intrusion. Figure 3 shows the effects produced by a dike located in the NW ridge, $2L = 2$ km, $U_3 = 1$ m, and $\delta = 80°$, extending from the mantle (about 25 km depth) to 2.5, 1.25, 0.5 and 0.25 km depth, respectively.

In both cases of volcanic intrusion, crustal deformations of approximately 1 cm could extend over large portions of the island, in particular for the case of the magma chamber. However significant

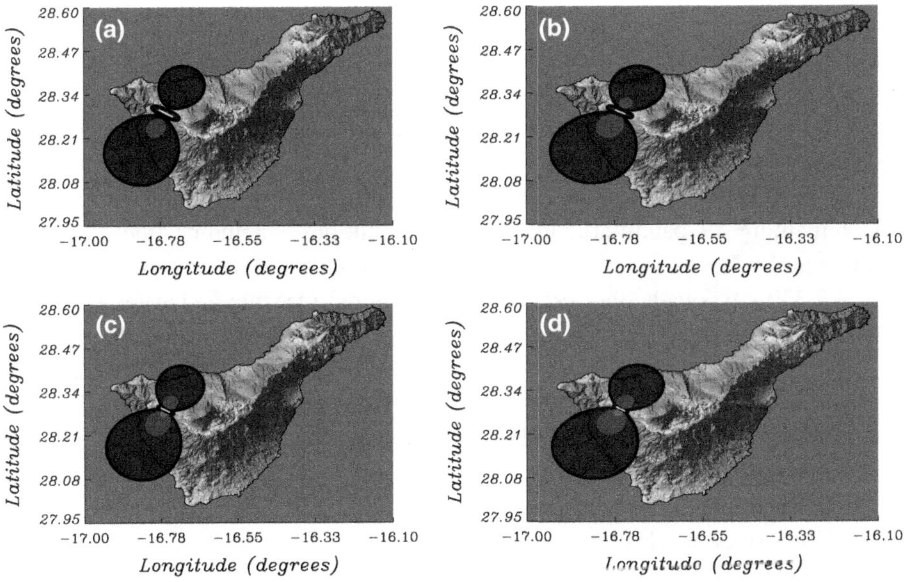

Figure 3
Areas affected by vertical displacement as the result of a nearly vertical ($\delta = 80°$) dike intrusion. The *three grey areas* (*dark to light grey*) represent crustal deformations of at least 1, 3 and 10 cm, respectively. The surface projection of the dike is represented by the white *solid line*. The dike extends from the mantle (about 25 km depth) to 2.5 km (*panel a*), 1.25 km (*panel b*), 0.5 km (*panel c*) and 0.25 km (*panel d*)

deformations larger than 3 cm are constrained to small spots around the surface projection of the dike or the magma chamber. Since the precise location of future eruptions is largely unknown, global observations of ground deformations, like those provided by DInSAR, are required.

3. *Data and Methodology*

In this work, three different data sets are analysed and compared to establish the potential of meteorological models to reduce the effect of tropospheric water vapour on DInSAR measurements of ground deformation. The three data sets correspond, on one hand to precipitable water vapour (PWV) measured by MODIS and calculated by the model WRF and, on the other hand, to phase delays measured by DInSAR.

3.1. Precipitable Water Vapour from MODIS

Various precipitable water products have been developed and are available operationally for assessing the state of the atmosphere with respect to the magnitude of the moisture and its transport. In this work, data provided by the Moderate Resolution Imaging Spectroradiometer (MODIS) sensor onboard Terra and Aqua satellites were employed. The MODIS instrument is a passive imaging spectroradiometer providing high radiometric sensitivity (12 bit) in 36 spectral bands ranging in wavelength from 0.4 to 14.4 μm (Fig. 4). Two bands are imaged at a nominal resolution of 250 m at nadir, with five bands at 500 m and the remaining 29 bands at 1 km. A ±55° scanning pattern at the Terra (or Aqua) orbit of 705 km achieves a 2,330 km swath and provides global coverage every 1–2 days (Nishihama *et al.* 1997).

In order to obtain precipitable water, MOD05 product (Gao and Kaufman, 2003) was selected. This product provides, among others, the near-infrared total precipitable water parameter, which is supplied for each daylight satellite overpass. This parameter is derived from the attenuation by water vapour of near-IR solar radiation. Techniques employing ratios of MODIS water-vapour-absorbing channels 17, 18, and 19 with the atmospheric window channels 2 and 5 are used. The ratios remove partially the effects of variation of surface reflectance with wavelength and result in the atmospheric water-vapour transmittances. From this, the column-water-vapour amounts are derived from the transmittances based on theoretical radiative-transfer calculations and using look-up-table procedures. The algorithm is applied over clear land areas of the globe and above the clouds over both land and ocean.

However, several undesired artifacts might show up in MODIS water vapour images. First, as previously mentioned, for those pixels identified as clouds from the MODIS cloud algorithms, MOD05 product provides total precipitable water over the cloud layer, hence the information below is not considered (Gao and Kaufman, 2003) and the water vapour content is underestimated. Moreover, the presence of brightness spots near the coast can be observed. These anomalous values can be attributed to errors in the algorithm as a consequence of mixed land–ocean pixels. In addition, although the nominal spatial resolution of this product is 1 km at nadir, in those pixels near the edge of the satellite track, this resolution is reduced significantly. In topographically complex regions, such as the Canary Islands, this can lead to erroneous values in the retrieved precipitable water.

3.2. Precipitable Water Vapour from WRF

The Weather Research and Forecasting (WRF, version 3.1) model is an up to date mesoscale modeling system (Skamarock and Klemp, 2008) developed as a collaborative effort among different institutions, including the NCAR Mesoscale and Microscale Meteorology (MMM) Division, the National Oceanic and Atmospheric Administration's (NOAA) National Centers for Environmental Prediction (NCEP), Forecast System Laboratory (FSL), the Department of Defense's Air Force Weather Agency (AFWA), Naval Research Laboratory (NRL), the Center for Analysis and Prediction of Storms (CAPS) at the University of Oklahoma, and the Federal Aviation Administration (FAA). The WRF is a community model suitable for both research and forecasting. The dynamical core of WRF is based on the fully compressible, non-hydrostatic Euler equations, with terrain following Eta coordinate

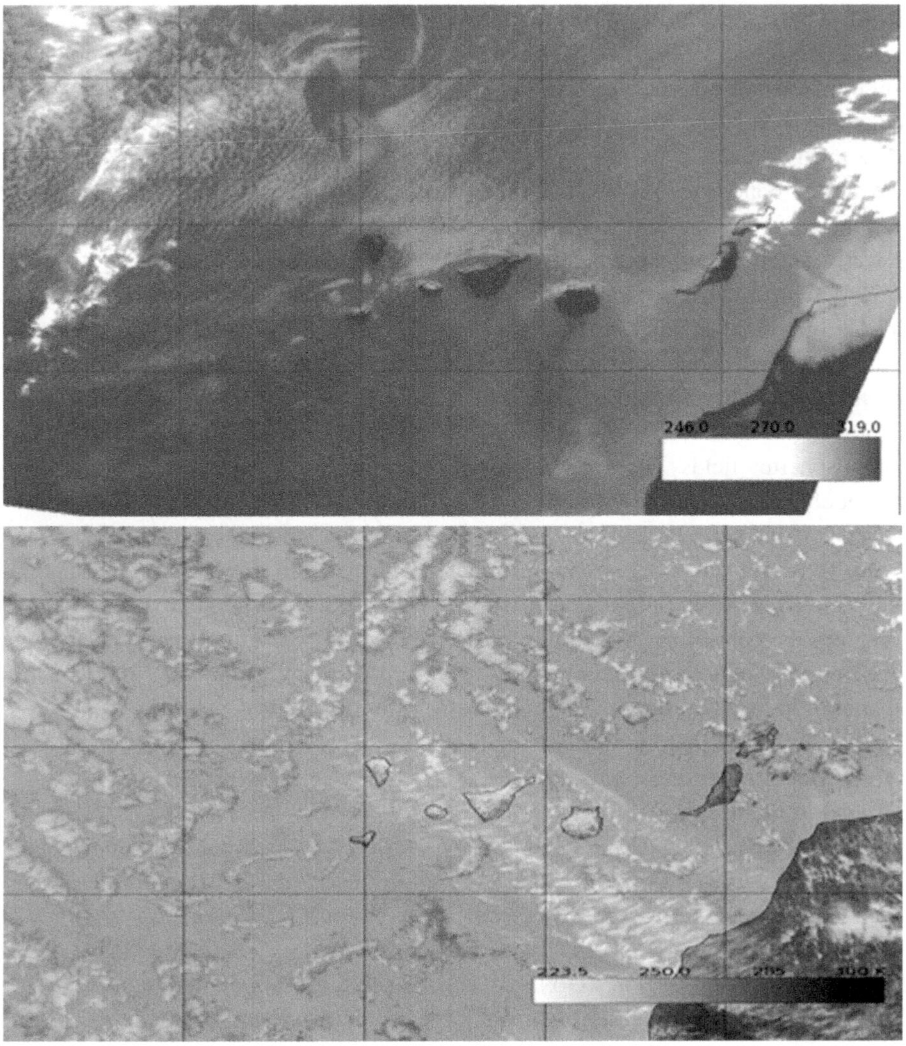

Figure 4
Calibrated brightness temperatures from MODIS infrared channel 31 (10.8 μm) for the Canary Islands region collected on 1 June 2007 (*upper panel*) and 27 November 2007 (*lower panel*)

which is well suited for studies of high-resolution simulations with the capability to explicitly resolve convections and cloudiness. This model introduces several improvements over its predecessor MM5, especially in the advanced numerics and dynamical core it uses, making it better suitable for high spatial resolution (1–10 km) simulations.

The WRF model has a multiple nest capability so that one domain can be forced both by the model output from coarse-resolution domain to account for large scale effects, so as by the model output from the finer-resolution domain (2-way nesting) which accounts for local effects. In this work, we defined an experimental setup with four nested domains, centered in the Canary Islands, where the horizontal resolution of the outer domains are 27, 9 and 3 km and their grids sizes are 88 × 52, 121 × 79 and 121 × 91, respectively. The fourth domain is centered in Tenerife, having a horizontal resolution of 1 km and a grid size of 91 × 121. The spatial resolution could be improved increasing the computational cost of the numerical calculations; however, this goes well beyond the scope of this work. For the described experimental setup, the total computation time is approximately 4 h per simulated day on a workstation with two Xeon E5320 quadcore (2 GHz) and 4 GB of RAM. The computation time is significantly reduced to 1.5 h per simulated day when the higher resolution domains are not simulated, obtaining a final horizontal resolution of 3 km, which corresponds to the third domain.

Initial and boundary conditions (ICBCs) for the large-scale atmospheric fields, sea surface temperature (SST), as well as initial soil parameters (soil water, moisture and temperature) are given by the 1° × 1° NCEP Global Final Analysis. Boundary conditions at the specified zone are determined entirely by temporal interpolation from the 6 h NCEP Final Analysis data.

Simulated atmospheric fields were generated using the WRF model. The simulations were initialized at 12:00 UTC on the day before the InSAR measurement, and then run for 48 h in the four defined domains with 28 vertical output levels. As water vapour content is controlled by cloud and precipitation processes through condensation, transportation and evaporation, it is important to simulate correctly these processes to adequately estimate total precipitable water. To resolve these processes at subgrid scales, nested Regional Mesoscale numerical meteorological models (RCMs) such as WRF employ several sets of parameterizations to this aim. Exhaustive sensitive studies have been carried out to assess the skill of WRF to predict precipitable water amounts in the Canary Islands region (González *et al.*, 2010b). These authors computed the root mean square differences between the hourly estimations of precipitable water vapour calculated through the WRF and those derived from GPS data for the year 2009. They found differences lower than 2 mm for the three GPS stations used in the study, located in different islands and at different altitudes (ranging from 197 to 2417 m a.s.l.). Hence, we have assumed that 2 mm correspond to the accuracy of the precipitable water vapour estimates through the WRF model for the Canary Islands.

3.3. DInSAR and Phase Delays

We used three radar images acquired from descending orbits by the European Remote Sensing satellites ERS-1/2, namely on 5 August 2005 (slave image) and on 1 June 2007 and 27 November 2007 (master images). The differential interferograms were computed by performing a complex average (multi-look) operation with four range looks and 20 azimuth looks, resulting in a pixel size of approximately 90 × 90 m (González *et al.* 2010a).

For the same dates, we obtained the PWV from MODIS observations and WRF calculations. All these results were degraded to a spatial resolution of 1 × 1 km, corresponding to the spatial resolution of the WRF data. Hereafter, the data obtained from MODIS on 5 August 2005, 1 June 2007 and 27 November 2007 will be referred as MODIS-A, MODIS-B and MODIS-C, respectively; the same labelling will be sued for WRF and DInSAR data. Once the PWV data are obtained, it is necessary to calculate the spurious signal that water vapour introduces in the crustal deformation maps obtained by DInSAR. The slant phase delay induced by the presence of atmospheric water vapour may be approximated following the equation proposed by Hanssen (2001):

$$\phi_{p,q} = \frac{4\pi\Pi^{-1}\Delta_{\mathrm{PWV}}}{\lambda\cos(\theta_{\mathrm{inc}})} \tag{1}$$

where $\Phi_{p,q}$ (radians) represents the predicted interferometric phase difference between pixel p and q, Π is a constant factor, Δ_{PWV} (mm) the precipitable water column obtained by MODIS or WRF, λ (mm) is the radar wavelength and $\cos(\theta_{\mathrm{inc}})$ is the cosine of the incidence angle of the satellite taking the radar images. Considering a typical parameter set of $\Pi \approx 0.15$ (Bevis *et al.*, 1996) and $\theta_{\mathrm{inc}} = 23°$, it is expected that up to 7.5 Δ_{PWV} (mm) of the signal in the interferometric images are due to the phase delay resulting from the presence of water vapour in the troposphere. In any case, the approximation expressed by Eq. (1), assumes a linear relation between DInSAR data and the PWV estimates given by MODIS and WRF.

4. Results

The PWV estimates obtained by MODIS and WRF for the 3 days used in this work present a pyramid-like shape (Figs. 5, 6), being the PWV larger at the coast and lower at the summit areas of the centre of the island. This shape has also been found in the yearly and seasonal averages of PWV for the island of Tenerife (EFF-DARWICH *et al.* 2009). Largest values of PWV in both MODIS and WRF data do not exceed 25 mm, this value being compatible with the yearly and seasonal variations of PWV studied by EFF-DARWICH *et al.* (2009). In this sense, the data used in this work are representative of the spatial distribution of PWV on the island of Tenerife.

Although both MODIS and WRF give similar overall regional results for the spatial distribution of PWV, there are significant differences in the spatial coverage of the MODIS and WRF data. The MODIS cloud mask algorithm (the cloud mask) removes areas that might be covered by clouds with at least 90% of confidence level (see Fig. 5). In the case of MODIS-C, that corresponds to a cloudy day (Fig. 4), the cloud mask algorithm indicates that the PWV estimates are only reliable in the eastern part of the island and a small area in the centre of the island. Moreover, MODIS-C data were collected nearly at the edge of the satellite pass, hence the spatial resolution of the PWV estimates is highly degraded.

In order to assess the overall differences between MODIS and WRF data, we analysed the variations of the PWV estimates of MODIS as a function of the WRF estimates (Fig. 7). For the data obtained on 5 August 2005 and 1 June 2007, the slope of the linear fit between MODIS and WRF data reaches 0.992 for PWV lower than 12 mm. If we consider all data with PWV lower than 25 mm, the slopes of the linear fits became 0.77 for the data collected on 5 August 2005 and 0.89 for the data collected on 1 June 2007. In this sense, the WRF underestimates the PWV content relative to the MODIS data for larger PWV contents, e.g. near the coast. In any case, for these 2 days, the root mean square (RMS) difference between WRF and MODIS data is approximately 6 mm. The MODIS data obtained on 23 November 2007 are severely affected by the presence of clouds, hence there is a poor correlation between MODIS and WRF estimates, particularly when the PWV content exceeds 10 mm.

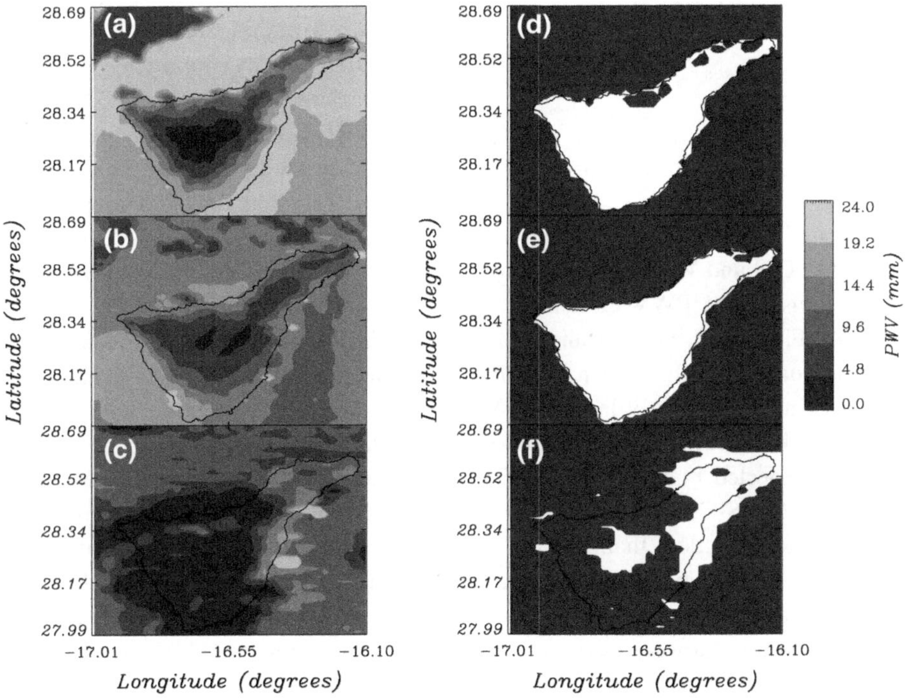

Figure 5
Spatial distribution of PWV content (mm) obtained from MODIS-A, MODIS-B and MODIS-C (*panels a, b and c*, respectively). For the corresponding dates, the MODIS cloud masks are presented in *panels d, e and f*. The *dark grey filled* areas are assigned by the MODIS cloud algorithm as being covered by clouds with a 90% confidence level, and hence these areas are not considered in the PWV analysis, since they give erroneous estimates

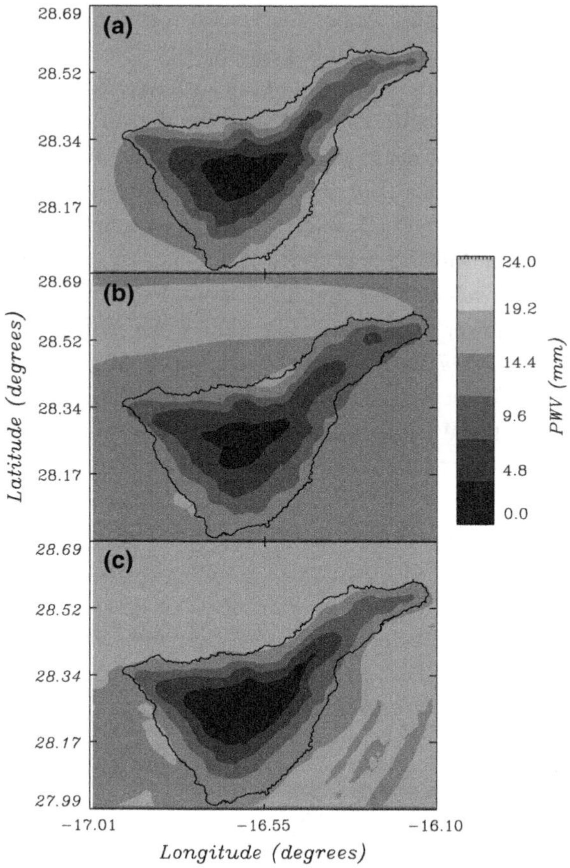

Figure 6
Spatial distribution of PWV content (mm) obtained from WRF-A, WRF-B and WRF-C (*panels a, b and c*, respectively)

In an attempt to understand in more detail the differences between MODIS and WRF estimates, we carried out cross-sections of the PWV content at a longitude of 16.66°W (Figs. 8, 9), hence transecting El Teide volcano from north to south. In both MODIS and WRF, we did not find significantly larger PWV contents in the northern slope relative to the southern slope. We did find an expected result for the PWV content of WRF-C and MODIS-C, being significantly lower (particularly at high altitudes) than the PWV content of the other sets. This is because the epoch associated with WRF-C and MODIS-C corresponds to the winter, when PWV is lower, whereas the epochs associated to the rest of the sets correspond to the summer, when the PWV content is highest.

If the analysis of the variation of PWV obtained by WRF as a function of the PWV calculated by MODIS is carried out in the transect at a longitude of 16.66°W (Fig. 10), it is found that the PWV content calculated by WRF in the northern slope is underestimated relative to the PWV content obtained by MODIS. It is not yet possible to explain this effect, since larger data sets (not only 3 days), spanning months to years are necessary to assess the statistical significance of this result.

We have shown that in cloud free conditions both MODIS and WRF provide compatible estimates of the PWV content, being the RMS difference approximately 6 mm. If clouds are present during the observations the WRF methodology is superior to MODIS, because either MODIS estimates are not available (restrictions of the cloud mask) or these estimates contain spurious signals. Moreover, the WRF data might be obtained anytime during the day or night, whereas MODIS data are limited to satellite overpass time, hence sparse repeat times. However, it is still necessary to test the efficiency of WRF to reduce the effect of PWV on DInSAR measurements of ground deformation. In this sense, two differential interferograms were calculated substracting DInSAR-A (slave) from DInSAR-B and DInSAR-C (masters). The two differential images will be called DInSAR-BA and DInSAR-CA. Following the same procedure, we have calculated MODIS-BA, MODIS-CA, WRF-BA and WRF-CA. The MODIS and WRF differential data have to be multiplied by 7.5 to be converted into phase delays, following Eq. (1). The spatial distribution of the phase delays associated to DInSAR-BA (Fig. 11) is closely reproduced by MODIS-BA and WRF-BA. This means that most of the features found in the DInSAR data are the result of atmospheric path delays induced by the presence of water vapour and hence, they are not associated with geologically induced crustal deformations on the island. DInSAR-CA (Fig. 11) has to be severely affected by the presence of clouds, since the interferogram shows an unrealistic subsidence in the central part of the island (more than 5 cm of subsidence in 5 months) that have not been detected by other conventional geodetic techniques or GPS measurements. Moreover, this subsidence is not related to anomalies in the PWV content, since WRF-CA does not significantly differ from WRF-BA (Fig. 11).

If we substract WRF-BA or MODIS-BA from DInSAR-BA, and WRF-CA or MODIS-CA from

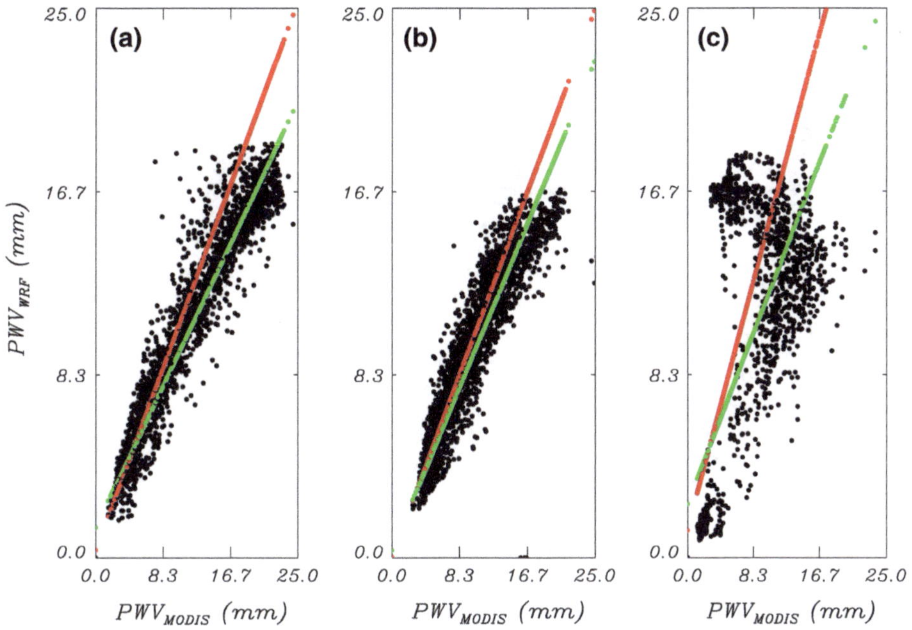

Figure 7
PWV obtained from MODIS as a function of the PWV obtained by WRF. Data obtained on 5 August 2005, 1 June 2007 and 23 November 2007 are presented in *panels a, b and c*, respectively. The *red dots* correspond to the best linear fit if we only consider PWV < 12 mm, whereas the *green dots* correspond to the best linear fit considering all data where PWV < 25 mm

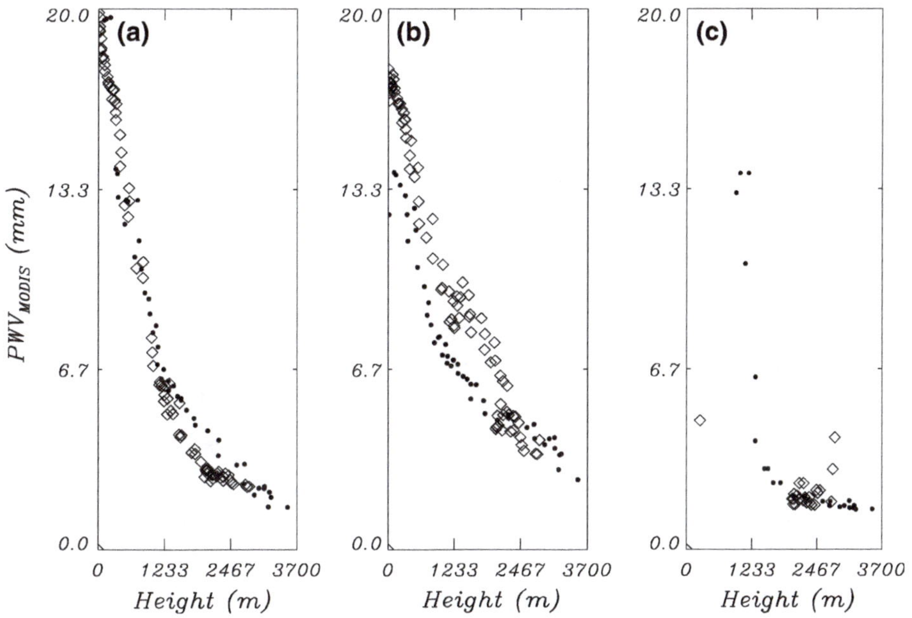

Figure 8
PWV (mm) obtained from MODIS as a function of height and at a longitude of 16.66°W. Results from MODIS-A, MODIS-B and MODIS-C are presented in *panels a, b and c*, respectively. *Diamonds* correspond to the southern slope of the island, whereas *black dots* represent the PWV on the northern slope of the island

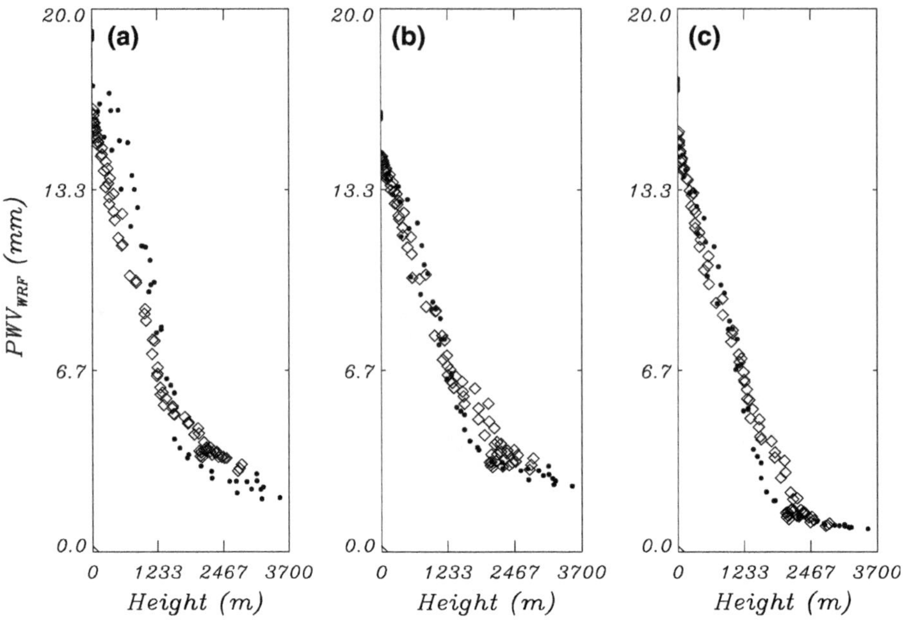

Figure 9
As in Fig. 8 but for WRF estimates

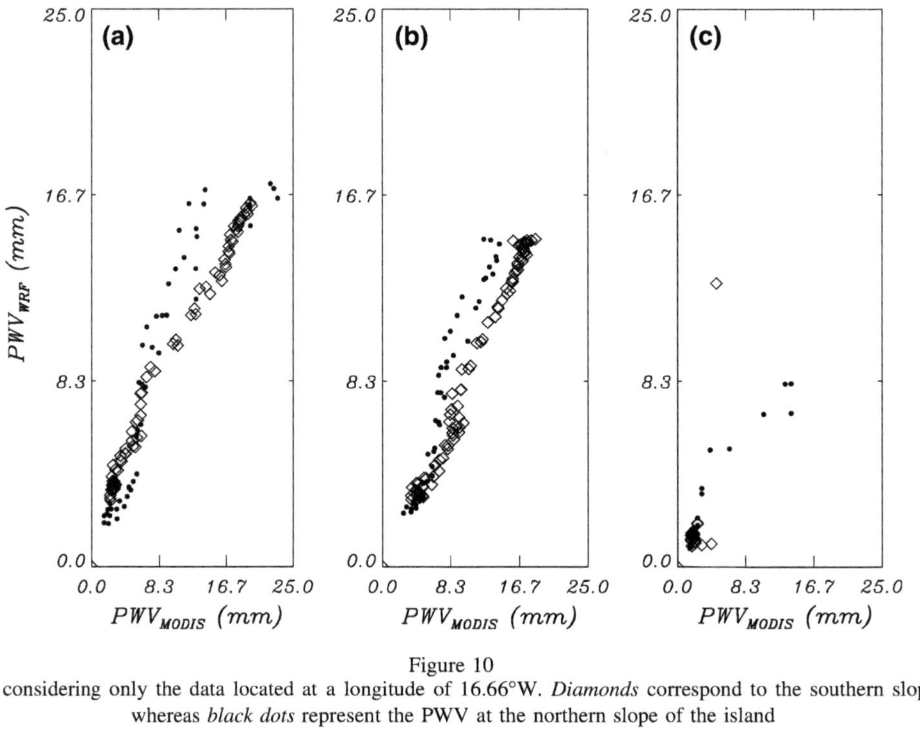

Figure 10
As in Fig. 7, but considering only the data located at a longitude of 16.66°W. *Diamonds* correspond to the southern slope of the island, whereas *black dots* represent the PWV at the northern slope of the island

DInSAR-CA, we will obtain interferograms where the effect of PWV have been significantly reduced (Fig. 12). Phase delays in DInSAR-BA and DInSAR-CA range from −50 to 50 mm, approximately; however, when the effect of PWV is removed, the range of values of the DInSAR phase delays is

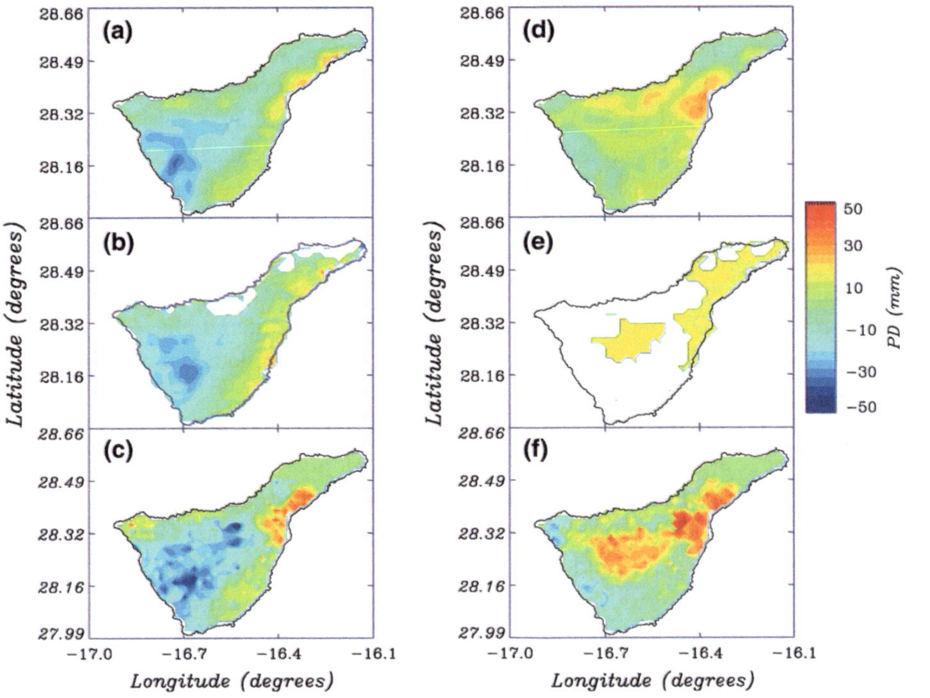

Figure 11
Spatial distribution of Phase Delays (PD, in mm) corresponding to WRF-BA and WRF-CA (*panels a and d*, respectively), MODIS-BA and MODIS-CA (*panels b and e*, respectively), and DInSAR-BA and DInSAR-CA (*panels c and f*, respectively)

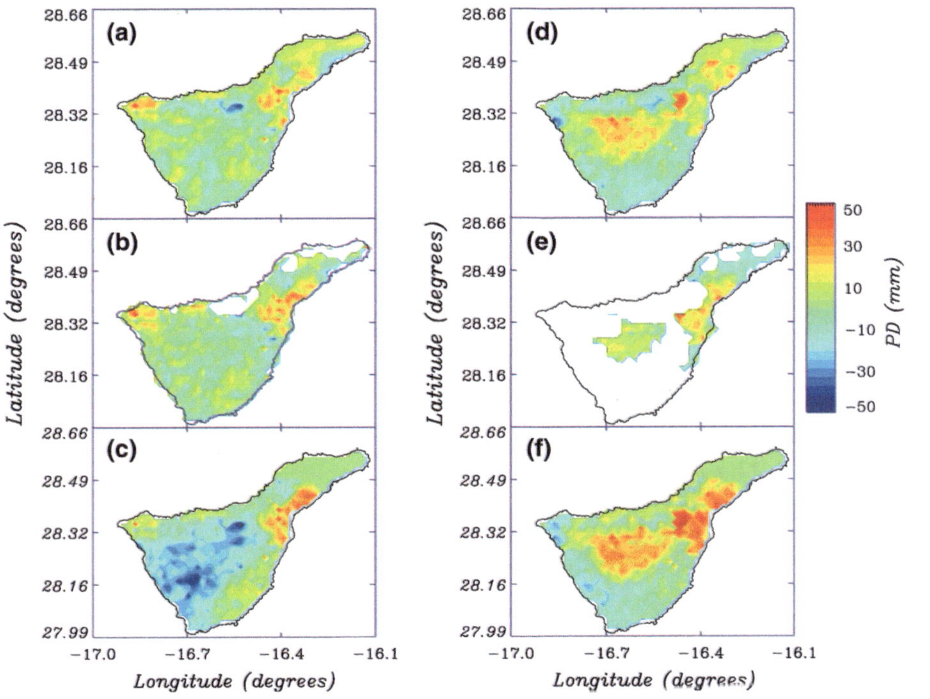

Figure 12
Spatial distribution of Phase Delays (PD, in mm) corresponding to DInSAR-BA after the substraction of WRF-BA (*panel a*) and MODIS-BA (*panel b*). *Panels d and e* correspond to DInSAR-CA after the substraction of WRF-CA and MODIS-CA, respectively. *Panels c and f* for DInSAR-BA and DInSAR-CA are left as in Fig. 11

reduced to approximately −17 to 17 mm. The potential of a WRF-based filtering technique for DInSAR data is shown in Fig. 13, where a detailed view of the largest subsidence found in the island is presented (see Fig. 1 for details). Since our data spans approximately 2 years in time (5 August 2005 to 23 November 2007), the subsidence should reach nearly 3 cm in depth. It is not possible to follow the evolution of this feature with MODIS, since the cloud mask removes the results (MODIS-CA) for that portion of the island. WRF provides data for the two epochs when the differential interferometric images were calculated. The subsidence is visible in the two WRF-filtered images (panels a and d). The significance in the detection of the subsidence will increase if more than 3 days of data were included in the analysis and hence, statistical analysis could be carried out.

In an attempt to analyze the effect of the filtering technique devised in this work, when the effect of densely vegetated areas is removed, we calculated the phase delays along the transect shown in panel c) of Fig. 13. This transect passes through sparsely vegetated areas, where the large subsidences a and b (see Fig. 1) are located. Both subsidences are visible in the phase delays retrieved along the transect (Fig. 14), particularly a; however, the data dispersion and the presence of spurious signals are significantly reduced in the WRF-filtered phase delays. In this sense, the subsidences are better constrained in the WRF-filtered signal than in the raw DInSAR profile. The MODIS-filtered phase delays were only retrieved for one of the data sets, due to the lack of data on MODIS-C.

We have found promising results (Figs. 11, 12, 13, 14); however, it is necessary to include more data sets to reduce even more the spurious signals in the interferometric data, enhancing the geologically induced crustal deformations. DInSAR statistical filtering techniques, such as SBAS (Berardino *et al.*, 2002), require a large number of data sets to remove atmospheric effects. The methodology we are introducing here already works with two data sets (master and slave), since we are directly substracting the

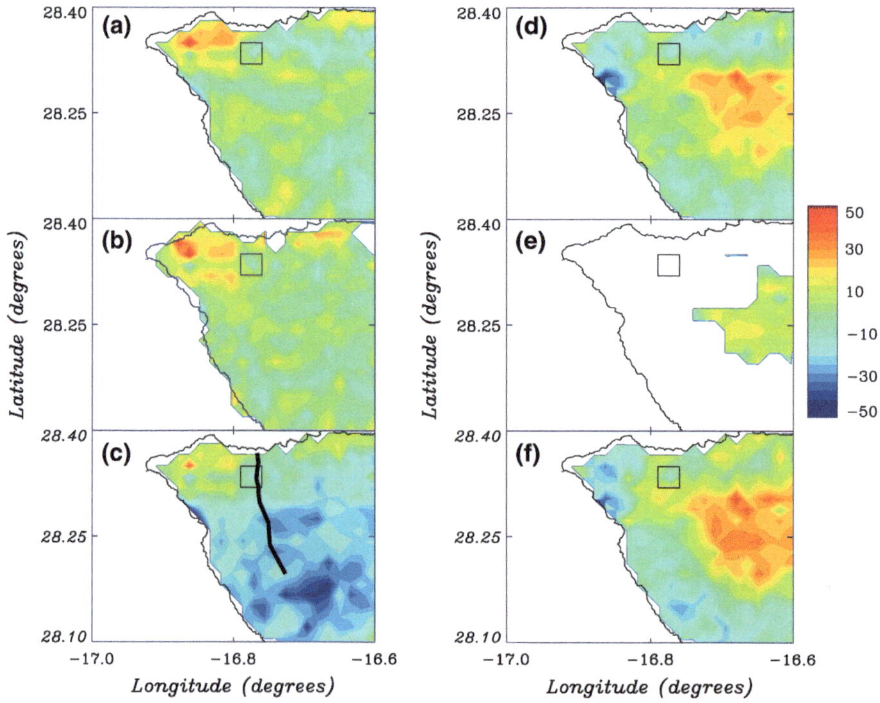

Figure 13
As in Fig. 12, but with a detailed view of the northwestern portion of the island. The *black square* locates the largest subsidence (marked as *a* in Fig. 1) detected in Tenerife (see text for details). *Black continuous line in panel c*) corresponds to the path chosen to represent the phase delay as a function of distance that is presented in Fig. 14

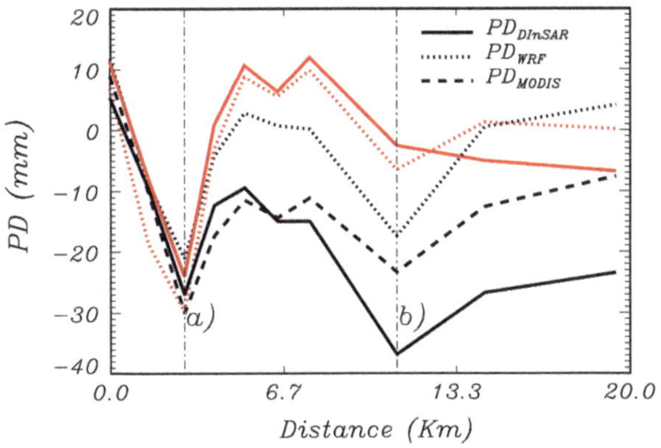

Figure 14
Phase delays as a function of distance for the profile shown in *panel c)* of Fig. 13. *Solid, dotted and dashed black lines* represent the phase delays corresponding to DInSAR-BA, DInSAR-BA after the substraction of WRF-BA and DInSAR-BA after the substraction MODIS-BA, respectively. *Solid and dotted red lines* represent the phase delays corresponding to DInSAR-CA and DInSAR-CA after the substraction of WRF-CA, respectively. The *dotted-dashed vertical lines* (labelled as *a* and *b*) mark the positions of the subsidences *a* and *b* of Fig. 1

atmospheric contribution to the phase delay obtained from DInSAR. However, the performance of the method will improve analysing a series of data sets (not only 3 days of data).

5. Conclusions

We have proposed a methodology, based on WRF mesoscale numerical meteorological models, to correct the atmospheric effects in single DInSAR interferograms that will be used as crustal deformation data for geological monitoring on the volcanic island of Tenerife, Canary Islands. The spatial distributions of water vapour above Tenerife, obtained by WRF and by the space platform MODIS, are converted into interferometric phase delays that are subsequently substracted from the raw DInSAR images. The WRF methodology attempts to fulfil the limitations of MODIS-based corrections of DInSAR data, namely, the sparse repeat times of the satellite, the presence of spurious signal (or even the absence of data) in cloudy conditions and/or the spatial deg radation of the PWV estimates if the data are collected at the edges of the satellite pass.

Our results show that WRF models are promising tools to filter out atmospheric effects in the phase delays obtained by DInSAR on the island of Tenerife. Phase delays in raw DInSAR range from −50 to 50 mm, approximately; however, when the effect of PWV is removed through WRF-based data, the range of values of the DInSAR phase delays is reduced to approximately −17 to 17 mm. The range of correction would be improved if more that 3 days were used in the analysis, through the averaging of a sample of interferometric sets.

Acknowledgments

This work was partially funded by the Canarian ACIISI projects SOLSUBC200801000243 and proID20100121. Research by JF and PJG has been supported by the Spanish MICINN projects: GEOMOD (CGL2005-05500-C02), PEL2G (CGL2008-06426-C01-01/BTE) and CGL2010-21366-C04-01, and has been also done in the frame of the Moncloa Campus of International Excellence (UCM-UPM, CSIC). PJG was partially funded by a UCM PhD Thesis Research Fellowship. The authors thank the University of La Laguna, the Agencia Canaria de Investigación, Innovación y Sociedad de la Información and the Fondo Europeo de Desarrollo Regional (FEDER) for partial support (Proyectos estructurantes de la ULL). The boundary conditions for WRF simulations were obtained from the Research Data Archive (RDA) which is maintained by the Computational and Information Systems Laboratory (CISL) at the National Center for Atmospheric Research (NCAR). NCAR is sponsored by the National Science Foundation (NSF). The original data are available from the RDA (http://dss.ucar.edu) in data set number ds083.2. We thank the two referees for helping to greatly improve the quality of this work.

References

Ablay, G., and Hurlimann, M. (2000), *Evolution of the north flank of Tenerife by recurrent giant landslides*, J. Volcanol. Geotherm. Res. *103*, 135–139.

Albert-Beltrán, J. F., Araña, V., Diez, J. L., and Valentín, A. (1990), *Physical–chemical conditions of the Teide volcanic system (Tenerife, Canary Islands)*, J. Volcanol. Geotherm. Res. *43*, 321–332.

Ancochea, E., Huertas, M. J., Cantagrel, J. M., Coello, J., Fúster, J. M., Arnaud, N., and Ibarrola, E. (1999), *Evolution of the Cañadas edifice and its implications for the origin of the Cañadas Caldera (Tenerife, Canary Islands)*, J. Volcanol. Geotherm. Res. *88*, 177–199.

Araña, V., Aparicio, A., García Cacho, L., and García García R. (1989), Mezcla de magmas en la región central del Teide. En Los volcanes y la caldera del parque nacional del Teide (Tenerife, Islas Canarias). In: V. Araña y J. Coello (eds), 269–299.

Araña, V., Camacho, A. G., García, A., Montesinos, F. G., Blanco, I., Vieira, R., and Felpeto, A. (2000), *Internal structure of Tenerife (Canary Islandas) based on gravity, aeromagnetic and volcanological data*, J. Volcanol. Geotherm. Res *103*, 43–64.

Berardino, P., Fornaro, A., Lanari, R., and Sansosti, E. (2002), *A new algorithm for surface deformation monitoring based on small baseline differential SAR interferograms*, IEEE Trans. Geosci. Remote Sens. *40*(11), 2375–2383.

Bevis, M., Chiswell, S., Businger, S., Herring, T. and Bock, Y. (1996), *Estimating wet delays using numerical weather analysis and predictions*, Radio Science *31*(3), 447–487.

Biggs, J., Amelung, F., Gourmelen, N., Dixon, T. H. and Kim, S. W. (2009), *InSAR observations of 2007 Tanzania rifting episode reveal mixedfault and dyke extension in an immature continental rift*, Geophys. J. Int. *179*, 549–558.

Cabrera, M. P., and Hernández-Pacheco, A. (1987), *Las erupciones históricas de Tenerife (Canarias) en sus aspectos vulcanológico, petrológico y geoquímica*, Rev. Mat. Proc. Geol. *V*, 143–182.

Carracedo, J. C., Paterne, M., Guillou, H., Perez Torrado, F. J., Paris, R., Rodriguez Badiola, E., and Hansen, A. (2003), *Dataciones radiometricas (C14 y K/Ar) del. Teide y el rift noroeste, Tenerife, Islas Canarias*, Estudios Geol. *59*, 15–29.

Díez, J. L., and Albert, J. F. (1989), Modelo termodinámico de la cámara magmática del Teide. En Los volcanes y la caldera del parque nacional del Teide (Tenerife, Islas Canarias). In: V. Araña y J. Coello (eds), 335–343.

Dzurisin D. (2007) Volcano deformation: New geodetic monitoring techniques, Springer-Verlach, Berlin Heilderberg New York, pp. 51–68 and pp. 89–95, ISBN 3-540-42642-6.

Eff-Darwich, A., Grassin, O. and Fernández, J. (2008a), *An Upper Limit to Ground Deformation in the Island of Tenerife, Canary Islands, for the Period 1997 2006*, Pure and Applied Geophysics, *165*, 1049.

Eff-Darwich, A., Coello, J., Viñas, R., Soler, V., Martin-Luis, M. C., Farrujia, I., Quesada, M. L. and De La Nuez, J. (2008b) *Underground Temperature Measurements as a Tool for Volcanic Activity Monitoring in the Island of Tenerife, Canary Islands*, Pure and Applied Geophysics, *165*, 135.

Eff-Darwich, A., García-Lorenzo, B., Pérez-Darias, J. C., González, A., Fernández, J., González, P., (2009), *Characterization of the distribution of water vapour for DInSAR studies on the volcanic island of Tenerife, Canary Islands*. Proceedings of the SPIE, Volume *7478*, pp. 747804–747808.

Feigl, K. L., and Dupré, E. (1999), *RNGCHN: a program to calculate displacement components from dislocations in an elastic half-space with applications for modeling geodetic measurements of crustal deformation*, Computers and Geosciences *25*, 695–704.

Fernández, J., and Rundle, J. B., (1994), *Gravity changes and deformation due to a magmatic intrusion in a two-layered crustal model*. Journal of Geophysical Research, *99*, 2737–2746.

Fernández, J., Rundle, J. B., Granell, R. D. R, Yu, T.-T., (1997). *Programs to compute deformation due to a magma intrusion in elastic-gravitational layered Earth models*. Computers & Geosciences, *23*, 231–249.

Fernández, J., Yu, T.-T., Rodríguez-Velasco, G., González-Matesanz, J., Romero, R., Rodríguez, G., Quirós, R., Dalda, A., Aparicio, A., and Blanco, M. J. (2003), *New geodetic monitoring system in the volcanic island of Tenerife, Canaries, Spain. Combination of InSAR and GPS techniques*, J. Volcanol. Geotherm. Res. *124*/3–4, 241–253.

Fernández, J., González-Matesanz, F. J., Prieto, J. F., Staller, A., Rodríguez-Velasco, G., Alonso-Medina, A., and Charco, M. (2004). *GPS Monitoring in the N-W part of the Volcanic Island of Tenerife, Canaries, Spain. Strategy and results*. Pure and Applied Geophysics, vol *161*, no 7, 1359–1377, doi:10.1007/s00024-004-2509-2.

Fernández, J., Romero, R.; Carrasco, D., Tiampo, K. F., Rodríguez-Velasco, G., Aparicio, A., Araña, V., González-Matesanz, F. J., (2005), *Detection of displacements in Tenerife Island, Canaries, using radar interferometry*. Geophysical Journal International, *160*, 33–45. doi:10.1111/j.1365-246X.2005.02487.x.

Fernández, J. Charco, M., Rundle, J. B., Tiampo, K. F., (2006), *A revision of the FORTRAN codes GRAVW to compute deformation produced by a point magma intrusion in elastic-gravitational layered Earth models*. Computers & Geosciences, *32*/2, 275–281. doi:10.1016/j.cageo.2005.06.015.

Fernández, J., Tizzani, P., Manzo, M., Borgia, A., González, P. J., Martí, J., Pepe, A., Camacho, A. G., Casu, F., Berardino, P., Prieto, J. F. and Lanary, J., (2009), *Gravity-driven deformation of Tenerife measured by InSAR time series analysis*, Geophys. Res. Lett., *36*, L04306. doi:10.1029/2008GL036920.

Foster, J., Brooks, B., Cherubini, T., Shacat, C., Businger, S., and Werner, C. L., (2006), *Mitigating atmospheric noise for InSAR using a high resolution weather model*, Geophys. Res. Lett., *33*, L16304, doi:10.1029/2006GL026781, 2006.

Gao, B. C., and Kaufman, Y. J., (2003) *Water vapour retrievals using Moderate Resolution Imaging Spectroradiometer (MODIS) near-infrared channels*, J. Geophys. Res., Vol. *108*, No. D13, 4389.

González, P. J., Samsonov, S., Manzo, M., Prieto, J. F., Tiampo, K. F., Tizziani, P., Casu, F., Pepe, A., Berardino, P., Camacho, A. G., Lanari, R., and Fernández, J. (2010a). *3D volcanic deformation fields at Tenerife Island: integration of GPS and Time Series of DInSAR (SBAS)*, Cahiers du Centre Européen de Géodynamique et de Séismologie. (in press).

González, A., Expósito, F. J., Pérez, J. C., Díaz J. P., and Taima, D. (2010b), *Assessment of WRF skills to infer precipitable water in orographic complex areas*, Submitted to Journal of Geophysical Research.

Grell, G. A., Dudhia, J., and Stauffer P. J., (1995), *A description of the fifth generation Penn State/NCAR Mesoscale Model (MM/5), NCAR Tech.Note 398*, 122 pp., Natl. Cent. For Atmos. Res., Boulder, Colo.

Guillou, H., Carracedo, J. C., Paris, R. and Pérez Torrado, F. J. (2004), *Implications for the early shield-stage evolution of Tenerife from K/Ar ages and magnetic stratigraphy*, Earth Planet. Sci. Lett. *222*, 599–614.

Hanssen, R. (2001), Radar Interferometry: Data Interpretation and Error Analysis, Vol. 2, Kluwer Academic Publishers. Dordrecht.

Irwan, M., Kimata, F., Fujii, N., Nakao, S., Watanabe, H., Sakai, S., Ukawa, M., Fujita, E. and Kawai, K. (2003), *Rapid Ground deformation of the Miyakejima volcano on 26-27 June 2000 detected by kinematic GPS analysis*, Earth Planets Space *55*, 13–16.

Li, Z., Fielding, E. J., Cross, P., Preusker, R. (2009), *Advanced InSAR atmospheric correction: MERIS/MODIS combination and stacked water vapour models*, International Journal of Remote Sensing, 30, *13*, 3343–3363.

Martí, J., Mitjavila, J., and Araña, V. (1994), *Stratigraphy, structure and geomorphology of the Las Cañadas Caldera (Tenerife, Canary Islands)*, Geol. Mag. *131*, 715–727.

Massonnet, D., Feigl, K. L., Rossi, M., Adragna, F. (1994) *Radar interferometry mapping of deformation in the year after the Landers earthquake*, Nature *369*:227–230.

Miranda, P. M. A., and James, I. N., (1992), *Non-linear three-dimensional effects on gravity wave drag: Splitting flow and breaking waves*. Quarterly Journal of the Royal Meteorological Society, vol *118*, 1057–1082.

Nishihama, M., Wolfe, R. and Solomon, D. (1997), MODIS level 1A earth location: Algorithm Theoretical Basis Document version 3.0. Available online at http://modis.gsfc.nasa.gov.

Okada, Y. (1985), *Surface deformation due to shear and tensile faults in a half-space*, Bull. Seismol. Soc. Am. *75*, 1135–1154.

Puysségur, B., Michel, R., and Avouac, J. -P. (2007), *Tropospheric phase delay in interferometric. synthetic aperture radar estimated from meteorological model and multispectral imagery*, J. Geophys. Res., *112*, B05419, doi:10.1029/2006JB004352.

Samsonov, S., Tiampo, K. F., González, P. J., Prieto, J. F., Camacho, A. G., Fernández, J. (2008), Surface deformation studies of Tenerife Island, Spain, from joint GPS-DInSAR observations. 2008 Second Workshop on Use of Remote Sensing Techniques for Monitoring Volcanoes and Seismogeneic Areas. IEEE Catalog Number: CFP0858E-DVD, ISBN: 978-1-4244-2547-1. 6 pp.

Skamarock, W. and, Klemp, J., (2008), *A time-split nonhydrostatic atmospheric model for weather research and forecasting applications*, Journal of Computational Physics, *227*, 3465–3485.

Wadge, G.; Webley, P. W., Stevens, N. F. (2004), Correcting In-SAR Data for Tropospheric Path Effects over Volcanoes using Dynamic Atmospheric Models, in Proceedings of the FRINGE 2003 Workshop (ESA SP-550). 1–5 December 2003, ESA/ES-RIN, Frascati, Italy. In: H. Lacoste (ed).

Yu, T. T., Fernández, J., Tseng, C. L., Sevilla, M. J., and Araña, V. (2000), *Sensitivity test of the geodetic network in Las Cañadas Caldera, Tenerife, for volcano monitoring*, J. Volcanol. Geotherm. Res. *103*, 393–407.

(Received June 28, 2010, revised January 4, 2011, accepted June 10, 2011, Published online August 31, 2011)

Pure Appl. Geophys. 169 (2012), 1443–1456
© 2011 Springer Basel AG
DOI 10.1007/s00024-011-0402-3

Pure and Applied Geophysics

An Elliptical Model for Deformation Due to Groundwater Fluctuations

Kristy F. Tiampo,[1] Francois-Alexis Ouegnin,[2,3] Sreeram Valluri,[2] Sergey Samsonov,[1] José Fernández,[4] and Garrett Kapp[1]

Abstract—Historically, surface subsidence as a result of subsurface groundwater fluctuations have produced important and, at times, catastrophic effects, whether natural or anthropogenic. Over the past 30 years, numerical and analytical techniques for the modeling of this surface deformation, based upon elastic and poroelastic theory, have been remarkably successful in predicting the magnitude of that deformation (Le Mouélic and Adragna in Geophys Res Lett 29:1853, 2002). In this work we have extended the formula for a circular-shaped aquifer (Geertsma in J Petroleum Tech 25:734–744, 1973) to a more realistic elliptical shape. We have improved the accuracy of the approximation by making use of the cross terms of the expansion for the elliptic coordinates in terms of the eccentricity, *e*, and the mean anomaly angle, *M*, widely used in astronomy. Results of a number of simulations, in terms of *e* and *M* developed from the transcendental Kepler equation, are encouraging, giving realistic values for the elliptical approximation of the vertical deformation due to groundwater change. Finally, we have applied the algorithm to modeling of groundwater in southern California.

Key words: Deformation, numerical techniques, groundwater hydrology, subsidence, inversion, geodesy.

1. Introduction

In many places today, anthropogenic surface deformation due to the evacuation of groundwater from underground aquifers presents a significant hazard to the surrounding communities and structures (Amelung *et al.*, 1999; Dixon *et al.*, 2006; Le Mouélic and Adragna, 2002; Tesauro *et al.*, 2000; Tomás *et al.*, 2005; Anderssohn *et al.*, 2008; Bell *et al.*, 2008; Motagh *et al.*, 2008). In addition, even in those areas where the potential damage is minimal, knowledge of the relationship between pumping rates and the associated subsidence can provide important information on the subsurface structure and nature of the associated aquifer.

Geertsma's (1973) pioneering work on the numerical modeling of land subsidence studied the causes of subsidence above hydrocarbon producing reservoirs. Here, a simple model was presented for the estimation of the reservoir compaction and the accompanying subsidence. This model is based on a mathematical formulation of the land subsidence above a disc-shaped oil and gas reservoir, and results in a circular approximation to the surface deformation caused by oil or groundwater pumping. The simplicity of this approach and its general applicability resulted in its widespread implementation.

In this paper we perform an expansion in terms of the eccentricity *e* and mean anomaly angle *M* of an ellipse in order to derive a second-order approximation to the ground deformation in an elliptic-shaped reservoir. We also exploit some interesting properties of the Bessel and Legendre special functions to find better estimates for that displacement.

Our work is organized as follows. First, we present a brief background discussion of the initial circular derivation and its application in Sect. 2. In the following Sect. 3 we present the derivation for the elliptical expansion and give examples of the induced deformation. We attempt to improve the estimates of the subsidence and their computation using a higher order approximation of the elliptic expansion. Section 4 demonstrates its applicability to the modeling of surface deformation due to groundwater fluctuations in southern California. Section 5 presents the conclusions.

[1] Department of Earth Sciences, University of Western Ontario, London, ON N6A-5B7, Canada. E-mail: ktiampo@uwo.ca
[2] Departments of Applied Math and Physics and Astronomy, University of Western Ontario, London, ON N6A 5B7, Canada.
[3] International University of Grand Bassam (IUGB), Grand Bassam, Côte d'Ivoire.
[4] Instituto de Geociencias (CSIC-UCM), Facultad de Ciencias Matemáticas, Ciudad Universitaria, Plaza de Ciencias 3, 28040 Madrid, Spain.

2. *Initial Deformation Model*

GEERTSMA (1973) invoked poroelasticity and thermoelasticity theory and assumed that both the reservoir and its surroundings were homogenous with respect to its compaction and permeability properties in order to derive an analytical formulation for the deformation induced by the removal of fluids from underground strata or reservoir. In that case, variables such as the uniaxial compaction coefficient, c_m, and the Poisson's ratio, v, remain fixed in an isotropic medium, and the derived surface deformation pattern is circular. The vertical (u_z) and radial (u_r) displacements at the surface ($z = 0$) then are obtained in terms of the product of two Bessel functions:

$$u_z(r,0) = -2c_m(1-v)\Delta pHR \int_0^\infty e^{-D\alpha} J_1(\alpha R) J_0(\alpha r) d\alpha \tag{1}$$

$$u_r(r,0) = +2c_m(1-v)\Delta pHR \int_0^\infty e^{-D\alpha} J_1(\alpha R) J_1(\alpha r) d\alpha \tag{2}$$

in which J_0 and J_1 are Bessel functions of the zero and first order, respectively, Δp is a uniform reservoir pressure reduction throughout the reservoir, H is the thickness of a disc-shaped reservoir, R is the radius of the reservoir, and D is the burial depth of that reservoir. Here z is positive in the up-direction; r is positive outward from the origin. The integration is performed over all possible nuclei of strain, α (GEERTSMA, 1973).

Note that, while GEERTSMA (1973) originally formulated his analysis in order to derive the subsidence induced by the pumping of oil from underground strata, the model can be applied to the pumping of groundwater from aquifers or, inversely, the injection of fluids into porous underground layers resulting in local uplift. Subsequent studies used a similar framework for the detection of ground deformation in a variety of regions. Applications can be found in the literature for cities such as Napoli, New Orleans, and Las Vegas (AMELUNG *et al.*, 1999; DIXON *et al.*, 2006; TESAURO *et al.*, 2000).

In a recent study of the city of Paris, France, LE MOUÉLIC and ADRAGNA (2002) investigated surface uplift due to the infiltration of groundwater in the area of the Saint Lazare Railway. They used a remote sensing technique, interferometric synthetic aperture radar (InSAR) (MASSONNET and FEIGL, 1998) to identify the displacement, and the analytical model above (GEERTSMA, 1973), to estimate the ground deformation. They concluded that the groundwater deformation that occurred could be modeled successfully using this approach, and consisted of an elastic reversible deformation.

The main objective of this model is to describe the land subsidence due to the modification of the pressure in an underlying circular reservoir. Here, again, it is assumed that the medium is uniform and isotropic, such that the permeability in all directions is a constant. LE MOUÉLIC and ADRAGNA (2002) defined the following simple framework: an elastic earth treated as an elastic half-space with a Poisson ratio v, a disc-shaped reservoir of radius R and thickness h, buried at a depth D, and a groundwater height variation due to the pressure drop that is denoted dh. The vertical displacement of the surface $u_z(r)$ depends on the radial distance r, where the origin above the center of the reservoir is determined by the location of the pumping or injection operation.

The formula for the displacement is then:

$$u_z(r) = +(2v-2)dhR \int_0^\infty e^{-D\alpha} J_1(\alpha R) J_0(\alpha r) d\alpha \tag{3}$$

$$u_r(r) = -(2v-2)dhR \int_0^\infty e^{-D\alpha} J_1(\alpha R) J_1(\alpha r) d\alpha \tag{4}$$

here dh is linked to the reservoir thickness H, and the increase of pressure Δp by the relation $dh = c_m H\Delta p = \frac{H\Delta p}{E}$ (LANDAU and LIFSHITZ, 1986; LE MOUÉLIC and ADRAGNA, 2002).

3. *Elliptical Approximation*

In natural underground aquifers, the medium rarely is isotropic. While variations can occur on many scales, the predominant case seen in nature is where there are differing permeabilities, or hydraulic conductivity, in two different, primarily orthogonal directions, so that permeability varies with azimuthal direction (BURBEY, 2006; FREEZE and CHERRY, 1979; KIM, 2005). This anisotropy can be caused by differences in directional porosity that are a result of grain size variations, depositional environment, fracture patterns, and pressure. This causes a slower rate of groundwater evacuation in one direction than the other, and a smaller total amount of extraction in that direction and a shorter distance over which that extraction occurs. The result is an elliptical shape for the surface height change, as seen from above. Here we attempt to provide a better approximation to the actual deformation seen in nature through the derivation of a formula for surface deformation that varies elliptically with the angle of rotation of the radius r around the center of the aquifer. This formula is obtained via a mathematical approximation of the transcendental Kepler equation that has been used to relate the eccentric anomaly angle E in terms of the mean anomaly M and the orbital eccentricity e for planetary motion and other astronomical applications.

In order to obtain an accurate elliptical orbit that relates the orbital eccentricity e and the mean anomaly angle M to the eccentric anomaly E, VALLURI *et al.* (2006) introduced the following modifications to the Cartesian coordinates x and y of a circular orbital motion,

$$x = a(\cos E - e) \tag{5}$$

$$y = a\left(\sqrt{1-e^2}\right)\sin E \tag{6}$$

$$E \approx M + \left(e - \frac{e^3}{8}\right)\sin M + \frac{1}{2}e^2 \sin 2M + \frac{3}{8}e^3 \sin 3M + O(e^4) \tag{7}$$

where E and M are the eccentric and mean anomaly angles widely used in astronomy and astronautics, etc., x and y are the Cartesian coordinates that describe the motion of the stellar object, and a and e stand for the radius of the circular orbit and the eccentricity of the elliptical orbit.

Equation (7) is obtained from an iterative method of the transcendental Kepler equation:

$$E_{i+1} = M + e\sin E_i, \quad i = 0, 1, 2\ldots \tag{8}$$

The trigonometric formulae for $\sin(A+B)$ and the series expansion of $\sin x$ and $\cos x$ for small x are also used (MURRAY and DERMOTT, 1999). The final series for $E - M$ is in fact a Fourier sine series in the mean anomaly M, with the coefficients of the series $C_S(e)$ of the form:

$$C_S(e) = \frac{2}{s}J_s(se), \quad s = 1, 2, 3, \cdots \tag{9}$$

where $J_s(se)$ is the Bessel function with running index s and argument se.

It is of interest to note that a more generic analysis of the Kepler equation displays the connection of the singularities in the complex M plane to those of the eccentricity e (HAGIHARA, 1970). The singularities are obtained by the solution in terms of the Lambert W function. The Lambert W function has seen a renaissance not only among physicists and mathematicians but also researchers in many diverse fields during recent years (CORLESS *et al.*, 1996). This function has shown its ubiquity in innumerable applications. Also of special interest for this present paper is the application of the Lambert W function for the calculation of wave water heights due to refraction and friction (VALLURI *et al.*, 2000) and the movement of water in the soil from the Richards equation in hydrology (BARRY *et al.*, 1993). The application of the Lambert W function and the Lambert hyperfunction in connection with the transcendental Kepler equation and the curves of Hippias (CORLESS *et al.*, 1996) warrants a separate study in its own right.

Different levels of accuracy can be obtained by taking into consideration the cross terms that result from the expression for the radial coordinate r in terms of x and y. We choose, for brevity, to study only two cases, the case without the cross terms, which we call the zero order accuracy approximation, and the case with the cross terms included, which we designate as the higher order accuracy approximation.

3.1. Zero Order Accuracy Approximation

We insert the series for E in Eq. (7), above, into the formulas for x and y in Eqs. (5) and (6), and omit the cross terms in the trigonometric functions to obtain a zero-order approximation of the following form:

$$x = -\frac{3ae}{2} + a\left(1 - \frac{3e^2}{8}\right)\cos(M) + \frac{1}{2}ae\cos(2M) + \frac{3}{8}ae^2\cos(3M) \quad (10)$$

$$y = a\left(1 - \frac{3e^2}{8}\right)\sin(M) - \frac{1}{4}ae^2\sin(M) + \frac{1}{2}ae\sin(2M) + \frac{3}{8}ae^2\sin(3M). \quad (11)$$

The formulae in Eqs. (10) and (11) take into account the eccentricity and the mean anomaly. New radii $r(e)$ and $R(e)$ can be defined to approximate the elliptic shape taking the eccentricity into consideration. After simplification, each new radius has the following approximate form:

$$r(e) = \left\{a^2\left(1 - \frac{3}{8}e^2\right)\left[1 + \frac{160\,e^2 + (20 - 2\cos 2M)e^4}{\left(1 - \frac{3}{8}e^2\right)}\right]\right\}^{\frac{1}{2}} \quad (12)$$

The eccentricity is chosen to be greater than 0 and less than 1 in order to arrive at an ellipse, whereas the angle remains between $-\pi$ and π. The description of the eccentric and mean anomaly and their computation can be found in Logsdon (1997). Therefore, the formula of the vertical displacement becomes

$$u_{ze}(r_e) = +(2v - 2)dhR_e\int_0^\infty \mathrm{e}^{-D\alpha}J_1(\alpha R_e)J_0(\alpha r_e)\mathrm{d}\alpha \quad (13)$$

$$u_{re}(r_e) = -(2v - 2)dhR_e\int_0^\infty \mathrm{e}^{-D\alpha}J_1(\alpha R_e)J_1(\alpha r_e)\mathrm{d}\alpha \quad (14)$$

where u_e, r_e, and R_e are a function of the eccentricity, e.

Here we have not revised the original analytical computation for the deformation, based upon elasticity theory. We have approximated the variation in shape resulting from bidirectional variations in the hydraulic conductivity, or permeability, via a modification to the allowable spatial extent of the deformation that is controlled by the extent of the anisotropy in the permeability. In Fig. 1 we plot the results for this approximation for the vertical deformation $u_{ze}(r_e)$ for the case in which $e = 0$, or the circular case, again using the values $H = 23$ m, $D = 25$ m, $v = 0.3$, $dh = 1.01$ cm, and $R = 400$ m. The contour shown is coincident with the upper corner value at the aquifer edge. Note that these results are identical to those of the circular formulation given in Eq. (3).

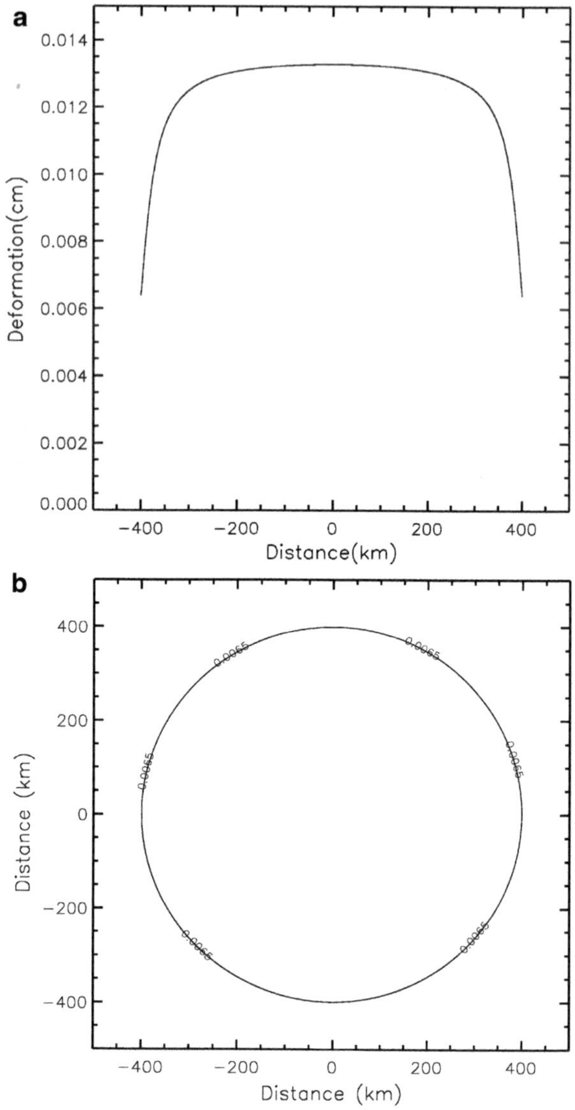

Figure 1
Results for change in surface height, $u_z(r)$, for $H = 23$ m, $D = 25$ m, $v = 0.3$, $dh = 1.01$ cm, and $R = 400$ m, for a circular aquifer, using the elliptical approximation of Eq. 13 (*solid line*). **a** Shows a cross section across the centerline of the aquifer, while **b** shows the map view along the outside edge of the aquifer. Note that these results are indistinguishable from those of Eq. 3 (Le Mouélic and Adragna, 2002)

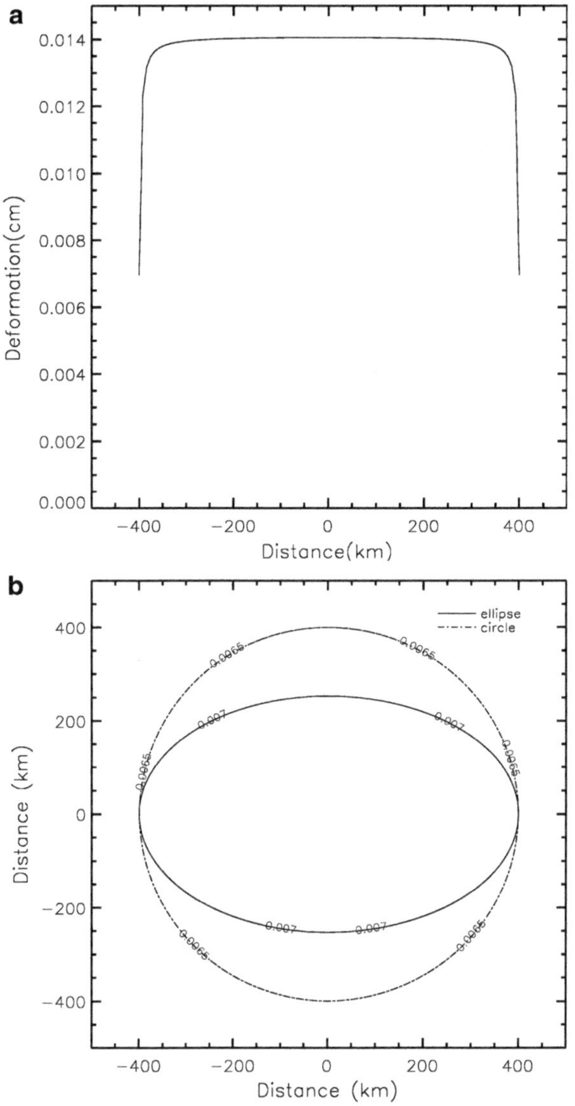

Figure 2
Results for change in surface height, $u_z(r_e)$, for $H = 23$ m, $D = 25$ m, $v = 0.3$, $dh = 1.01$ cm, and $R = 400$ m, for an aquifer with eccentricity = 0.6, using the elliptical approximation of Eq. 13 (*solid line*). **a** and **b** are the same as for Fig. 1 above. Included in **b**, for reference, is the circular approximation of Eq. 3 (*dashed line*)

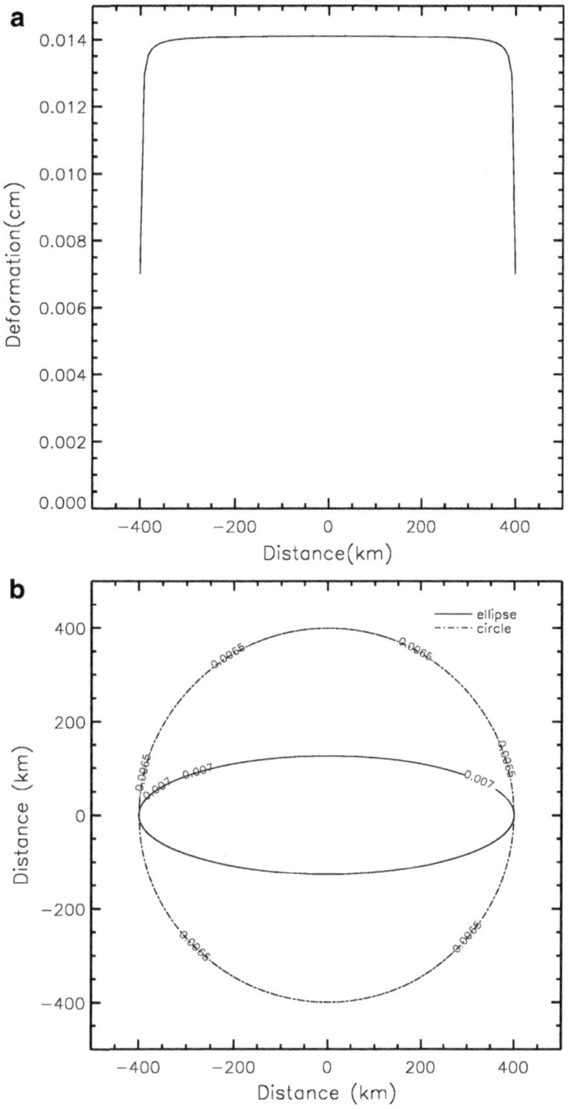

Figure 3
Results for change in surface height, $u_z(r_e)$, for $H = 23$ m, $D = 25$ m, $v = 0.3$, $dh = 1.01$ cm, and $R = 400$ m, for an aquifer with eccentricity = 0.9, using the elliptical approximation of Eq. 13 (*solid line*). **a** and **b** are the same as for Fig. 1 above. Included in **b**, for reference, is the circular approximation of Eq. 3 (*dashed line*)

As additional examples, we focus on two cases where the eccentricity differs from zero. We chose the values 0.6 and 0.9 and we provide their plots in Figs. 2 and 3, respectively. Included with each is the circular approximation for comparison.

Figure 4 shows a cross section of the vertical deformation along the long axis of the elliptical aquifer, using the zero order approximation of Eq. (13). Note that the maximum values of the estimates from our elliptical approximation increase gradually with increasing eccentricity, and that the edges of the profiles become sharper and steeper as well.

3.2. Higher Order Accuracy Approximation

A higher order accuracy approximation for $r_h(e)$ can be obtained by including the cross terms of the corrections presented in (Valluri *et al.*, 2006). The following expression is obtained using the properties of Bessel functions and their associated integrals (Gradshteyn and Ryzhik, 1965; Wang and Guo, 1989):

$$r_h(e) = \left\{ \begin{array}{r} a^2\left(1 - \frac{3}{8}e^2 + \left(2 + \frac{13}{32}e^2\right)e^2\right) \\ \times\left[1 + \frac{e^2\left(\cos(M)\left(\frac{e^2-2}{e}\right) - e\cos(3M)\right) + e^4\left(-\cos(2M)\left(\frac{1+e^2}{2e^2}\right) + \frac{3\cos(4M)}{32}\right)}{\left(1 - \frac{3}{8}e^2 + \left(2 + \frac{13}{32}e^2\right)e^2\right)}\right] \end{array} \right\}^{\frac{1}{2}} \quad (15)$$

where $r_h(e)$ is the new radius, M is the mean anomaly as above, and a and e again are the radius of the circular orbit and the eccentricity. Again, the new radii $r_h(e)$ and $R_h(e)$ can be defined to approximate the elliptic shape, using Eq. (15). The new equations for deformation are then

$$u_{zh}(r_h) = +(2v - 2)dhR_h \int_0^\infty e^{-D\alpha} J_1(\alpha R_h) J_0(\alpha r_h) d\alpha \quad (16)$$

$$u_{rh}(r_h) = -(2v - 2)dhR_h \int_0^\infty e^{-D\alpha} J_1(\alpha R_h) J_1(\alpha r_h) d\alpha. \quad (17)$$

The higher order approximation of Eq. (16) is calculated for varying radii and the parameters used in Sect. 3.1, above. As before, Fig. 5 shows the calculated vertical deformation for an eccentricity of 0.6, and Fig. 6 shows the same calculation for an eccentricity of 0.9.

Figure 7 is a cross section of the vertical deformation across the long axis of the aquifier, for varying eccentricities and the higher order approximation of Eq. (16). Note that there are significant differences between the vertical displacement derived with the higher order approximation and the zero order approximation shown in Fig. 4. The zero order approximation had a more square shape, as noted above, while we observe a more bell-shape when we add the cross terms in the higher order approximation. The second point to remark is that the maximum value of the vertical deformation does not increase uniformly with increasing ellipticity, so that the deformation is not accommodated by a uniform volumetric increase.

4. Modeling Groundwater Deformation in Southern California

4.1. Background

The northern region of southern California's Los Angeles basin is bounded to the north by the San Gabriel Mountains and the Puente Hills and Montebello Hills to the south. This area, known as the San Gabriel Valley (Fig. 8), has a significant compressional tectonic strain component oriented approximately north–south. However, accurate estimation of the amplitude of the tectonic rate of strain has been difficult due to contamination from other geophysical sources, particularly anthropogenic signals associated with groundwater pumping in the region (Argus and Gordon, 2001; Bawden *et al.*, 2001; Watson *et al.*, 2002; Lanari *et al.*, 2004; Argus *et al.*, 2005; King *et al.*, 2007).

Dense population and rapid industrial development make this southern California a prime location for geodetic studies of earthquake hazard. However, the most rapid movements in this region are nontectonic deformations due to groundwater and oil extraction. For example, the 40-km-long Santa Ana basin subsides at the steady rate of 12 mm/year with the seasonal fluctuation of 55 mm in the vertical direction and 7 mm in the horizontal direction

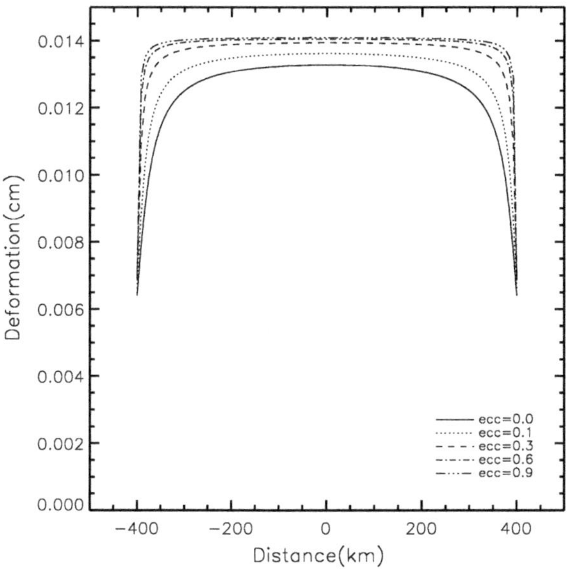

Figure 4
Cross section of the change in surface height, $u_z(r_e)$, through the long axis of the aquifer for varying eccentricity using the elliptical approximation of Eq. 13. As above, $H = 23$ m, $D = 25$ m, $\nu = 0.3$, $dh = 1.01$ cm, and $R = 400$ m

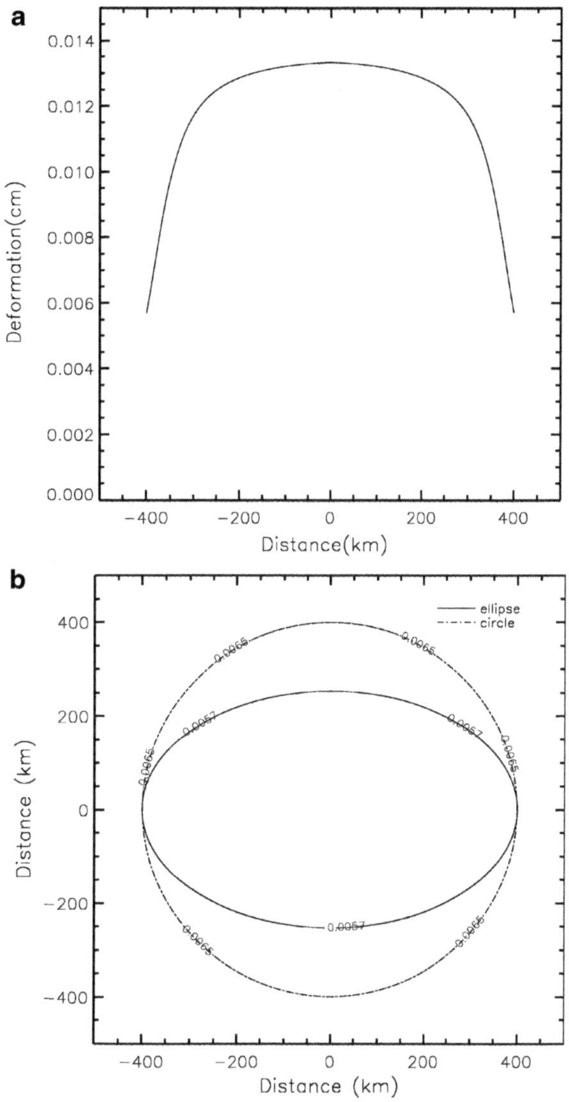

Figure 5
Results for change in surface height, $u_z(r_h)$, for $H = 23$ m, $D = 25$ m, $\nu = 0.3$, $dh = 1.01$ cm, and $R = 400$ m, for an aquifer with eccentricity = 0.6, using the elliptical approximation of Eq. 16 (*solid line*). **a** and **b** are the same as for Fig. 1 above. Included in **b**, for reference, is the circular approximation of Eq. 3 (*dashed line*)

[Watson *et al.*, 2002]. The horizontal movement is governed mostly by plate tectonics ranging from 1.8 cm/year (south) to 0.9 cm/year (north) in N–S direction and from −4 cm/year (southwest) to −3.2 (northeast) in the E–W direction. However, Bawden *et al.* (2001) estimated that after removing anthropogenic signals the uniaxial contraction across the Los Angeles basin can be observed at a rate of 0.44 cm/year in the northeast direction perpendicular to the major strike-slip faults in this area. Argus *et al.* (2005) estimated anthropogenic horizontal velocities to be more than 1 mm/year at more than 1/3 of the GPS stations in the region, and shortening (contraction) to the south of the San Gabriel mountains at 4.5 ± 1 mm/year.

4.2. Data

From previous investigations, it is evident that differential interferograms carry significant amounts of useful data in the studies of phenomena such as surface deformations due to seismic events (Jacobs *et al.*, 2002; Massonnet *et al.*, 1993, Massonnet and Feigl, 1998) and volcanic activities (Fernández *et al.*, 2005), mining subsidence (Gourmelen *et al.*, 2007) or groundwater extraction (Schmidt and Bürgmann, 2003; Bawden *et al.*, 2001, Watson *et al.* 2002). Different types of errors are also present in the interferograms and these need to be estimated and, if possible, corrected.

Recently, Samsonov *et al.* (2007) computed three-dimensional surface velocity maps for that part of

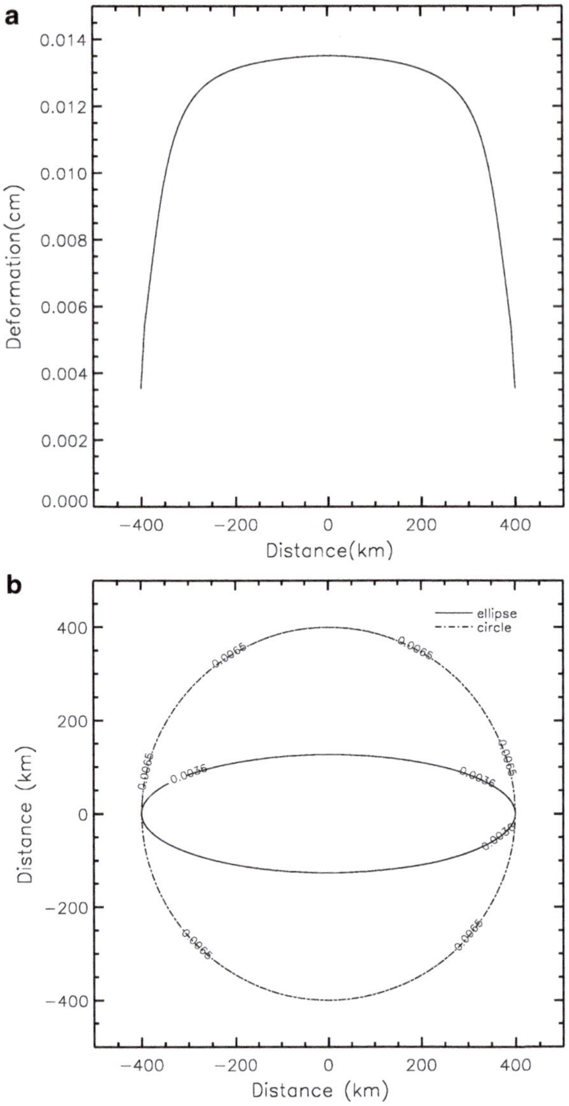

Figure 6
Results for change in surface height, $u_z(r_e)$, for $H = 23$ m, $D = 25$ m, $v = 0.3$, $dh = 1.01$ cm, and $R = 400$ m, for an aquifer with eccentricity = 0.9, using the elliptical approximation of Eq. 16 (*solid line*). **a** and **b** are the same as for Fig. 1 above. Included in **b**, for reference, is the circular approximation of Eq. 3 (*dashed line*)

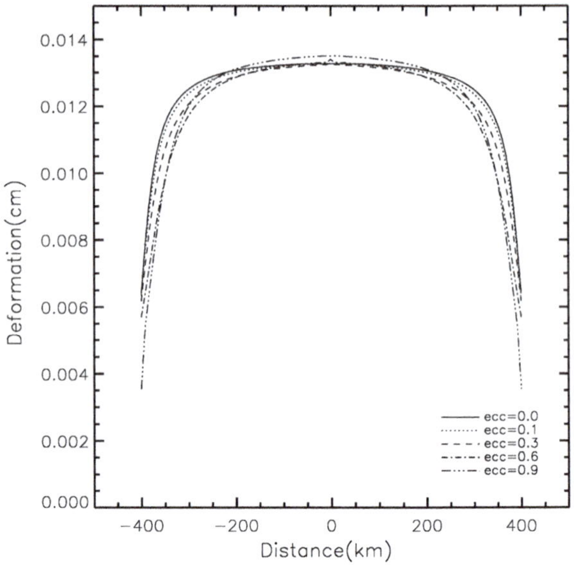

Figure 7
Cross section of the change in surface height, $u_z(r_e)$, through the long axis of the aquifer for varying eccentricity using the elliptical approximation of Eq. 16. As above, $H = 23$ m, $D = 25$ m, $v = 0.3$, $dh = 1.01$ cm, and $R = 400$ m

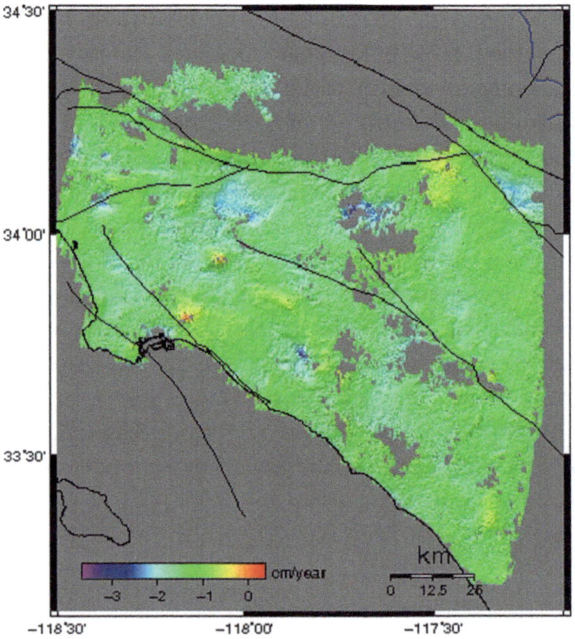

Figure 8
DInSAR interferogram, ERS2 September 2, 2000 to October 23, 2003

southern California between the Los Angeles basin and the San Gabriel Mountains. They developed a method based on random field theory within a Bayesian statistical framework to derive three-dimensional surface motion maps from sparse GPS measurements and differential interferometric synthetic aperture radar (DInSAR) interferograms in the southern California region. In this method, minimization of the Gibbs energy function is performed analytically (Samsonov and Tiampo, 2006). The problem is well posed and the solution is unique and stable and not biased by the continuity condition. The technique produces a three-dimensional field containing estimates of surface motion on the spatial scale of the DInSAR image, over a given time period, complete with error estimates. Significant improvement in the accuracy of the vertical component and moderate improvement in the accuracy of the horizontal components of velocity are achieved in comparison with the GPS data alone.

In this work we used the DInSAR interferogram shown in Fig. 8 with a time span of 2.97 year (02 September 2000—23 August 2003). This interferogram was processed from ERS-2 level 0 (raw data) using the JPL/Caltech Repeat Orbit Interferometry Package (*ROI PAC*, version 2.3) (Rosen *et al.*, 2004). To reduce water vapour effects on the interferogram, the GPS/MODIS integrated water vapour correction model demonstrated in (Li *et al.*, 2005) was applied. This processing involved the usual steps of image coregistration, interferogram computation, baseline estimation from Delft precise orbits and interferogram flattening and removal of topographic phase by using a 30 m SRTM DEM. At this stage, a Zenith Path Delay Difference Map (ZPDDM) derived from GPS-calibrated MODIS near IR water vapour fields was inserted into the interferometric processing sequence, followed by phase unwrapping and baseline refinement. In order to obtain the unwrapped water vapor corrected interferogram, a new simulated interferogram was created using the refined baseline and topography, and was subtracted from the unwrapped phase (including orbital ramp) with the water vapor model removed (Li *et al.*, 2005). Since InSAR has no absolute reference datum, the unwrapped phase has been shifted by the mean difference between InSAR and GPS range changes when compared to GPS-derived range changes in the LOS direction.

The GPS-DInSAR integration was performed using preprocessed time series from 140 GPS stations from the Southern California integrated GPS network Southern California Integrated GPS Network (SGICN) (Hudnut *et al.*, 2002). This data was reprocessed in order to remove outliers and offsets. Three components (north, east and up) of the velocity vector then were calculated by applying linear regression to the time span of the differential interferogram. Finally, ordinary kriging is used in order to interpolate GPS data at the intermediate locations.

The GPS-DInSAR optimization results are shown in Fig. 9. The appearance of the northern image practically does not change due to the orientation of the satellite in the current coordinate system. The eastern component of the velocity, however, has more significant changes, especially in the southern part. The strong velocity gradient in both the north and east components is visible, which suggests that this analysis can be used to improve the accuracy of the estimation of shortening across the Los Angeles basin. The vertical component shows the combination of signals from both GPS and DInSAR measurements. A few areas are of particular interest. A strong uplift with the approximate velocity of 0.6–0.8 cm/year is observed in the southern part of the Los Angeles basin (34.8N, −118.2E), due to seasonal changes and/or rebound in the local groundwater levels. A few areas of subsidence are presented in the northern part of the image. The middle region (around 34N, −117.6E) is the rapid subsidence which is presented on many other interferograms starting from the mid-1990s (Watson *et al.*, 2002). The top right subsidence (34.1N, −117.2E) that is originally presented only on the DInSAR interferogram is amplified on the final image and contours the fault. It is the groundwater uplift signal seen at −118.3, 33.8 that we will attempt to model here.

4.3. Inversion

The velocity change seen in Fig. 9 was converted to deformation over a smaller region centered on the vertical component of the groundwater deformation noted above, displayed in Fig. 10. Figure 10 shows the net deformation over this time period of almost 3 years, with the plate velocities removed. Deformation is shown in the east (Fig. 10a), north (Fig. 10b) and up (Fig. 10c) directions. While the vertical deformation is clearly defined, with a pseudo-elliptical shape, and a maximum deformation of approximately 4 cm, the deformation in the east and north direction is significantly smaller (note the difference in scale). However, there is some deformation above the background, particularly in the east component.

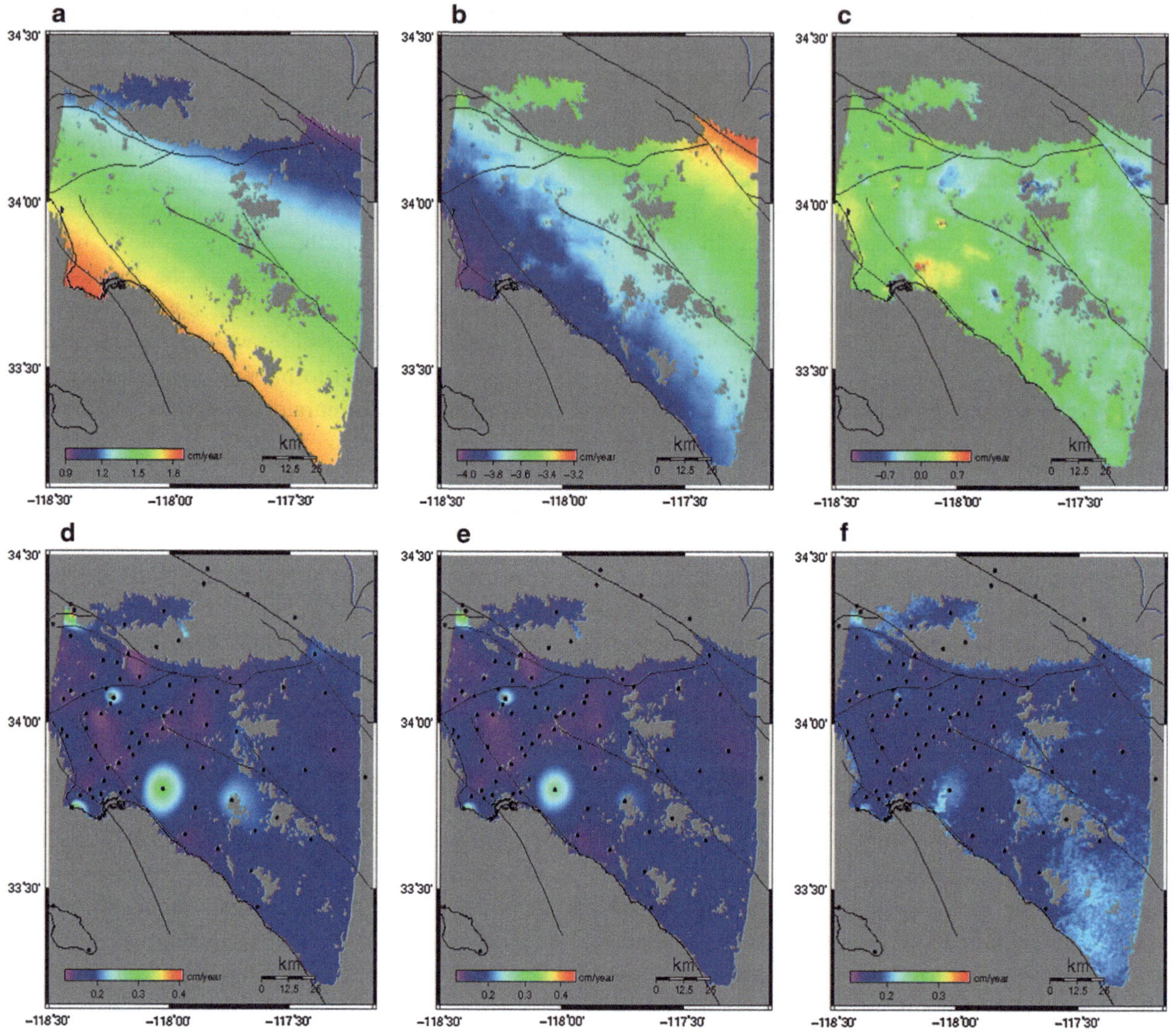

Figure 9
Optimized DInSAR/CGPS images of velocity for the period September 2, 2000 to October 23, 2003: **a** north component, **b** east component, **c** up component, **d** calculated error for the north component, **e** calculated error for the east component, **f** calculated error for the up component

Initial attempts to model the vertical deformation included a genetic algorithm inversion technique, details of which can be found in Tiampo *et al.* (2004), using initial parameter ranges for the aquifer from Poland *et al.* (1109) and Mills *et al.* (1999), and the circular groundwater deformation model from Le Mouélic and Adragna (2002). The depth to the top of the aquifer was fixed at 30 m and Poisson's ratio was set to 0.25. The inversion solved for the head change, *dh*, the radius of the aquifer (*R*), and the center location of the aquifer, *x* and *y*. Results for this inversion are shown in Fig. 11. Note that while the reduced chi-square value for this inversion was ~30, the reconstructed model in Fig. 11c does not fit the actual deformation pattern particularly well. This is because the genetic algorithm is attempting to find the best fit solution for a model that does not correspond to pattern particularly well, and therefore is averaging over the remaining points as best as possible. As a result, it significantly underestimates the maximum deformation. In addition, the pattern of horizontal deformation produced by the model does not resemble the actual deformation seen in Fig. 10. The best value for head change is 1.55 cm, while the optimal aquifer radius is 5,977 m.

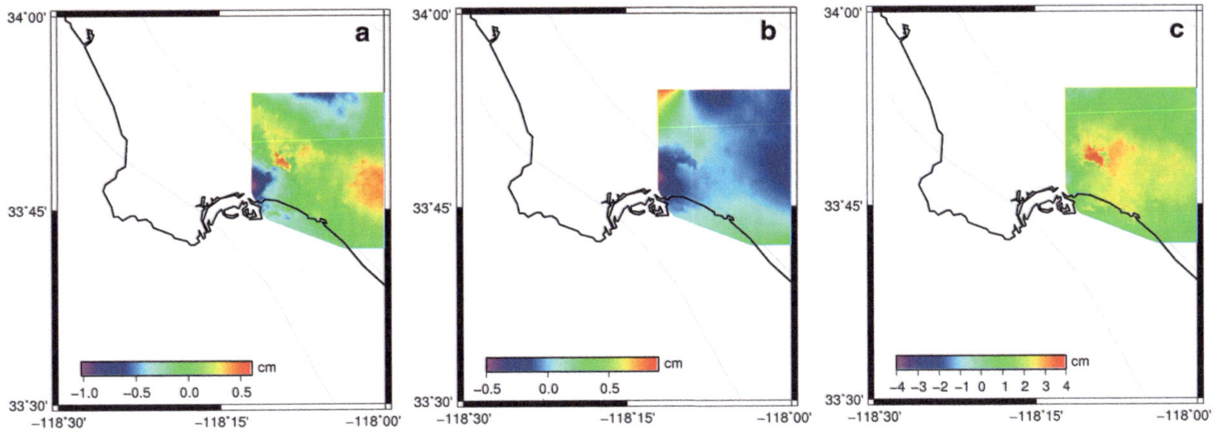

Figure 10
Optimized deformation images for the period September 2, 2000 to October 23, 2003: **a** east component, **b** north component and, **c** up component

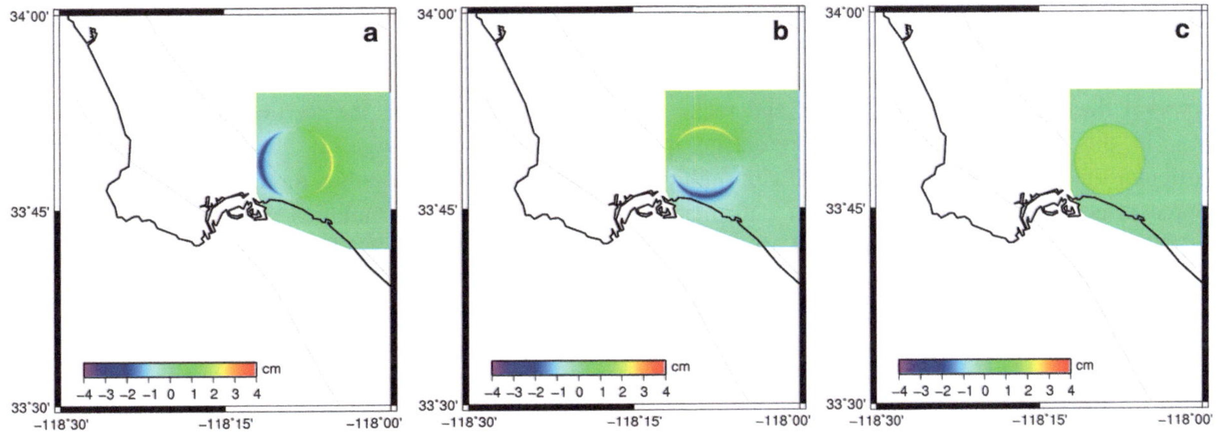

Figure 11
Deformation from circular source of Le Mouélic and Adragna (2002) for **a** east component, **b** north component and **c** up component and the time period September 2, 2000 to October 23, 2003

For comparison, we employed the elliptical deformation model developed above. The same center location was used, with a head change of 1.75 cm and an aquifer radius of 6,000 m, and an eccentricity of 0.5 and a rotation angle of 30° clockwise from horizontal. Results for the vertical deformation are shown in Fig. 12. Here we can see that while this model provides a better fit for the vertical deformation, it still underestimates the maximum uplift. It simplifies the variation in the uplift pattern across the aquifer, and there is a sharp an edge to the deformation at the aquifer edges, without the gradual changes seen by others (Argus *et al.*, 2005). In addition, the horizontal deformation is significantly larger than that of shown in Fig. 10. However, further examination suggests that modeling two separate elliptical sources of varying sizes and pumping volumes, offset and with slightly different rotations, could provide a significantly better fit to both the horizontal and vertical deformation patterns. Further modeling using a more complete inversion technique is necessary to determine the optimal relationship between radius, eccentricity and rotation.

The uplift in this area also differs in sign from that of the long-term regional trends determined by Argus *et al.* (2005). This result is more striking because

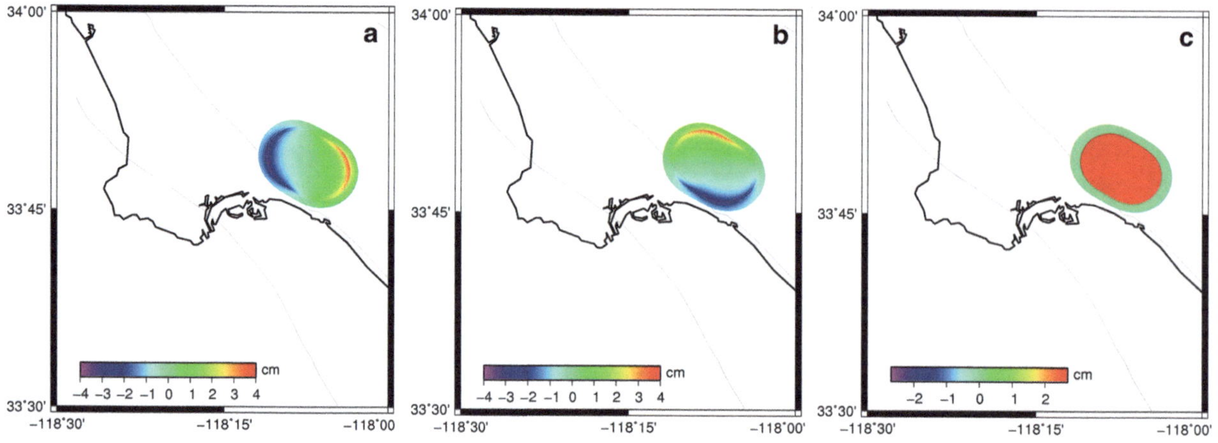

Figure 12
Deformation in the up-direction modeled from the elliptical adaption of the model shown in Fig. 11

other regions of known groundwater subsidence are visible in Fig. 9, in the same sense as observed by others (Watson *et al.*, 2002). Several explanations are possible. The most likely possibility is that temporal fluctuations in the discharge and recharge rates in this particular aquifer region during this particular time period, perhaps coupled with multiple source effects, are modulating the net deformation and obscuring the long-term trend. Further work, including a detailed analysis of many interferograms from this time period and the resulting time series will be necessary to provide a better interpretation.

5. Conclusions

Substantial progress in the detection of land subsidence (and uplift) has been made in recent years with the implementation of modern geodetic techniques, including continuous GPS networks and InSAR (Le Mouélic and Adragna, 2002). Mathematical formulation of the associated surface deformation provides an invaluable technique for the modeling of this deformation and the derivation and analysis of the important constitutive parameters, such as permeability, groundwater height change, and aquifer depth and thickness (Geertsma 1973; Le Mouélic and Adragna, 2002).

In this work we have extended the formula for a circular-shaped reservoir (Geertsma, 1973) to an elliptical-shaped reservoir (Figs. 2, 3, 4) and improved the accuracy of the approximation by making use of the cross terms of the expansion (Figs. 5, 6, 7). Results of the various simulations are encouraging, giving realistic values for the elliptical approximation of the vertical deformation due to groundwater change. It should be noted that, in addition, it is possible to develop a more generic mathematical formulation for an elliptic cylinder using the Mathieu-Hill differential equation and its associated solutions in terms of Mathieu functions (Abramowitz and Stegun, 1970; Valluri *et al.*, 1999, 2005; Vasconcelos *et al.*, 1991). Finally, we have applied the technique to groundwater deformation in southern California and show that model fits the pattern and magnitude of the vertical component of the elliptical signal better than a circular source, although additional studies in other regions that display a clearer elliptical pattern is ongoing in order to establish its universal applicability. The model is generalizable, and further tests will be conducted in order to determine the applicability of the horizontal deformation modeling, including a more detailed inversion for multiple sources.

Acknowledgments

K. F. Tiampo and S. R. Valluri gratefully acknowledge Discovery Grants from National Science Engineering Research Council Canada (NSERC), Human Resources Development Canada (HRDC) and the UWO Work Study Program for support of F. Ouegnin in this research project. The ERS data were supplied under ESA ENVISAT data grant AO ID = 853 (HAZARDMAP). Technical support for this work has been provided by the POLARIS network. Research by J. Fernández has been supported by Spanish MICINN projects: GEOMOD (CGL2005-05500-C02), PEL2G (CGL2008-06426-C01-01/BTE) and GEOSIR (AYA2010-17448). This research has been partially funded in the frame of the Moncloa Campus of International Excellence (UCM-UPM, CSIC). The DInSAR data was processed by the Repeat Orbit Interferometry Package (ROI PAC) developed at Caltech/Jet Propulsion Laboratory. The DEM data was provided by USGS. The images were plotted with the help of GMT software developed and supported by Paul Wessel and Walter H. F. Smith. The authors would like to thank S. Le Mouélic for providing the original IDL code for the circular deformation model and Dr. D. Argus and Dr. M. Motagh for thorough and helpful reviews.

References

Abramowitz, M. and Stegun, I.A., Handbook of Mathematical Functions (Dover, New York 1970).

Amelung, F., Galloway, D., Bell, J., Zebker, H., and Laczniak, R. (1999), *Sensing the ups and downs of Las Vegas: InSAR reveals structural control of land subsidence and aquifer-system deformation*, Geology *27*, 483-486.

Anderssohn, J., Wetzel, H.U., Walter, T.R., Motagh, M., Djamour, Y., and Kaufmann, H. (2008), *Land subsidence pattern controlled by old alpine basement faults in the Kashmar Valley, northeast Iran: results from InSAR and levelling*. Geophysical Journal International *174*, 287-294.

Argus, D.F., Heflin, M.B., Peltzer, G., Webb, F.H., and Crampe, F. (2005), *Interseismic strain accumulation and anthropogenic motion in metropolitan Los Angeles*, Journal of Geophysical Research *101*, doi:10.1029/2003JB002934.

Argus, D.F., and Gordon, R.G., (2001), *Present tectonic motion across the Coast Ranges and the San Andreas fault system in central California*, Geol. Soc. Am. Bull. *113*, 1580–1592.

Barry, D.A., Parlange, J.Y., Sander, G.C., and Sivaplan, M. (1993), *A class of exact solutions for Richards equation*, J. Hydrology *142*, 29-46.

Bawden, G., Thatcher, W., Stein, R., Hudnut, K., and Peltzer, G., (2001), *Tectonic contraction across Los Angeles after removal of groundwater pumping effects*, Nature *412*, 812-815.

Bell, J.W., Amelung, F., Ferretti, A., Bianchi, M., and Novali, F. (2008), *Permanent scatterer InSAR reveals seasonal and long-term aquifer-system response to groundwater pumping and artificial recharge*. Water Resources Research, *44*, 2 W02407. doi:10.1029/2007WR006152.

Burbey, T.J. (2006), *Three-dimensional deformation and strain induced by municipal pumping, Part 2: Numerical analysis*, J. of Hydrology *330*, 422-434.

Corless, R.M., Gonnet, G.H., Hare, D.E., Jeffrey, D.J., and Knuth, D.E. (1996), *On the Lambert W function*, Adv. Comput. Maths *5*, 329-359.

Dixon, T.H., Amelung, F., Ferretti, A., Novali, F., Rocca, F., Dokka, R., Sella, G., Kim, S.W., Wdowinski, S., and Whitman, D. (2006), *Subsidence and flooding in New Orleans*, Nature *441*, 587-588.

Fernández, J., Romero, R., Carrasco, D., Tiampo, K.F., Rodríguez-Velasco, G., Aparicio, A., Araña, V., and González-Matesan, F.J. (2005), *Detection of displacements in Tenerife Island, Canaries, using radar interferometry*. Geophysical Journal International *160*, 33-45, doi:10.1111/j.1365-246X.2005.02487.x.

Freeze, R.A., and Cherry, J.A., Groundwater (Prentice-Hall, NJ, 1979).

Geertsma, J. (1973), *Land Subsidence above Compacting Oil and Gas Reservoirs*, J. of Petroleum. Tech. *25*, 734-744.

Gourmelen, N., Amelung, F., Francesco, C., Mariarosaria, M., and Lanari, R. (2007), *Mining-related ground deformation in Crescent Valley, Nevada: Implications for sparse GPS networks*, Geophys. Res. Ltrs. *34*, L09309, doi:10.1029/2007GL029427.

Gradshteyn I.S., and Ryzhik, I.M., Table of Integrals Series and Products (Academic Press, NY, 1965).

Hagihara, Y., Celestial Mechanics, v. 1, (MIT Press, 1970).

Hudnut, K., Bock, Y., Galetzka, J., Webb, F., and Young, W. (2002), The Southern California Integrated GPS Network (SCIGN), Seismotectonics in Convergent Plate Boundaries, eds. Y. Fujinawa and A. Yosghida, Terrapub, 167-189 (http://www.scign.org/).

Jacobs, A., Sandwell, D., Fialko, Y., and Sichoix, L. (2002), *The 1997 (M 7.1) Hector Mine, California, earthquake: Near-field postseismic deformation from ERS interferometry*, Bull. Seis. Soc. Am. 92, 4, 1433–1442.

King, N.E., Argus, D., Langbein, J., Agnew, D.C., Bawden, G., Dollar, R.S., Liu, Z., Galloway, D., Reichard, E., Yong, A., Webb, F.H., Bock, Y., Stark, K., and Barseghian, D. (2007), *Space geodetic observation of expansion of the San Gabriel Valley, California, aquifer system, during heavy rainfall in winter 2004–2005*, Journal of Geophysical Research *112*, B03409, doi:10.1029/2006JB004448.

Kim, J. (2005), *Three-dimensional numerical simulation of fully coupled groundwater flow and land deformation in unsaturated true anisotropic aquifers due to groundwater pumping*, Water Resources Res., *41*, doi:10.1029/2003WR002941.

Lanari, R., Lundgren, P., Mariarosaria, M., and Casu, F. (2004), *Satellite radar interferometry time series analysis of surface deformation for Los Angeles, California*, Geophys. Res. Ltrs, *31*, L23613, doi:10.1029/2004GL021294, (2004).

Landau L.D., and Lifshitz, E.M. Theory of Elasticity, 3rd edition (Butterworth-Heinemann, Oxford, 1986).

Le Mouélic S., and Adragna, F. (2002), *Ground Uplift in the city of Paris (France) detected by Satellite Radar Interferometry*. Geophys. Res. Letters *29*, 1853, doi:10.1029/2002GL015630.

Li, Z., Muller, J.-P., Cross, P., and Fielding, E.J. (2005), *Interferometric synthetic aperture radar (InSAR) atmospheric correction: GPS, Moderate Resolution Imaging Spectroradiometer (MODIS), and InSAR integration*, J. Geophys. Res. *110*, B03410.

Logsdon, T. Orbital Mechanics: Theory and Applications (Wiley-Interscience, New York, 1997).

Massonnet, D., and Feigl, K.L. (1998), *Radar interferometry and its application to changes in the Earth's surface*, Rev. Geophys., *4*, 441-494.

Massonnet, D., Rossi, M., Carmona, C., Adragna, F., Peltzer, G., and Feigl, K. (1993), *The displacement field of the Landers earthquake mapped by radar interferometry*. Nature *364*, 138–142.

Mills, W., et al. Orange County Water District, Master Plan Report. 371 (Orange County Water District, Fountain Valley, California, 1999).

Motagh, M., Walter, T.R., Sharifi, M.A., Fielding, E., Schenk, A., Anderssohn, J., and Zschau, J. (2008), *Land subsidence in Iran caused by widespread water reservoir overexploitation*, Geophys. Res. Lett. *35*, L16403.

Murray C.D., and Dermott, S.F. Solar System Dynamics (Cambridge University Press, Cambridge, 1999).

Poland, J.F., and Piper, A.M., Ground-water geology of the coastal zone, Long Beach ± Santa Ana area, California, US Geol. Surv. Water-Supply Paper W1109 162 (US Geol. Survey, Reston, Virginia, 1956).

Rosen, P., Hensley, S., Peltzer, G., and Simons, M. (2004), *Updated repeat orbit interferometry package released*, EOS Transactions *85*, 47.

Samsonov, S. and Tiampo, K. (2006), *Analytical optimization of DInSAR and GPS dataset for derivation of three-dimensional surface motion*, Geoscience and remote sensing letters *3*, 107–111.

Samsonov, S., Tiampo, K.F., Rundle, J.B., and Li, Z. (2007), *Application of DInSAR-GPS Optimization for Derivation of Fine-Scale Surface Motion Maps of Southern California*, IEEE Transactions on Geoscience and Remote Sensing *45*, 2, doi: 10.1109/TGRS.2006.887166.

Schmidt, D., and Bürgmann, R. (2003), *Time-dependent land uplift and subsidence in the Santa Clara valley, California, from a large interferometric synthetic aperture radar data set*, Journal of Geophysical Research *108*, doi:10.1029/2002JB002267.

Tesauro, M., Berardino, P., Lanari, R., Sansisti, E., Fornaro, G., and Franschetti, G. (2000), *Urban subsidence inside the city of Napoli (Italy) observed by satellite radar interferometry*, Geophysical Research Letters *27*, 1961-1964.

Tiampo, K.F., Fernández, J., Jentzsch, G., Charco, M., and Rundle, J.B., (2004), *Inverting for the parameters of a volcanic source using a genetic algorithm and a model for magmatic intrusion in elastic-gravitational layered earth models*, Computers and Geosciences *30*, 9, 985-1001.

Tomás, R., Márquez, Y., Lopez-Sanchez, J.M., Delgado, J., Blanco, P., Mallorquí, J.J., Martínez, M., Herrera, G., and Mulas, J. (2005), *Mapping ground subsidence induced by aquifer overexploitation using advanced Differential SAR Interferometry: Vega Media of the Segura River (SE Spain) case study*, Rem. Sens. of Env., *98*, 269-283.

Valluri, S.R., Biggs, R.G., Harper W.L., and Wilson, C. (1999), *The significance of the Mathieu-Hill differential equation for Newton's apsidal precession theorem*, Can. J. Phys. *77*, 393-407.

Valluri, S.R., Chishtie, F.A., and Vajda, A. (2006), *The gravitational wave pulsar signal with Jovian and lunar perturbations and orbital eccentricity corrections*, Class. Quantum. Grav. *23*, 3323-3332.

Valluri, S.R., Jeffrey, D.J., and Corless, R.M. (2000), *Some applications of the Lambert W function to physics*, Can. J. Phys. *78*, 823-831.

Valluri, S.R., Yu, P., Smith, G.E., and Wiegert, P.A. (2005), *An Extension of Newton's Apsidal Precession Theorem*, Mon. Not. R. Astron. Soc. *358*, 1273-1284.

Vasconcelos, E.P., de Oliviera, N.T., and Farias, G.A. (1991), *Surface polaritons on an elliptic cylinder*, Phys Rev B *44*, 24, 13740.

Wang, Z.X., and Guo, D.R. Special Functions (World Scientific, Singapore, 1989).

Watson, K., Bock, Y., and Sandwell, D. (2002), *Satellite interferometric observation of displacements associated with seasonal groundwater in Los Angeles Basin*, J. Geophys. Res. *107*, 2074. doi:10.1029/2001JB000470.

(Received March 1, 2010, revised January 26, 2011, accepted July 10, 2011, Published online September 9, 2011)

Pure Appl. Geophys. 169 (2012), 1457–1462

DOI 10.1007/s00024-011-0419-7

Pure and Applied Geophysics

The Transition from Three-Dimensional Embedding to Two-Dimensional Euler-Lagrange Deformation Tensor of the Second Kind: Variation of Curvature Measures

Erik W. Grafarend[1]

Abstract—Based on the Stein formulation of changes of curvature parameters, we describe the changes of Riemann twodimensional manifolds of surface geometries. The three-dimensional left and right Euclidean manifolds with the curvature parameters "geodetic curvature, normal curvature, geodetic torsion" characterize the embedding "Riemann to Euclid" or 2d into 3d. The variation in time of the Euler-Lagrange deformation tensor of the second kind, in short "curvature variation" is studied.

Key words: Continuum mechanics, strain analysis, space-time Geodesy.

1. Introduction

Based on plate and shell theory within continuum mechanics we develop formulae for the tensor of change of curvature (TCC) based on surface theory (two-dimensional Riemann manifold) embedded into the ambient space (three-dimensional Euclidean space) at various time instants. Indeed, the curvature tensor is responsible for the detection of vertical displacements as proved by Moghtased-Azar and Grafarend (2009); Grafarend and Voosoghi (2003); Xu and Grafarend (1996). Here we take advantage of the second fundamental forms of surface geometry in the Lagrange-Eulerian versus the Euclidean version within ambient three-dimensional Euclidean space, regularized and graded. We shall present a set of invariants both in the Riemann space as well as in the ambient three-dimensional Euclidean space based on the variational formulation of Stein (1980) of TCC. In terms of differential geometry we take advantage of the Darboux reference frame in terms of {geodetic curvature, normal curvature, geodetic torsion} and the Gauss-Weingarten derivation equations in terms of Gaussian curvature and mean curvature. Our contribution generalizes the work of *Vanicel et al.* Here we enjoy taking reference to modern textbooks on continuum mechanics, for instance, Holzapfel (2000); Hutter and Joehnk (2004); Ernst (1981); Eschenauer and Schnell (1993); Simo and Hughes (1998); Libai and Simmonds (1976); Mushtari and Galimov (1961) and Naghdi (1972). The topic of simultaneous diagonalization of two quadratic forms is treated by Aravind (1988).

2. *Mapping from the Left Riemann Manifold $\mathbb{M}^2_\ell$ to the Right Riemann manifold $\mathbb{M}^2_r$*

Let us start from the left two-dimensional Riemann manifold $\{\mathbb{M}^2_\ell, \mathbf{G}_{\Lambda,\Phi}\}$ as the undeformed manifold embedded in the ambient three-dimensional Euclidean manifold and the right two-dimensional Riemann manifold $\{\mathbb{M}^2_r, \mathbf{g}_{\lambda,\phi}\}$ as the deformed manifold embedded in the ambient three-dimensional Euclidean manifold. In order to represent the two Riemann manifolds in a proper geodetic reference frame, we represent its metric in a Gauss coordinate system which represents the GPS coordinate system: A surface point on the Earth's topography is mapped orthogonally onto the reference ellipsoid by geodetic longitude and geodetic latitude. The third coordinate is defined by the Euclidean distance between the topographic point and the footprint of the reference ellipsoid by an orthogonal projection.

Let us define the reference ellipsoid-of-revolution by the semi-major axis A_1 and the semi-minor axis A_2 leading to the "first eccentricity squared", $E^2 := (A_1^2 - A_2^2)/A_1^2$, at the undeformed manifold as well as

[1] Geodaetisches Institut, Geschwister-Scholl-Str. 24D, 70174 Stuttgart, Germany. E-mail: grafarend@gis.uni-stuttgart.de

the reference ellipsoid-of-revolution by the semi-major axis a_1 and the semi-minor axis a_2 leading to the "first eccentricity squared", $e^2 := (a_1^2 - a_2^2)/a_1^2$, at the deformed manifold. We use the ellipsoidal gauge $A_1 = a_1, A_2 = a_2, E^2 = e^2$. In surface geometry we take advantage of the ellipsoidal height representations $H(\Lambda, \Phi)$ and $h(\lambda, \phi)$ introduced in terms of ellipsoidal functions by GRAFAREND and ENGELS (1992) ("amplitude-modified spherical harmonic functions" and "phase-modified spherical harmonic functions").

$$\mathbf{X}(\Lambda, \Phi, H) \in \{\mathbb{R}^3, \mathbf{G}_{\Lambda,\Phi,H}\} \quad \text{or} \quad \mathbf{X}(\Lambda, \Phi, H(\Lambda, \Phi)) \in \{\text{RIEMANN}\}$$

versus

$$\mathbf{x}(\lambda, \phi, h) \in \{\mathbb{R}^3, \mathbf{g}_{\lambda,\phi,h}\} \quad \text{or} \quad \mathbf{x}(\lambda, \phi, h(\lambda, \phi)) \in \{\text{RIEMANN}\}$$

$$\mathbf{X}(\Lambda, \Phi, H) = \mathbf{J}_1 X^1(\Lambda, \Phi, H) + \mathbf{J}_2 X^2(\Lambda, \Phi, H) + \mathbf{J}_3 X^3(\Lambda, \Phi, H)$$

$$\mathbf{X}(\Lambda, \Phi) = [\mathbf{J}_1, \mathbf{J}_2, \mathbf{J}_3] \times \begin{bmatrix} \left[\frac{\sim_1}{\sqrt{1-\mathrm{E}^2\sin^2\Phi}} + \mathrm{H}(\Lambda,\Phi)\right]\cos\Phi\cos\Lambda \\ \left[\frac{\sim_1}{\sqrt{1-\mathrm{E}^2\sin^2\Phi}} + \mathrm{H}(\Lambda,\Phi)\right]\cos\Phi\sin\Lambda \\ \left[\frac{\sim_1(1-\mathrm{E}^2)}{\sqrt{1-\mathrm{E}^2\sin^2\Phi}} + \mathrm{H}(\Lambda,\Phi)\right]\sin\Phi \end{bmatrix}$$

versus

$$\mathbf{x}(\lambda, \phi, h) = \mathbf{j}_1 x^1(\lambda, \phi, h) + \mathbf{j}_2 x^2(\lambda, \phi, h) + \mathbf{j}_3 x^3(\lambda, \phi, h)$$

$$\mathbf{x}(\lambda, \phi) = [\mathbf{j}_1, \mathbf{j}_2, \mathbf{j}_3] \times \begin{bmatrix} \left[\frac{\mathrm{a}_1}{\sqrt{1-\mathrm{e}^2\sin^2\phi}} + \mathrm{h}(\lambda,\phi)\right]\cos\phi\cos\lambda \\ \left[\frac{\mathrm{a}_1}{\sqrt{1-\mathrm{e}^2\sin^2\phi}} + \mathrm{h}(\lambda,\phi)\right]\cos\phi\sin\lambda \\ \left[\frac{\mathrm{a}_1(1-\mathrm{e}^2)}{\sqrt{1-\mathrm{e}^2\sin^2\phi}} + \mathrm{h}(\lambda,\phi)\right]\sin\phi \end{bmatrix}$$

The minimal distance mapping of the topography to the reference ellipsoid has been analyzed in detail by BARTELME and MEISSL (1975); BENNING (1974); FROEHLICH and HANSEN (1976); GRAFAREND and LOHSE (1991); HECK (1987); HEIKKINEN (1982); PAUL (1973); PENEV (1978); PICK (1985); SUENKEL (1976); VINCENTY (1976); VINCENTY (1980).

An important restriction of the *Gauss reference frame* based on two tangent vectors, elements of the *tangent space* $\mathbb{T}_P(\Lambda, \Phi)$ as well as the tangent space $\mathbb{T}_p(\lambda, \phi)$ of surface geometry and the normal space $\mathbb{N}_P(\Lambda, \Phi)$ as well as $\mathbb{N}_p(\lambda, \phi)$, namely,

$$\left\{\mathbf{G}_1 := \frac{\partial \mathbf{X}}{\partial \Lambda}, \mathbf{G}_2 := \frac{\partial \mathbf{X}}{\partial \Phi}\right\} \quad \text{as well as} \quad \left\{\mathbf{g}_1 := \frac{\partial \mathbf{x}}{\partial \lambda}, \mathbf{g}_2 := \frac{\partial \mathbf{x}}{\partial \phi}\right\}$$

versus

$$\mathbf{G}_3 := \frac{\mathbf{G}_1 \times \mathbf{G}_2}{\|\mathbf{G}_1 \times \mathbf{G}_2\|} \quad \text{as well as} \quad \mathbf{g}_3 := \frac{\mathbf{g}_1 \times \mathbf{g}_2}{\|\mathbf{g}_1 \times \mathbf{g}_2\|}$$

is that, due to the parameterization, they become dependent of the ellipsoidal height functions $H(\Lambda, \Phi)$ as well as $h(\lambda, \phi)$ and in consequence also H_Λ, H_Φ as well as h_λ, h_ϕ. A conclusion is that the surface reference frames of type Gauss are no longer orthogonal nor normalized. We summarize by our representation the theory behind the Euler-Lagrange deformation tensor of the second kind as a function of the spatial displacement vector **u** and **w**. We collect results in Tables 1, 2.

For further details of the first, second and third fundamental forms, namely on the curvature matrix and its eigenspace, we refer to KLINGENBERG (1978): The Gauss curvatures $\det \mathbf{K}_\ell$ and $\det \mathbf{K}_r$ are the only curvature functions which do not change sign under the orientation-reversing isometrics or changes of variables. In the case of a differential map from the left to the right differential manifold and vice versa, we summarize various representations of the Euler-Lagrange deformation tensor of the second kind in terms of curvature measures.

3. *Curvature Forms in Gaussian Surface Geometry*

Deformation measures have been developed for changes of placement vectors of a left versus right Riemann manifold embedded into ambient three-dimensional Euclidean manifolds. They are related to changes in

$$\boxed{\textit{thefirstfundamentalformofsurfacegeometry}} \longleftrightarrow \boxed{\textit{thesecondfundamentalformofsurfacegeometry}}$$

Table 1

Left versus right curvature matrices

Left Riemann manifold surface curvature	Right Riemann manifold surface curvature
Weingarten Map	Weingarten Map
$\mathbf{dG}_3 = -[\mathrm{d}\Lambda, \mathrm{d}\Phi]\mathbf{H}_\ell \mathbf{G}_\ell^{-1} \begin{bmatrix} \mathbf{G}_1 \\ \mathbf{G}_2 \end{bmatrix}$	$\mathbf{dg}_3 = -[\mathrm{d}\lambda, \mathrm{d}\phi]\mathbf{H}_r \mathbf{G}_r^{-1} \begin{bmatrix} \mathbf{g}_1 \\ \mathbf{g}_2 \end{bmatrix}$
$\frac{\partial}{\partial[\Lambda,\Phi]} = -\mathbf{H}_\ell \mathbf{G}_\ell^{-1} \begin{bmatrix} \mathbf{G}_1 \\ \mathbf{G}_2 \end{bmatrix}$	$\frac{\partial}{\partial[\lambda,\phi]} = -\mathbf{H}_r \mathbf{G}_r^{-1} \begin{bmatrix} \mathbf{g}_1 \\ \mathbf{g}_2 \end{bmatrix}$
Left curvature matrix	Right curvature matrix
$\mathbf{K}_\ell := -\mathbf{H}_\ell \mathbf{G}_\ell^{-1}$	$\mathbf{K}_r := -\mathbf{H}_r \mathbf{G}_r^{-1}$
Left Gauss curvature, left mean curvature (trace, determinant)	Right Gauss curvature, right mean curvature (trace, determinant)
$h_\ell := -\frac{1}{2}\mathrm{tr}\mathbf{K}_\ell = \frac{1}{2}\left[\kappa_\ell^1 + \kappa_\ell^2\right]$	$h_r := -\frac{1}{2}\mathrm{tr}\mathbf{K}_r = \frac{1}{2}\left[\kappa_r^1 + \kappa_r^2\right]$
$k_\ell := \det\mathbf{K}_\ell = \kappa_\ell^1 \kappa_\ell^2$	$k_r := \det\mathbf{K}_r = \kappa_r^1 \kappa_r^2$
(C.F. Gauss)	(C.F. Gauss)
$\kappa_\ell^1, \kappa_\ell^2 \ldots$ left eigenvalues of $\mathbf{K}_\ell$	$\kappa_r^1, \kappa_r^2 \ldots$ right eigenvalues of $\mathbf{K}_r$

Table 2

Euler-Lagrange deformation tensor of the second kind, Gauss map

Left Euler-Lagrange deformation of the second kind	Right Euler-Lagrange deformation of the second kind
$\mathbb{I}_r - \mathbb{I}_\ell := k_{\lambda,\phi}\, \mathrm{d}q^\lambda\, \mathrm{d}q^\phi - K_{\Lambda,\Phi}\, \mathrm{d}Q^\Lambda\, \mathrm{d}Q^\Phi$ for $\Lambda, \Phi, \lambda, \phi \in \{1, 2\}$	$\mathbb{I}_r - \mathbb{I}_\ell := k_{\lambda,\phi}\, \mathrm{d}q^\lambda\, \mathrm{d}q^\phi - K_{\Lambda,\Phi}\, \mathrm{d}Q^\Lambda\, \mathrm{d}Q^\Phi$ for $\Lambda, \Phi, \lambda, \phi \in \{1, 2\}$
$\mathbb{I}_r - \mathbb{I}_\ell := \left(k_{\lambda,\phi} \frac{\partial q^\lambda}{\partial Q^\Lambda} \frac{\partial q^\phi}{\partial Q^\Phi} - K_{\Lambda,\Phi}\right) \mathrm{d}Q^\Lambda\, \mathrm{d}Q^\Phi$	$\mathbb{I}_r - \mathbb{I}_\ell := \left(k_{\lambda,\phi} - K_{\Lambda,\Phi} \frac{\partial q^\Lambda}{\partial Q^\lambda} \frac{\partial q^\Phi}{\partial Q^\phi}\right) \mathrm{d}q^\Lambda\, \mathrm{d}q^\phi$
$\mathbb{I}_r - \mathbb{I}_\ell := \kappa_{\Lambda,\Phi}\, \mathrm{d}Q^\Lambda\, \mathrm{d}Q^\Phi$	$\mathbb{I}_r - \mathbb{I}_\ell := \kappa_{\lambda,\phi}\, \mathrm{d}q^\Lambda\, \mathrm{d}q^\phi$
subject to	*subject to*
$\kappa_{\Lambda,\Phi} := k_{\lambda,\phi} \frac{\partial q^\lambda}{\partial Q^\Lambda} \frac{\partial q^\phi}{\partial Q^\Phi} - K_{\Lambda,\Phi}$	$\kappa_{\lambda,\phi} := k_{\lambda,\phi} - K_{\Lambda,\Phi} \frac{\partial q^\Lambda}{\partial Q^\lambda} \frac{\partial q^\Phi}{\partial Q^\phi}$

Table 3

Euler-Lagrange deformation tensor of the second kind, Gauss map

Left representation	Right representation
$\kappa_{\Lambda,\Phi} = k_{\lambda,\phi} \frac{\partial q^\lambda}{\partial Q^\Lambda} \frac{\partial q^\phi}{\partial Q^\Phi} - K_{\Lambda,\Phi} \quad (2)$	$\kappa_{\lambda,\phi} = k_{\lambda,\phi} - K_{\Lambda,\Phi} \frac{\partial Q^\Lambda}{\partial q^\lambda} \frac{\partial Q^\Phi}{\partial q^\phi} \quad (6)$
$\kappa_{\Lambda,\Phi} =$	$\kappa_{\lambda,\phi} :=$
$:= -\frac{\partial q^\lambda}{\partial Q^\Lambda} \frac{\partial q^\phi}{\partial Q^\Phi} \left\langle \frac{\partial \mathbf{g}_3}{\partial q^\lambda}, \mathbf{g}\phi \right\rangle - K_{\Lambda,\Phi} \quad (3)$	$:= k_{\lambda,\phi} + \frac{\partial Q^\Lambda}{\partial q^\lambda} \frac{\partial Q^\Phi}{\partial q^\phi} \left\langle \frac{\partial \mathbf{G}_3}{\partial Q^\lambda}, \mathbf{G}\phi \right\rangle \quad (7)$
$\kappa_{\Lambda,\Phi} = -\frac{\partial q^\lambda}{\partial Q^\Lambda} \frac{\partial q^\phi}{\partial Q^\Phi} \cdot \cdot \left\langle \frac{\partial(\mathbf{w}+\mathbf{G}_3}{\partial q^\lambda}, \frac{\partial(\mathbf{X}+\mathbf{u})}{\partial q^\phi} \right\rangle - K_{\Lambda,\Phi} \quad (4)$	$\kappa_{\lambda,\phi} = k_{\lambda,\phi} - \frac{\partial Q^\Lambda}{\partial q^\lambda} \frac{\partial Q^\Phi}{\partial q^\phi} \cdot \cdot \left\langle \frac{\partial(\mathbf{w}-\mathbf{g}_3}{\partial Q^\Lambda}, \frac{\partial(\mathbf{x}-\mathbf{u})}{\partial Q^\Phi} \right\rangle \quad (8)$
$\kappa_{\Lambda,\Phi} =$	$\kappa_{\lambda,\phi} =$
$= -\langle \mathbf{w}_{1_\Lambda}, \mathbf{G}\Phi \rangle - \langle \mathbf{w}_{1_\Lambda}, \mathbf{u}_\Phi \rangle - -\langle \mathbf{u}_{1_\Phi}, \mathbf{G}_{3,\Lambda} \rangle \quad (5)$	$= -\langle \mathbf{w}_{1_\lambda}, \mathbf{g}_3 \rangle - \langle \mathbf{w}_{1_\lambda}, \mathbf{u}\phi \rangle - -\langle \mathbf{u}_{1_\phi}, \mathbf{g}_{3,\lambda} \rangle \quad (9)$

Here we concentrate on the tensor of change of the curvature (TCC) in terms of left and right coordinates as developed by PIETRASZKIEWICZ (1977); [STEIN (1980). Their theory forms the basis of plate and shell theory. In analogy to the Euler-Lagrange deformation of the first kind, they introduced the deformation tensor of the second kind or the TCC as functions of the spatial displacement of surface normal vectors in the left and right Riemann manifold, namely

$$\mathbf{w} := \mathbf{g}_3 - \mathbf{G}_3 \tag{1}$$

([Stein (1980)])

In our geodetic concept of TCC we have documented already that TCC measures the changes of vertical displacements of manifolds. In a set of formulae we collect the results of TCC theory, the Stein equations. We use the inner product notation of partial derivatives of the normal vector bundle based on $\mathbf{w}, \mathbf{g}_3$ and $\mathbf{G}_3$. Finally we are left with the Euler-Lagrange deformation tensor of the second kind as a function of spatial displacement vectors $\mathbf{u}$ and $\mathbf{w}$.

We have derived the Mean Curvature $h := \frac{1}{2}\mathrm{tr}\mathbf{K}$ and the Gauss Curvature $k := \det\mathbf{K}$ from the eigenvalue–eigenvector representation

$$\begin{aligned}
&\kappa\mathbf{I}\mathbf{v} = \mathbf{H}\mathbf{G}^{-1}\mathbf{v} \Leftrightarrow \det(\kappa\mathbf{I} - \mathbf{H}\mathbf{G}^{-1}) = 0\\
&\Leftrightarrow \det(\mathbf{H} - \kappa\mathbf{G}) = 0,\ \mathbf{G}\text{ positive-definite}\\
&\det(\kappa\mathbf{I} - \mathbf{H}\mathbf{G}^{-1}) = \kappa^2 - 2h\kappa + k = 0\\
&\kappa_{1,2} = h \pm \sqrt{h^2 - k^2}.
\end{aligned} \tag{10}$$

We are confronted with the problem of determining simultaneously diagonalize the pair of matrices $-\mathbf{H}_\ell\mathbf{G}_\ell^{-1}$ and $-\mathbf{H}_r\mathbf{G}_r^{-1}$ forming left and right curvature. Unfortunately the matrices of curvature "left and right" are not positive-definite, neither semi-definite, in general. Indeed, we cannot apply our result from Theorem 1.1 of Grafarend and Krumm (2006) page 1:

If $\mathbf{A} \in \mathbb{R}^{n\times n}$ is a symmetric matrix and $\mathbf{B} \in \mathbb{R}^{n\times n}$ is a symmetric positive definite matrix such that the product $\mathbf{A}\mathbf{B}^{-1}$ exists, then there exists a non-singular matrix $\mathbf{X}$ such that both following matrices are diagonal matrices, where $\mathbf{I}_n$ is the n-dimensional unique matrix.

$$\mathbf{X}^{\mathrm{T}}\mathbf{A}\mathbf{X} = \mathrm{diag}(\lambda_1, \ldots, \lambda_n), \tag{11}$$

$$\mathbf{X}^{\mathrm{T}}\mathbf{B}\mathbf{X} = \mathbf{I}_n = \mathrm{diag}(1, \ldots, 1). \tag{12}$$

We come back to this problem elsewhere.

4. *Curvature Forms in the Darboux Reference Frame in the Three-Dimensional Euclidean Space*

So far we have analyzed the Gauss curvature concept and introduced the Euler-Lagrange deformation tensor of the second kind according to surface geometry and left/right Riemann manifolds of change of curvature. Here we present the theory of the Darboux frame of reference in terms of

$$geodetic curvature, normal curvature, geodetic torsion$$

and study its changes in terms of the embedding into the ambient three-dimensional Euclidean space.

Let us first define the orthonormal triad or 3-leg of a surface curve of type Darboux, a special CARTAN frame of reference.

$$\begin{aligned}
\mathbf{d}_1 &:= \frac{\partial\mathbf{x}}{\partial\lambda}/\|\mathbf{x}_\lambda\|\\
\mathbf{d}_2 &:= \mathbf{g}_3 \times \mathbf{d}_1\\
\mathbf{d}_3 &:= \mathbf{g}_3
\end{aligned}$$

Next we study its embedding into the three-dimensional Euclidean space. We start from the representation

$$\mathbf{dd} = \Omega\mathbf{d},\quad \Omega = -\Omega^{\mathrm{T}},\quad \Omega := (\mathbf{dR})\mathbf{R}^{\mathrm{T}}$$

The antisymmetric connection matrix Ω has, therefore, only three components called

$$geodetic curvature, normal curvature, geodetic torsion$$

$$\begin{bmatrix}\mathbf{dd}_1\\ \mathbf{dd}_2\\ \mathbf{dd}_3\end{bmatrix} = \begin{bmatrix}0 & \omega_1^2 & \omega_1^3\\ -\omega_1^2 & 0 & \omega_2^3\\ -\omega_1^3 & -\omega_2^3 & 0\end{bmatrix}\begin{bmatrix}\mathbf{d}_1\\ \mathbf{d}_2\\ \mathbf{d}_3\end{bmatrix}$$

$$\kappa_g := \frac{\omega_1^2}{\sigma^1} \quad \text{"geodetic curvature"}$$

$$\kappa_n := \frac{\omega_1^3}{\sigma^1} \quad \text{"normal curvature"}$$

$$\tau_g := \frac{\omega_2^3}{\sigma^1} \quad \text{"geodetic torsion"}$$

$$\begin{aligned}
\mathbf{dd}_1 &= \sigma^1\kappa_g\mathbf{d}_2 + \sigma^1\kappa_n\mathbf{d}_3\\
\mathbf{dd}_2 &= -\sigma^1\kappa_g\mathbf{d}_1 + \sigma^1\tau_g\mathbf{d}_3\\
\mathbf{dd}_3 &= -\sigma^1\kappa_n\mathbf{d}_1 - \sigma^1\tau_g\mathbf{d}_2
\end{aligned}$$

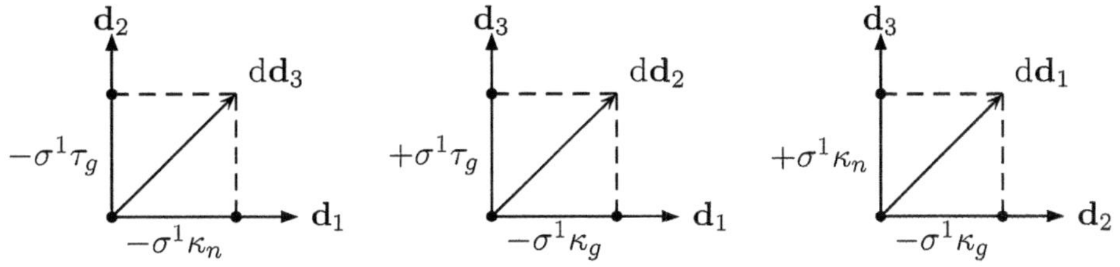

Figure 1
Projections of $\{\mathbf{d}_1, \mathbf{d}_2, \mathbf{d}_3\}$ on the tangential plane as well as on the horizontal plane

subject to $\sigma^1 := \|\mathbf{x}_\lambda\|$

Geodetic curvature κ_g measures the change of the tangent vector $\mathbf{d}_1$ and of the tangent vector $\mathbf{d}_2$ located on the local tangent plane. The normal curvature κ_n takes care of the change of the tangent vector $\mathbf{d}_1$ as well as the normal vector $\mathbf{d}_3$ while geodetic torsion τ_g measures the change of the surface normal vector projected on the tangent vector $\mathbf{d}_1$.

Figure 1 illustrates the various projections of $\mathbf{dd} = \Omega\mathbf{d}$.

Basic is the transformation of the Darboux frame of reference to a local Frenet frame of reference.

Lemma 1 *(J. Meusnier: Memoir sur la courbure des surfaces, Mem. des Savants etr.* 10 *(1785) 504)*

$$\kappa_g = \kappa_1 \sin\theta, \quad \kappa_n = \kappa_2 \cos\theta, \quad \tau_g = \kappa_2 + \frac{d\theta}{\sigma^1}$$

$$\kappa_g^2 + \kappa_n^2 = \kappa_1^2$$

$(\kappa_g, \kappa_n, \tau_g)$ *are the elements of the connection of the Darboux 3-leg*, (κ_1, κ_2) are the descriptive elements of the connection of the Frenet 3-leg. Of course, Frenet 1 = Darboux 1 holds. θ denotes the angle between the base vectors Frenet 2 (principal normal vector) and the surface normal vector Darboux 3. How do we transfer this result to the Euler-Lagrange deformation tensor of surface geometry, a surface embedded into a three-dimensional Euclidean space? We consider the Stein equations $\mathbf{w} = \mathbf{g}_3 - \mathbf{G}_3$ and its derivatives as a model of changes of curvature parameters, namely Eqs. (1)–(10). They describe the changes of Riemann two-dimensional manifolds of surface geometries. The three-dimensional Euclidean manifolds with the curvature parameters "geodetic curvature, normal curvature, geodetic torsion" characterize the embedding Riemann→Euclid or 2d into 3d. Their variation in time is the basis of the Euler-Lagrange deformation tensor of the second kind, in short "curvature variation".

5. Conclusion

We have first represented the Euler-Lagrange deformation tensor of the second kind, the surface curvature tensor deformation in the left and right forms, as well as its transformation "left-right" by formula (6)–(9). In contrast, we next introduced the curvature variational forms in the Darboux reference frame in the three-dimensional Euclidean space, the embedding space illustrated in Fig. 1. Based on the Meusnier Lemma we presented the transformation from three-dimensional Euclidean space, namely its curvature parameters {geodetic curvature, normal curvature, geodetic torsion}, into the eigenvalues of the Gauss curvature tensor in surface geometry. In addition, there appears the angle θ between base vector Frenet 2 (principal normal vector) and the surface normal vector Darboux 3. We leave the question open to transform the variational equations for the Euler-Lagrange tensor of the second kind from the left to the right form and vice versa. We have two forms of the Meusnier Lemma, one for the left manifold and one for the right manifold. The variation in time of the manifolds, left and right, has to be calculated by

$$\begin{aligned}\kappa_g(\text{left}) - \kappa_g(\text{right}) &= f\{\kappa_1(\text{left}), \kappa_2(\text{left}), \theta(\text{left}),\\ &\qquad \kappa_1(\text{right}), \kappa_2(\text{right}), \theta(\text{right})\},\\ \kappa_n(\text{left}) - \kappa_n(\text{right}) &= f\{\ \},\\ \tau_g(\text{left}) - \tau_g(\text{right}) &= f\{\ \},\end{aligned}$$

subject to the condition

$$\kappa_g^2 + \kappa_n^2 = \kappa_1^2(\text{left}) \quad \text{and}$$
$$\kappa_g^2 + \kappa_n^2 = \kappa_1(\text{right}).$$

References

ARAVIND PK (1988) *Geometrical interpretation of the simultaneous diagonalization of two quadratic forms.* J Phys *57*:309–311

BARTELME N, MEISSL P (1975) *Ein einfaches, rasches und numerisch stabiles Verfahren zur Bestimmung des kuerzesten Abstandes eines Punktes von einem sphaeroidischen Rotationselipsoid.* Allg Vermessungsnachrichten *82*:436–439

BENNING W (1974) *Der kuerzeste Abstand eines in rechtwinkligen Koordinaten gegebenen Auenpunktes vom Ellipsoid.* Allg Vermessungsnachrichten *81*:429–433

ERNST LJ (1981) A geometrically nonlinear finite element shell theory. PhD thesis, Department of Mechanical Engineering, TU Delft

ESCHENAUER H, SCHNELL W (1993) Elastizitaetstheorie, 3rd edn. BI Wissenschaftsverlag, Mannheim-Leipzig

FROEHLICH H, HANSEN HH (1976) *Zur Lotfupunktberechnung bei rotationsellipsoidischer Bezugsflaeche.* Allg Vermessungsnachrichten *83*:175–179

GRAFAREND EW, ENGELS J (1992) *A global representation of ellipsoidal heights – geoidal undulations or topographic heights – in terms of orthonormal functions, part 1: "amplitude-modified" spherical harmonic functions.* manuscripta geodaetica *17*:52–58 and 59–62

GRAFAREND EW, KRUMM F (2006) Map projections, cartographic information systems. Springer Verlag, Berlin-Heidelberg

GRAFAREND EW, LOHSE P (1991) *The minimal distance mapping of the topographic surface onto the (reference) ellipsoid of revolution.* manuscripta geodaetica *16*

GRAFAREND EW, VOOSOGHI B (2003) *Intrinsic deformation analysis of the earth's surface based on displacement fields derived from space geodetic measurements. case studies: present-day deformation patterns of Europe and of the Mediterranean area (ITRF data sets).* J Geodesy *17*(5–6):303–326, doi:10.1007/s00190-003-0329-2

HECK B (1987) Rechenverfahren und Auswertemodelle der Landesvermessung. Wichmann-Verlag, Karlsruhe

HEIKKINEN M (1982) *Geschlossene Formeln zur Berechnung raeumlicher geodaetischer Koordinaten aus rechtwinkligen Koordinaten.* Zeitschrift fuer Vermessungswesen *107*:207–211

HOLZAPFEL GH (2000) Nonlinear solid mechanics. Colchester Weinheim

HUTTER K, JOEHNK K (2004) Continuum methods of physical modelling. Springer Verlag, Berlin-Heidelberg

KLINGENBERG W (1978) A course in differential geometry. Springer Verlag, New York–Heidelberg–Berlin

LIBAI A, SIMMONDS JG (1976) The Nonlinear Theory of Elastic Shells. Cambridge University Press

MOGHTASED-AZAR K, GRAFAREND EW (2009) *Surface deformation analysis of dense GPS networks based on intrinsic geometry – deterministic and stochastik aspects.* J Geodesy *83*:431–454

MUSHTARI KM, GALIMOV KZ (1961) The non-linear theory of elastic shells Israel Program Sci Transl (Translated from Russian)

NAGHDI PM (1972) The theory of shells and plates. Handbuch der Physik VI, A2. Springer, Berlin-Heidelberg-New York

PAUL MK (1973) *A note on computation of geodetic coordinates from geocentric (Cartesian) coordinates.* Bulletin Godsique *108*:135–139

PENEV P (1978) *Transformation of rectangular coordinates into geographical coordinates by closed formulas.* Mapping and Photogrammetry *20*:175–177

PICK M (1985) *Closed formulas for transformation of the Cartesian coordinate system into a system of geodetic coordinates.* Studia Geoph et Geod *29*:112–119

PIETRASZKIEWICZ W (1977) Introduction to the non-linear theory of shells. tech. rep. 10. Mitteilungen aus dem Institut fuer Mechanik

SIMO JC, HUGHES TJR (1998) Computational inelasticity. Springer Verlag, New York-Heidelberg

STEIN E (1980) Variational functionals in the geometrical nonlinear theory of thin shells and finite-element-discretzations with applications to stability problems. North-Holland, Amsterdam-New York, Tbilisi, pp 509–535, in: Theory of Shells, Proceedings of the 3rd IUTAM Symp. on Shell Theory

SUENKEL H (1976) *Ein nicht-iteratives Verfahren zur Transformation geodaetischer Koordinaten.* oesterreichische Zeitschrift fuer Vermessungswesen *64*:29–33

VINCENTY T (1976) *Ein Verfahren zur Bestimmung der geodaetischen Hoehe eines Punktes.* Allg Vermessungsnachrichten *83*:179

VINCENTY T (1980) *Zur raeumlich-ellipsoidischen Koordinaten-Transformation.* Zeitschrift fuer Vermessungswesen *105*:519–521

XU PL, GRAFAREND EW (1996) *Statistics and geometry of the eigenspectra of three-dimensional decond-rank symmetric random tensor.* Geophysical Journal International *127*(3):744–756, doi:10.1111/j.1365-246X.1996.tb04053.x

(Received November 11, 2010, revised March 29, 2011, accepted May 11, 2011, Published online November 12, 2011)

Pure Appl. Geophys. 169 (2012), 1463–1482

DOI 10.1007/s00024-011-0403-2

Pure and Applied Geophysics

A Quantitative Assessment of DInSAR Measurements of Interseismic Deformation: The Southern San Andreas Fault Case Study

Mariarosaria Manzo,[1] Yuri Fialko,[2] Francesco Casu,[1] Antonio Pepe,[1] and Riccardo Lanari[1]

Abstract—We investigate the capabilities and limitations of the Differential Interferometric Synthetic Aperture Radar (DInSAR) techniques, in particular of the Small BAseline Subset (SBAS) approach, to measure surface deformation in active seismogenetic areas. The DInSAR analysis of low-amplitude, long-wavelength deformation, such as that due to interseismic strain accumulation, is limited by intrinsic trade-offs between deformation signals and orbital uncertainties of SAR platforms in their contributions to the interferometric phases, the latter being typically well approximated by phase ramps. Such trade-offs can be substantially reduced by employing auxiliary measurements of the long-wavelength velocity field. We use continuous Global Positioning System (GPS) measurements from a properly distributed set of stations to perform a pre-filtering operation of the available DInSAR interferograms. In particular, the GPS measurements are used to estimate the secular velocity signal, approximated by a spatial ramp within the azimuth-range radar imaging plane; the phase ramps derived from the GPS data are then subtracted from the available set of DInSAR interferograms. This pre-filtering step allows us to compensate for the major component of the long-wavelength range change that, within the SBAS procedure, might be wrongly interpreted and filtered out as orbital phase ramps. With this correction, the final results are obtained by simply adding the pre-filtered long-wavelength deformation signal to the SBAS retrieved time series. The proposed approach has been applied to a set of ERS-1/2 SAR data acquired during the 1992–2006 time interval over a 200 × 200 km area around the Coachella Valley section of the San Andreas Fault in Southern California, USA. We present results of the comparison between the SBAS and the Line Of Sight (LOS)—projected GPS time series of the USGC/PBO network, as well as the mean LOS velocity fields derived using SBAS, GPS and stacking techniques. Our analysis demonstrates the effectiveness of the presented approach and provides a quantitative assessment of the accuracy of DInSAR measurements of interseismic deformation in a tectonically active area.

Key words: Deformation time series, differential SAR interferometry, Small BAseline Subset (SBAS), interseismic deformation, San Andreas Fault.

1. Introduction

Differential Synthetic Aperture Radar Interferometry (DInSAR) is a microwave remote sensing technique that allows measuring surface deformation with a centimeter to millimeter accuracy at high resolution (tens of meters) and large spatial coverage (Gabriel *et al.*, 1989). The DInSAR technique exploits the phase difference (interferogram) between two temporally separated SAR acquisitions to provide a measure of the ground deformation along the radar Line Of Sight (LOS).

Initially applied to characterize sizeable deformation events (Massonnet *et al.*, 1993, 1995; Peltzer and Rosen, 1995; Rignot, 1998; Amelung *et al.*, 1999; Fialko *et al.*, 2001), the DInSAR methodology has successively been adapted to analyze the temporal evolution of surface deformation via the generation of LOS displacement time series. For this purpose, the information available from each interferometric SAR data pair must be properly related to that contained in other pairs by generating and inverting an appropriate sequence of DInSAR interferograms. In this context, several advanced DInSAR approaches have been implemented; they can be grouped into two main categories: the Persistent Scatterer (PS) (Ferretti *et al.*, 2000; Werner *et al.*, 2003; Hooper *et al.*, 2004) and the Small Baseline (SB) (Berardino *et al.*, 2002; Mora *et al.*, 2003; Prati *et al.*, 2010) methods, although a solution that incorporates both the PS and SB approaches has also been recently proposed (Hooper, 2008). The first

[1] Istituto per il Rilevamento Elettromagnetico dell'Ambiente (IREA), National Council of Research (CNR), Via Diocleziano 328, 80124 Naples, Italy. E-mail: manzo.mr@irea.cnr.it

[2] Institute of Geophysics and Planetary Physics, Scripps Institution of Oceanography, University of California San Diego, La Jolla, CA 92093-0225, USA.

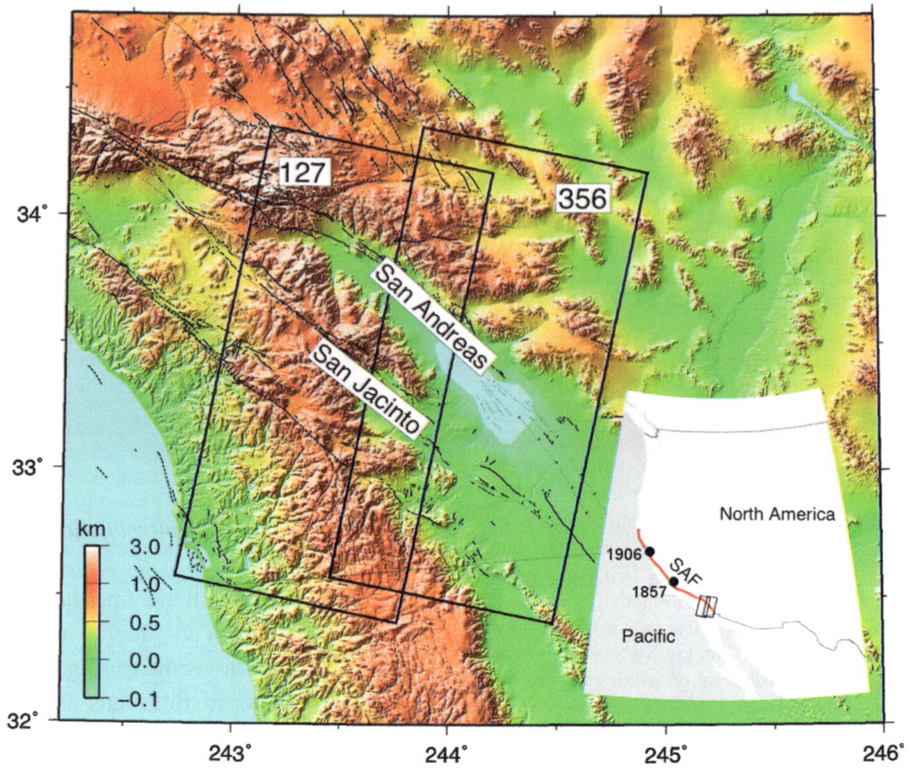

Figure 1
Shaded topography map of the study area. *Black wavy lines* denote active faults. *Black rectangles* represent ERS SAR tracks used in this study. The *red line* in the inset denotes the San Andreas Fault (SAF); moreover, the *black circles* indicate the locations of the 1857 and 1906 earthquakes

class of time-dependent DInSAR algorithms operates on single-look interferograms generated with respect to a selected common SAR image (usually referred to as "master image"), with no constraint on the spatial and temporal separation (baselines) of the SAR data pairs. The second approach considers a combination of SAR interferograms generated from an appropriate selection of the SAR data pairs characterized by relatively small baselines. In the latter case, both multi-look (Berardino *et al.*, 2002) and single-look (Lanari *et al.*, 2004a) interferograms can be analyzed. In this paper, we focus on the advanced SB-DInSAR technique referred to as Small BAseline Subset (SBAS) (Berardino *et al.*, 2002) approach that has previously been demonstrated to measure LOS velocities and displacements with an accuracy of about 1–2 mm/year and 5–10 mm, respectively (Lanari *et al.*, 2004b; Casu *et al.*, 2006; Manzo *et al.*, 2006; Lanari *et al.*, 2007a, b; Tizzani *et al.*, 2007; Neri *et al.*, 2009).

Deformation monitoring of seismogenetic areas is a key application of the DInSAR techniques, including both secular and transient deformation. However, in this case the presence of phase signal components within the generated differential interferograms, caused by inaccuracies in the SAR sensor orbit information, represents a key limitation. Orbital errors are typically well approximated by long-wavelength trends in the radar interferograms (Rosen *et al.*, 2000), usually referred to as orbital ramps. Distinguishing between deformation signals and phase patterns due to orbital uncertainties is critical for characterizing surface motions that result from interseismic deformation due to active faults. On the other hand, the estimation and subsequent removal of possible orbital phase artifacts remains a mandatory operation that, if neglected, can lead to significant errors in the estimate of the inter-seismic strain accumulation (Peltzer *et al.*, 2001; Wright *et al.*, 2004; Fialko, 2006; Biggs *et al.*, 2007) and the retrieved deformation time series (Burgmann *et al.*, 2006; Gourmelen *et al.*, 2010).

One way to mitigate this problem is to exploit independent data. For instance, measurements from the continuous Global Positioning System (GPS) stations in the study area can provide important information about the long-wavelength spatial characteristics as well as the time dependence of the surface velocity field (e.g., Burgmann *et al.*, 2006). Although the spatial distribution of continuous GPS is typically limited compared to the one provided by the DInSAR technology, GPS measurements may be readily used to estimate and subsequently remove possible orbital phase artifacts (Fialko *et al.*, 2006; Gourmelen *et al.*, 2010). The DInSAR approach exploited below is rather simple because it has minimum impact on the SBAS processing chain; indeed, it is based on carrying out a pre-filtering operation of the available DInSAR interferograms by using the measurements provided by a limited number of continuous GPS stations. These measurements are used to get an estimate of the long-term deformation signal, approximating a spatial ramp that is subsequently used to identify and filter out the orbital phase artifacts affecting the considered DInSAR interferograms. In this study, we apply this technique to analyze space geodetic data from a rather large (200 × 200 km) area around the San Andreas Fault (SAF) system in Southern California, USA (Fig. 1). This area is well suited for C-band radar interferometry, has a dense continuous GPS network (with some stations operating since early 1990 s) and has extensively been studied using space-geodetic methods (Johnson *et al.*, 1994; Bennett *et al.*, 1996; Fialko, 2006; Lundgren *et al.*, 2009). In this paper, we analyze a large set of ERS-1 and ERS-2 acquisitions spanning the 1992–2006 time interval and we generate mean LOS velocity maps and the corresponding deformation time series for each coherent pixel of the investigated area (i.e., for each pixel whose phase measurement is considered reliable). The derived time series are then compared to those provided by continuous GPS stations of the USGS/PBO networks (http://pasadena.wr.usgs.gov/scign/Analysis/) to identify secular and transient signals in surface velocities. Presented results considerably expand previously published LOS velocity data (Fialko, 2006; Lundgren *et al.*, 2009), both temporally (by adding 7 years of data from track 356) and spatially (by processing data from the adjacent track 127). In particular, the new data extend coverage to the central and northern sections of the San Jacinto fault (Fig. 1).

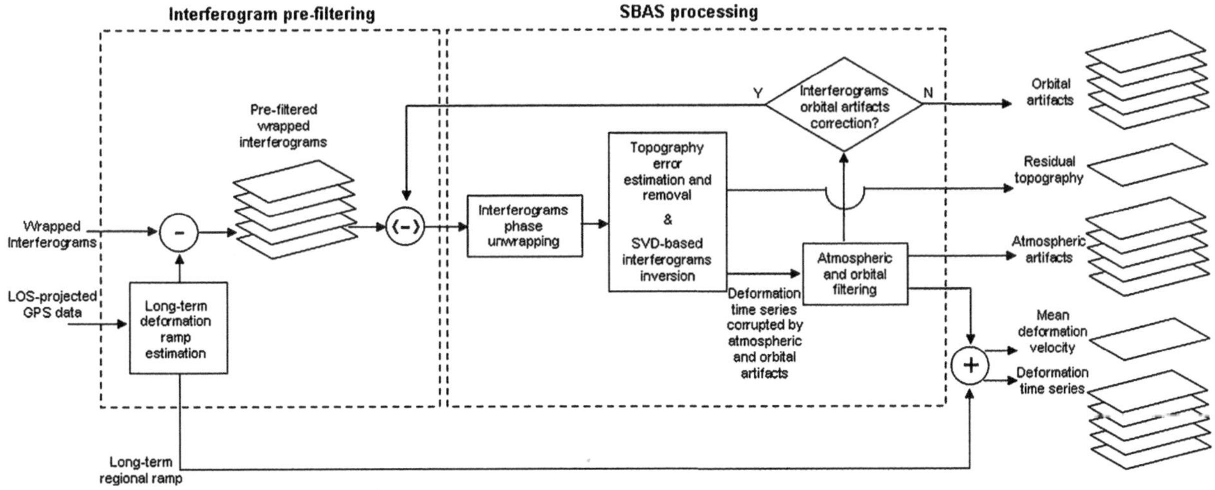

Figure 2
Block diagram of the exploited SBAS-DInSAR processing chain

Results presented in this paper highlight the necessity of a joint exploitation of SAR and GPS data in studies of interseismic deformation due to active faults in the Earth's crust. In particular, they provide detailed information about the temporal evolution and spatial distribution of the LOS velocities in the Southern SAF area. We present results of a comparison of the SBAS and LOS-projected GPS time series of the USGS/PBO network as well as mean LOS velocity fields derived using SBAS and stacking techniques. Our analysis shows a good agreement between these data sets and provides a quantitative assessment of the accuracy of the SBAS-DInSAR measurements in seismogenic areas.

The paper is organized as follows. First, we discuss the geological setting of the Southern SAF system. We then briefly describe the SBAS algorithm, highlighting the minor modifications required to make use of the GPS data. We then present and discuss the SBAS-DInSAR results obtained by jointly exploiting the radar and GPS measurements; in particular, we compare the retrieved DInSAR time series to the LOS-projected measurements of the available GPS network on selected sites and the mean deformation velocity map with that obtained

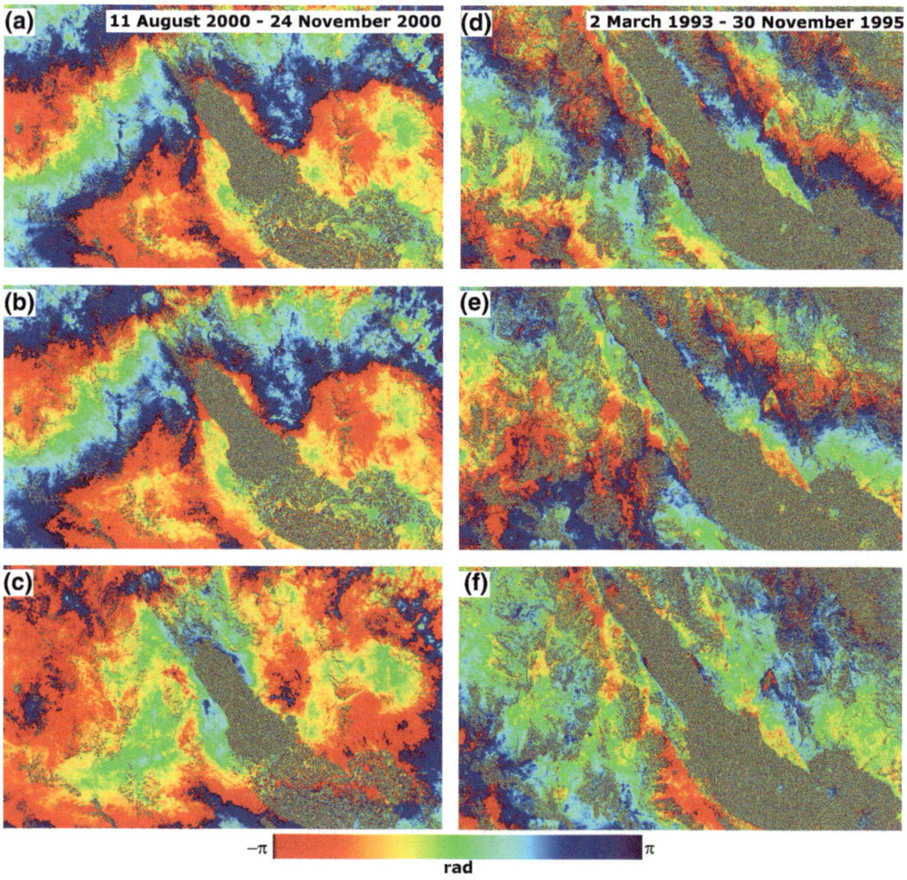

Figure 3
Impact of GPS driven pre-filtering versus orbital error correction for two selected ERS differential interferograms. **a** Original differential interferogram relevant to the 11 August 2000–24 November 2000 data pair; **b** differential interferogram compensated for the GPS-derived secular deformation pattern; **c** same as Fig. 3b, but also compensated for the orbital ramp. **d** Original differential interferogram relevant to the 2 March 1993–30 November 1995 data pair; **e** differential interferogram compensated for the GPS-derived secular deformation pattern; **f** same as Fig. 3e, but also compensated for the orbital ramp. Note that one full color cycle, say from *red to violet*, represents one interferometric phase fringe corresponding to a LOS-displacement of about 2.8 cm. We remark that the horizontal axis corresponds to the azimuth direction while the vertical one to the range

through the stacking technique and GPS. The last section is devoted to some conclusive remarks.

2. *Tectonic Setting*

Our study area is centered on the SAF system in Southern California (Fig. 1). This area is well suited for studies using DInSAR techniques. Arid climate and limited vegetation result in stability of reflective properties of the ground over time periods of the order of 10 years (Peltzer *et al.*, 2001; Lyons and Sandwell, 2003; Fialko, 2004, 2006). Extensive imaging of the area by the DInSAR-capable ERS-1/2 satellite missions of the European Space Agency, combined with a dense network of continuous GPS stations, gave rise to a geodetic dataset of excellent coverage and spatio-temporal resolution that can be exploited to characterize slow interseismic deformation due to the SAF system. The latter is a mature continental transform fault accommodating a significant fraction of the ~0.05 m/yr relative motion between the North American and Pacific plates (e.g., Thatcher and Lisowski, 1987; DeMets *et al.*, 1990). Except for the 100-km long fault section between Parkfield and San Juan Batista that undergoes a steady creep, the SAF exhibits a stick–slip behavior and is capable of producing great earthquakes. The two most recent great earthquakes on the SAF have ruptured its central and northern sections in 1857 and 1906, respectively (see the inset of Fig. 1). The southernmost section of the SAF has not produced major earthquakes in historic time (over more than 300 years), as it is currently believed to be late in the interseismic phase of the earthquake cycle (Working Group on California Earthquake Probabilities, 1995; Weldon *et al.*, 2005; Fialko, 2006). Estimates of seismic hazard on the SAF as well as on other major faults in Southern California critically depend on the present-day strain rates and the degree of fault locking in the seismogenic crust (i.e., the presence and extent of fault creep). Both factors can in principle be evaluated with help of precise spatially dense measurements of surface deformation. In this paper, we quantitatively assess and validate DInSAR

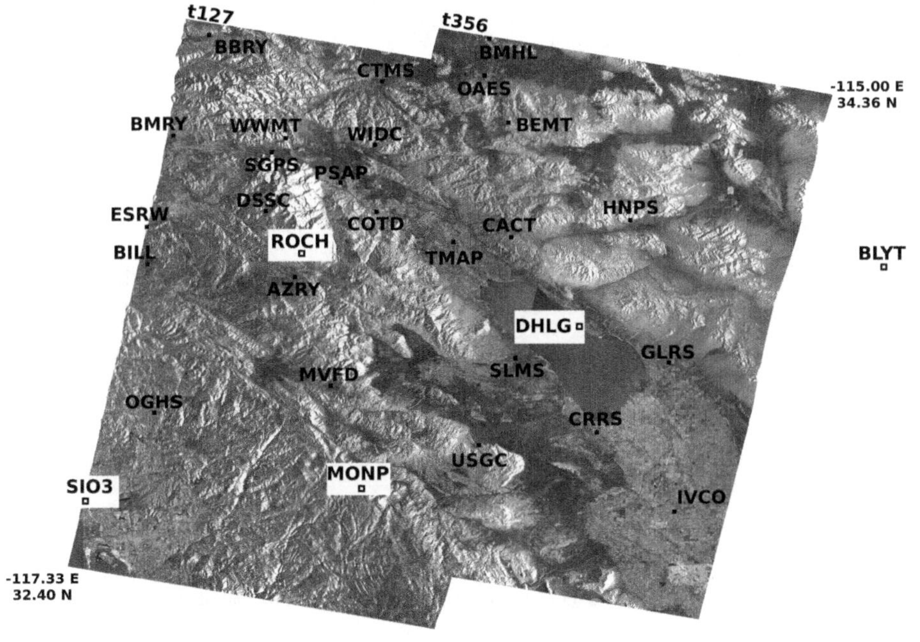

Figure 4

Mosaic of the multi-look SAR amplitude images of the investigated area for the SAR data belonging to 127 and 356 orbit tracks, that are presented in Fig. 5. The *black* and *white squares* mark the locations of the continuous GPS stations deployed in the area and belonging to the USGS/PBO network. Note that the *white ones* represent the five GPS stations selected to perform the estimate of the regional tectonic deformation rate

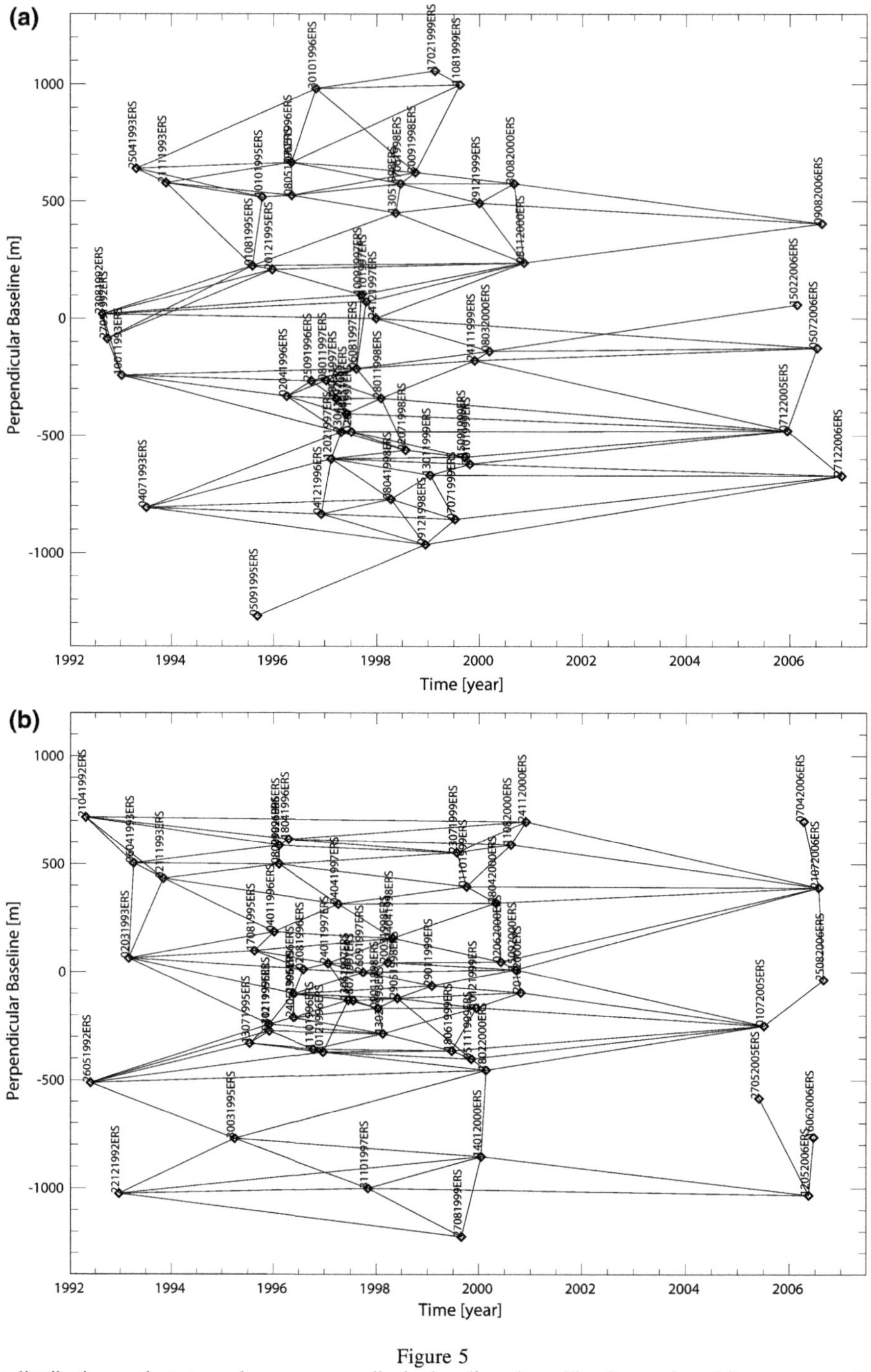

Figure 5
Interferometric data distribution on the temporal versus perpendicular baseline plane. The *diamonds* and *lines* represent the SAR acquisitions and data pairs, respectively. **a** SAR data distribution for track 127. **b** SAR data distribution for track 356

techniques commonly used for measuring subtle interseismic deformation and present analysis of DInSAR data from the Southern SAF collected between 1992 and 2006. The new results provide a much improved description of surface velocities; they can be used to constrain the geometry and kinematics of deep "roots" of faults comprising the Southern SAF system as well as to quantify the deformation rates in the overlying nominally locked seismogenic crust.

3. SBAS-DInSAR Technique

The SBAS technique is a DInSAR algorithm that allows us to detect the Earth surface deformation and to analyze its temporal evolution by generating mean LOS velocity maps and time series of LOS displacements. In particular, this technique relies on the use of multiple master multi-look interferograms (Rosen *et al.*, 2000) generated via an appropriate selection of SAR data pairs characterized by small [less than a selected threshold, see Lanari *et al.*, (2007b)] spatial and temporal baselines. The key objective of this data selection is to mitigate the noise (decorrelation) effects, thus maximizing the number of temporally coherent pixels [for definition of temporal coherence, see Pepe and Lanari, (2006)]. Moreover, this baseline selection may imply that the data pairs used to produce the interferograms are arranged in few SBASs separated by large baselines. In the latter case, because there is no suitable interferogram that connects elements belonging to different subsets, they turn out to be independent of each other. Therefore, an underdetermined problem has to be solved, for instance by using a Least Squares minimization based on the singular value decomposition method (Berardino *et al.*, 2002).

In this work, we exploit the extended version of the SBAS technique (Casu *et al.*, 2008) that allows us to detect and analyze deformation phenomena with large spatial wavelength. This is done by exploiting relatively low resolution multi-look DInSAR interferograms computed from long SAR image strips, which are obtained by jointly focusing contiguous ERS raw data frames (belonging to the same satellite track). Following the generation of suitable interferograms, we average them on a grid of about 120 × 120 m and invert via the conventional SBAS technique (Berardino *et al.*, 2002) to retrieve the mean LOS velocity and the corresponding LOS displacement time series for each coherent pixel.

We point out that the SBAS inversion limits the impact of possible topographic artifacts present in the DEM used to compute the differential interferograms and improves the performance of the phase unwrapping procedure by making use of both the spatial and temporal variations in the radar phase among multiple interferograms (Pepe and Lanari, 2006). Moreover, the SBAS approach includes a filtering operation for the atmospheric phase components, based on the observation that the atmospheric phase signal is highly correlated in space but poorly in time (Goldstein, 1995; Ferretti *et al.*, 2000; Berardino *et al.*, 2002).

In addition to the atmospheric artifacts correction, the SBAS technique also includes a step for the detection and removal of orbital fringes (Lanari *et al.*, 2004b) that, as mentioned before, are well approximated by phase ramps. In particular, an estimate of these orbital patterns is performed by searching for the best-fit ramp to the retrieved time series signal component. This step leads to the generation of the required deformation time series. However, we remark that the orbital phase artifacts are particularly relevant in seismic areas, because it is not an easy task distinguishing between orbital artifacts and deformation signals that may mimic the orbital phase patterns. Accordingly, we propose to exploit the measurements from a limited number of continuous GPS stations in order to mitigate this problem. The rationale of the approach is quite simple and is focused on exploiting the GPS measurements in order to identify the long-wavelength interseismic signal; the GPS data are used to pre-filter the original differential interferograms representing the input of the SBAS processing. In particular, we first search for the best-fit ramp to the mean deformation velocity values of the selected GPS data that have been projected on the SAR sensor LOS in order to be consistent with the radar observations. Subsequently, we estimate and subtract the GPS-derived deformation ramp from each interferogram, by taking into account the time span of the considered interferometric pair. At this stage, the overall standard SBAS processing chain is applied to the pre-processed interferograms, including the orbital artifacts detection and removal step. However, in this case the filtering operation is more robust, because we have drastically mitigated the contribution of long term displacements that may be wrongly interpreted as being due to orbital errors.

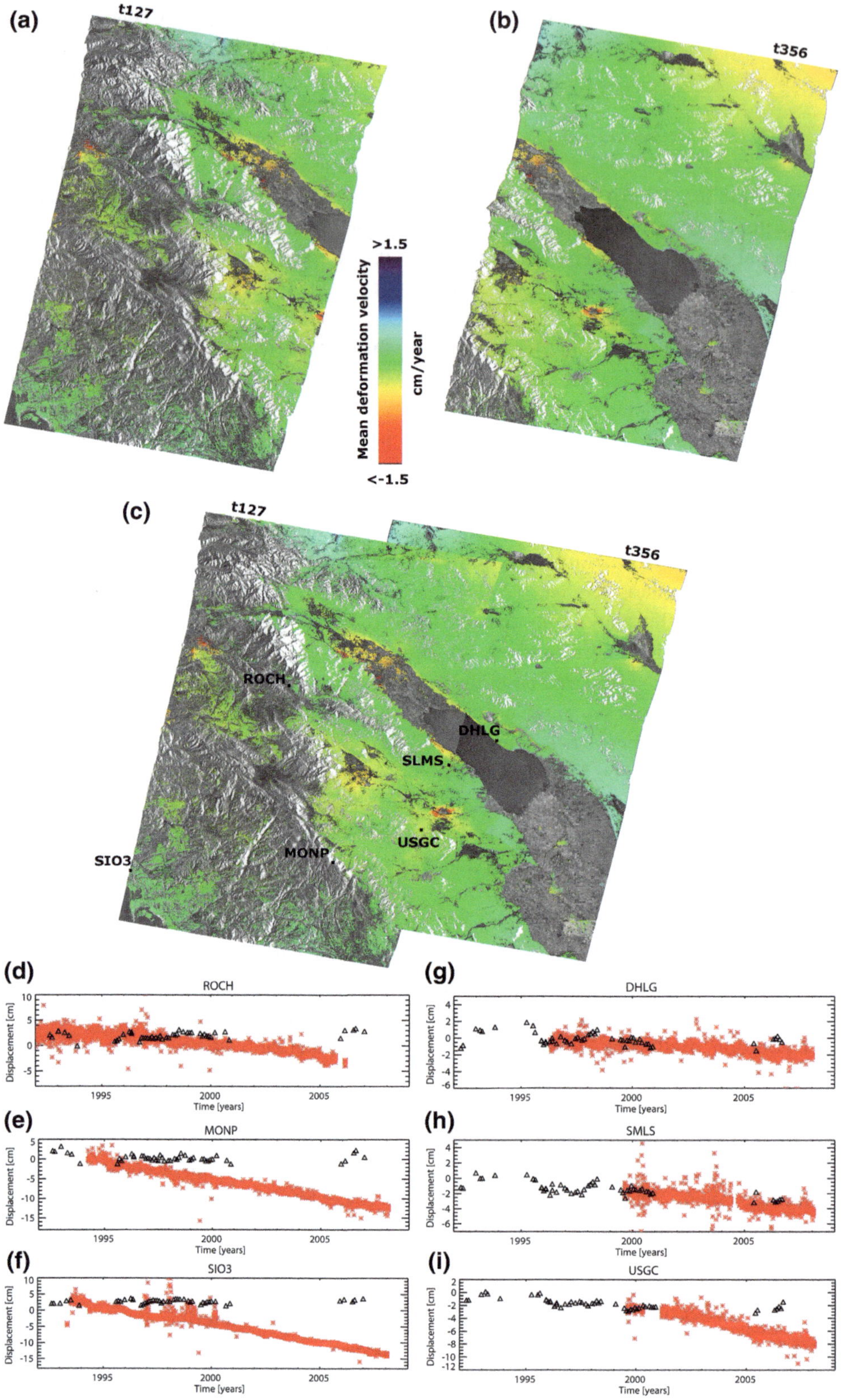
(a)
t127
(b)
t356
>1.5
Mean deformation velocity
cm/year
<-1.5
(c)
t127
t356
ROCH
DHLG
SLMS
USGC
MONP
SIO3
(d)
ROCH
(e)
MONP
(f)
SIO3
(g)
DHLG
(h)
SMLS
(i)
USGC
Displacement [cm]
Time [years]
1995
2000
2005

Figure 6
SBAS-DInSAR results obtained by applying the conventional SBAS algorithm, where the orbital corrections were carried out without exploiting the available GPS measurements. **a**, **b** Mean deformation velocity maps for track 127 and track 356, geocoded and superimposed on the SAR images of the area. **c** Mosaic of the two velocity maps shown in Fig. 6a, b. **d–i** Comparison between the retrieved DInSAR measurements (*black triangles*) and the LOS-projected GPS time series (*red stars*), the latter obtained by projecting the displacement vector onto the radar LOS; data are relevant to the locations of the continuous GPS sites labeled in Fig. 6c as ROCH (**d**), MONP (**e**) and SIO3 (**f**) for track 127, and DHLG (**g**), SLMS (**h**) and USGC (**i**) for track 356

Note that at this stage we have available the orbital correction time series, i.e., the orbital ramps we estimate for each acquisition with respect to the first image of the time series. Accordingly, we may use them to correct the pre-filtered interferograms without introducing any inconsistency among interferograms. This operation may be relevant, particularly in low coherent areas, because it reduces, together with the GPS-derived correction, the phase gradient components, thus simplifying the phase unwrapping operation (Pepe and Lanari, 2006). We typically carry out just one iteration of this correction as a compromise between computing time and results accuracy.

Finally, in order to retrieve the actual surface motion across the fault, we add the GPS-retrieved long-term regional trend to the computed deformation time series. To summarize the rationale of the applied processing chain, we show its block diagram in Fig. 2.

The GPS measurements have been used to correct the DInSAR results before (e.g., Burgmann *et al.*, 2006; Gourmelen *et al.*, 2010). In this study, instead of using independent measurements to carry out a final correction of our deformation estimates, we exploit them to pre-filter the data in order to possibly improve the performances of the SBAS processing steps, particularly of the phase unwrapping operation.

We finally note that, although the proposed pre-filtering solution has a minor impact on the SBAS processing chain, it may play a fundamental role in the generation of the DInSAR results. To better clarify the impact of the GPS driven pre-filtering step vs. the orbital error correction, two examples are shown in Fig. 3, based on exploiting the SAR data presented in the following experimental results. In particular, we present in Fig. 3a the ERS differential interferogram corresponding to a short time interval 11 August 2000–24 November 2000 where a phase ramp is clearly visible. We remark that in this case the effect of the interseismic deformation retrieved through the GPS measurements is negligible due to the short time span and its removal has nearly no impact on the pre-filtered interferogram, see Fig. 3b. Accordingly, the detected phase ramp is fully due to orbital errors; the corrected interferogram is depicted in Fig. 3c. In contrast, Fig. 3d shows the interferogram covering a longer time interval, 2 March 1993–30 November 1995. In this case, again a phase ramp is visible but now, due to the significantly longer time span, it can be partially attributed to interseismic deformation. By compensating for this component, a significant correction is achieved (see Fig. 3e) and we may finally identify and compensate for the remaining orbital artifacts, as shown in Fig. 3f.

This simple example clearly highlights that in areas of active tectonic deformation a straightforward correction of orbital phase ramps may wrongly map ground motion into orbital errors. Accordingly, more advanced processing schemes are needed, particularly for deformation time series retrieval, as shown through the experiments presented in the next section.

4. *SBAS-DInSAR Results*

The SBAS-DInSAR processing chain summarized before and depicted in Fig. 2 has been applied to a set of ERS-1/2 SAR data acquired from two adjacent satellite tracks spanning the southern segment of the SAF system over the 1992–2006 time interval. Figure 4 shows the mosaic of the SAR amplitude images of the investigated area with the highlighted locations (black and white squares) of the continuous GPS stations belonging to the USGS/PBO network (http://pasadena.wr.usgs.gov/scign/Analysis/). The white squares denote five continuous GPS sites used to estimate the regional tectonic trend (note that the station named BLYT is slightly outside the radar swaths, located east of the studied area). These sites were chosen as they provide measurements since

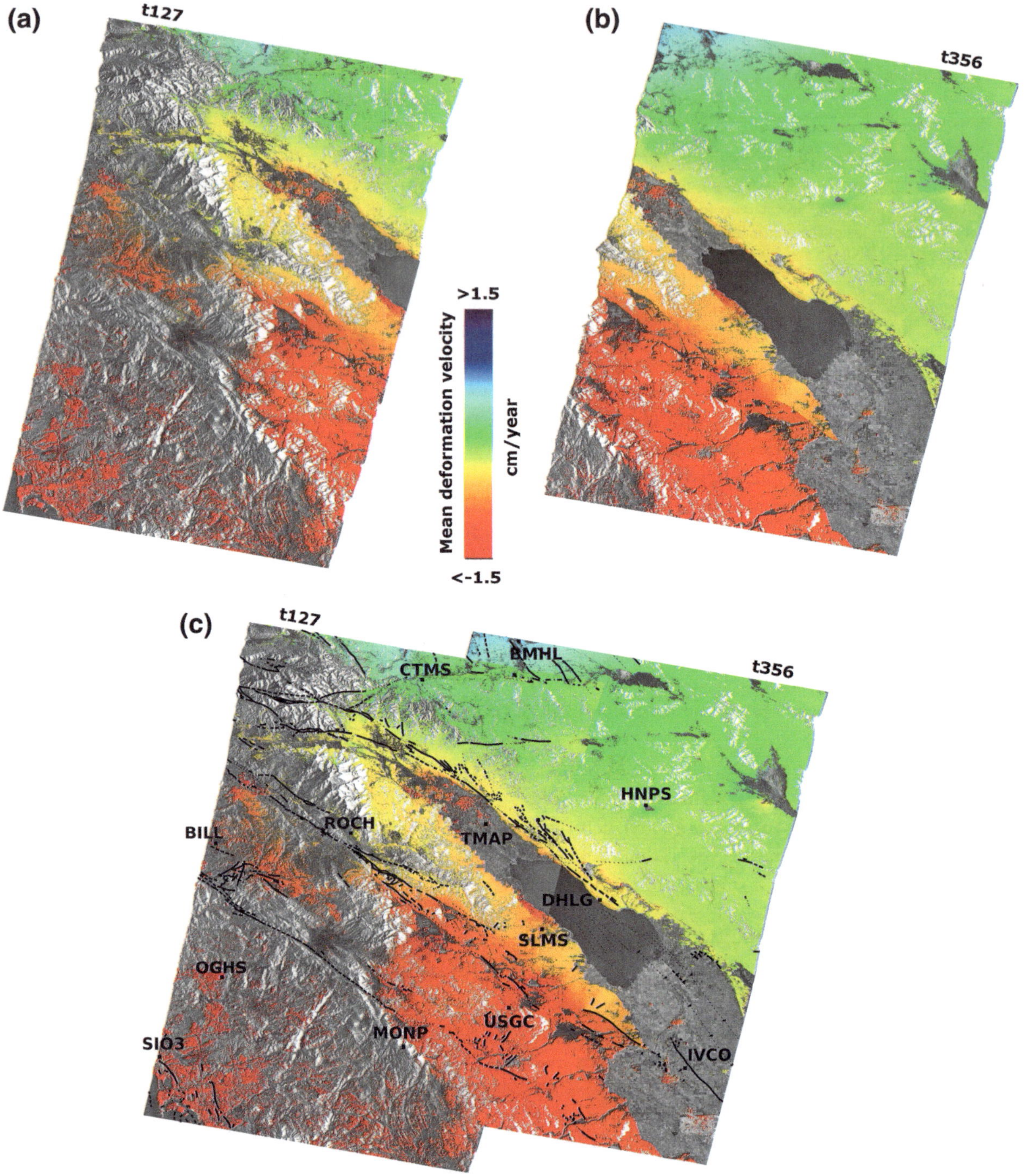

Figure 7

SBAS-DInSAR results obtained by applying the modified version of the SBAS algorithm, shown in Fig. 2, where the orbital corrections were carried out by exploiting the available GPS measurements. **a–b** Mean deformation velocity maps for track 127 (**a**) and track 356 (**b**). **c** Mosaic of the two velocity maps shown in Fig. 7a, b. The trace of the SAF system and the locations of the GPS sites relevant to the plots shown in Fig. 8 are indicated by the black lines and squares, respectively

1996 or earlier, thus ensuring at least 10 years of temporal overlap with the available SAR data.

We made use of 49 ERS acquisitions from the descending track 127 (frames: 2925–2943) to generate a set of 129 interferograms with a spatial resolution of about 120 × 120 m. We also used 53 ERS acquisitions from the descending track 356 (frames: 2925–2943) to generate 141 interferograms with the same spatial resolution. We chose the interferometric SAR pairs with a perpendicular baseline smaller than 450 m and a maximum Doppler Centroid separation of about 1100 Hz. This Doppler constraint is relevant since the ERS-2 SAR data acquired after year 2000, i.e., following the ERS-2 gyroscope failures (Miranda *et al.,* 2003), may give rise to large Doppler centroids. In our processing of SAR data we took advantage of precise satellite orbits provided by the Delft University and a 3-arcsec Digital Elevation Model (DEM) made available by the Shuttle Radar Topography Mission DEM (SRTM) (Rosen *et al.*, 2001). In Fig. 5a, b a sketch of the exploited SAR images and of the implemented selection of DInSAR interferograms is shown for the data of the 127 and 356 tracks, respectively. All the SBAS results presented below (i.e., velocity maps and displacement time series) were computed with respect to a reference pixel chosen at the location of the BEMT continuous GPS station (see Fig. 4). The coregistered SAR images used to generate the interferograms were also used to compute the multi-look SAR amplitude images, as shown in Fig. 4.

To highlight the importance of a joint analysis of SAR and GPS data, we first present results obtained by applying the conventional SBAS algorithm, wherein the orbital corrections were carried out without exploiting the available GPS measurements.

Figure 6 shows the mean LOS velocity maps for tracks 127 and 356 (Fig. 6a and b, respectively), as well as the mosaic of the two maps (Fig. 6c), geocoded and superimposed on the SAR images of the area (see Fig. 4). Areas where the measurement accuracy is compromised by decorrelation are represented by the radar amplitude only. Note that results shown in Fig. 6 represent a time-averaged velocity of the Earth's surface in the satellite LOS. As one can see in Fig. 6, there exist some localized deformation anomalies, but the area is devoid of a long-wavelength deformation due to active faults. This is in contrast to the previously published results obtained using stacking (Fialko, 2006) and time series (Lundgren *et al.*, 2009) techniques. SBAS-DInSAR results shown in Fig. 6 are wrong because a significant component of the long-wavelength deformation signal was misinterpreted as orbital ramps and filtered out; this clearly shows that a straightforward phase ramps filtering is inappropriate for tectonically active areas, such as Southern California. The impact of these errors is further demonstrated by a comparison between the retrieved LOS displacement time series at the locations of the continuous GPS sites and the GPS time series obtained by projecting the displacement vector onto the radar LOS (Fig. 6d–i).

Next, we present the results obtained by using the modified version of the SBAS-DInSAR algorithm described in Sect. 3 and sketched in Fig. 2. In particular, we show in Fig. 7 the mean LOS velocity maps for tracks 127 and 356 (Fig. 7a, b, respectively), as well as the mosaic of the two maps (Fig. 7c). In this case, one can clearly see a regional deformation pattern due to interseismic strain accumulation on major faults (in particular, the San Andreas and San Jacinto faults, see Fig. 1). In order to further investigate the inferred velocity field, we carried out an extensive comparison between the DInSAR-derived and the LOS-projected GPS time series. As previously, all displacements and velocities (SAR- and GPS-derived) were computed with respect to the reference BEMT site (Fig. 4). While our analysis involves measurements from all the GPS stations located in areas that are coherent within the DInSAR velocity maps, for illustration purposes we limited the comparison to 13 sites, as shown in Fig. 8. For each of the selected sites (whose locations are denoted by black squares in Fig. 7c) we compare the DInSAR results (black triangles) with the respective LOS-projected GPS displacements (red stars). The LOS projection of the GPS data used local radar incidence angles. Inspection of Fig. 8 reveals a generally good agreement between the LOS displacement time series derived from DInSAR and GPS data. The overall agreement is perhaps not surprising for stations ROCH, MONP, DHLG and SIO3 that were used to estimate the

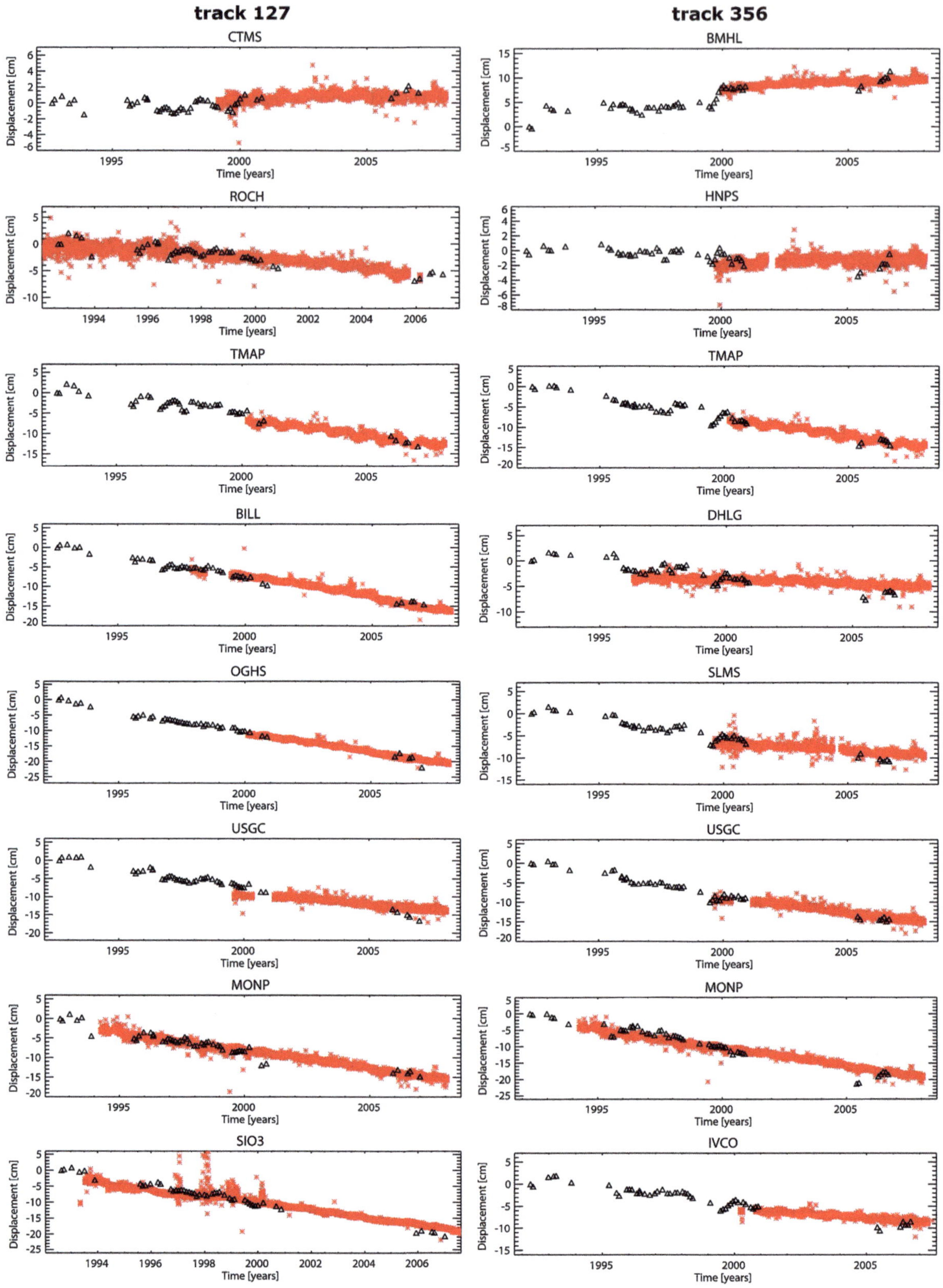
track 127
track 356
CTMS
BMHL
ROCH
HNPS
TMAP
TMAP
BILL
DHLG
OGHS
SLMS
USGC
USGC
MONP
MONP
SIO3
IVCO
Displacement [cm]
Time [years]

Figure 8
Comparison between the retrieved DInSAR deformation measurements (*black triangles*) and the LOS-projected GPS time series (*red stars*) for the continuous GPS sites identified in Fig. 7c; DInSAR time series versus GPS measurements for the sites of track 127 are shown on left side, while those for the sites of track 356 are on the right

orbital ramps, as discussed in Sect. 3. Importantly, a rather good agreement is also observed for other stations (Fig. 8), although minor discrepancies persist in some cases. We evaluated the accuracy of the DInSAR results by computing the standard deviation of the differences between the DInSAR and the LOS-projected GPS displacements for all the sites identified in Fig. 4 by the black and white squares (except for the BLYT station where no DInSAR measurements were available); the obtained results are summarized in Tables 1 and 2 for data from tracks 127 and 356, respectively. Based on these measurements, we computed the average standard deviation (σ) value representative of the differences between the DInSAR and LOS-projected GPS data; for both tracks we obtained $\sigma \approx 1.2$ cm. This result is in good agreement with that previously reported by Casu *et al.* (2006) and Tizzani *et al.* (2007). It clearly confirms the need to account for long-wavelength deformation in the SBAS-DInSAR analysis, using auxiliary geodetic data to effectively filter out orbital artifacts. In order to further clarify the impact of the carried out correction of the DInSAR results, we also present in Tables 3 and 4 the standard deviations of the differences between the LOS-projected GPS displacements (also in this case for all the sites identified in Fig. 4 by the black and white squares) and the DInSAR time series without any correction (left column) and with the correction of the orbital artifacts only (right column) for data from tracks 127 and 356, respectively. By considering Tables 3 and 4 it is evident that the orbital artifacts correction can be very relevant as for the case of track 127 for which the impact of the correction of the DInSAR time series exceeds 1 cm on average; this exceeds by a factor of two the effect of filtering of atmospheric artifacts (Tables 1 and 2).

To provide a visual interpretation of the impact of the orbital artifacts, in Fig. 9 we present the DInSAR

Table 1

Results of the comparison between LOS-projected GPS measurements and DInSAR deformation time series relevant to track 127

GPS stations	Standard deviation of the difference between LOS-projected GPS and DInSAR measurements (cm)
AZRY	1.2
BBRY	1.9
BILL	0.9
BMRY	1.3
CACT	0.8
COTD	1.8
CTMS	0.8
DSSC	1.4
ESRW	0.6
MONP	1.0
MVFD	1.3
OAES	1.0
OGHS	1.1
PSAP	1.5
ROCH	1.1
SGPS	0.8
SIO3	2.4
SLMS	1.9
TMAP	0.7
USGC	2.4
WIDC	1.3
WWMT	0.2
Average value	1.2

Table 2

Results of the comparison between LOS-projected GPS measurements and DInSAR deformation time series relevant to track 356

GPS stations	Standard deviation of the difference between LOS-projected GPS and DInSAR measurements (cm)
BMHL	0.8
CACT	0.4
CRRS	1.1
DHLG	1.5
GLRS	1.2
HNPS	1.3
IVCO	1.5
MONP	1.7
OAES	1.6
SLMS	1.5
TMAP	0.7
USGC	0.9
Average value	1.2

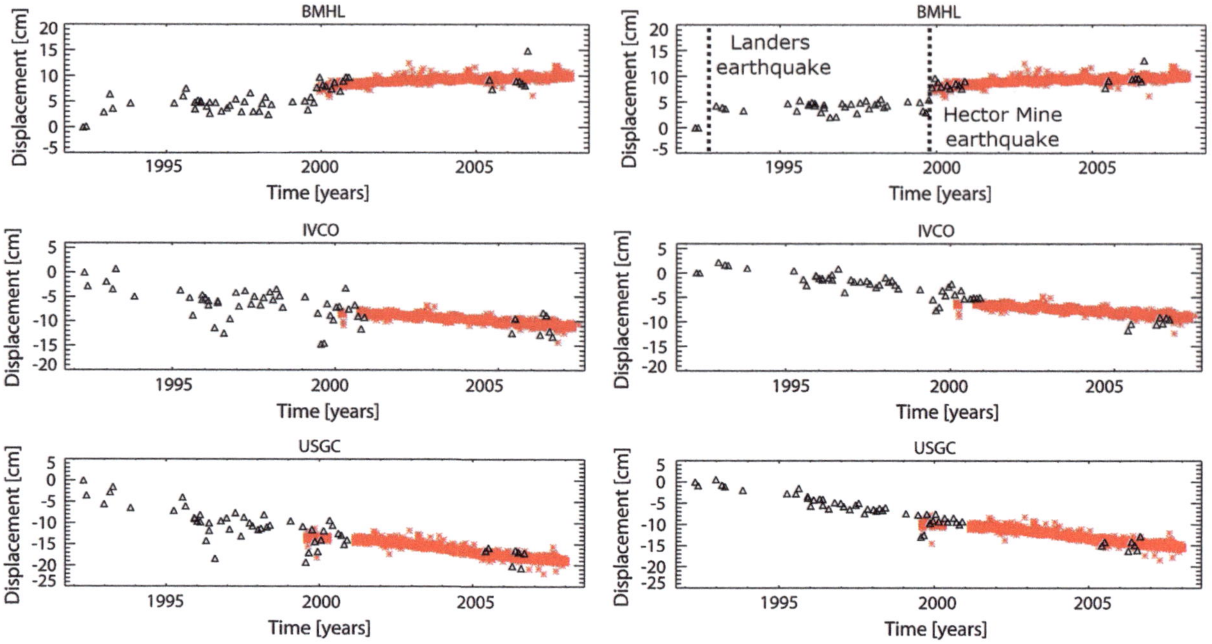

Figure 9
Comparison between the retrieved DInSAR deformation time series for data belonging to track 356 (*black triangles*) and the LOS-projected GPS measurements (*red stars*) at the locations of the continuous GPS sites labeled as BMHL, IVCO and USGC (Fig. 4). *Left*: the DInSAR time series are computed without orbital artifact correction; *right*: the DInSAR time series are computed with the orbital ramp correction. Note that the *vertical dashed lines* identify the dates of the Landers and Hector Mine earthquakes

time series for track 356 without (left column) and with (right column) orbital correction for three selected GPS sites labeled as BMHL, IVCO and USGC. The relevance of this correction is evident, particularly for the BMHL site, where it allows us to clearly identify, even without any atmospheric filtering, the effect of the Hector Mine earthquake (16 October 1999) and, particularly, of the Landers earthquake (28 June 1992). Therefore, although the impact of our correction for track 356 is not very prominent on average (a few mm, see Table 4), it may play an important role for the investigation of specific seismic events.

For sake of completeness, we also report similar results for track 127, see Fig. 10, where we selected the sites BILL, OGHS and USGC. The success of the correction for BILL and OGHS is evident, as already expected from Table 3. We also report the less favorable case of the USGC site, where our analysis reveals that the computed standard deviation exceeds 2 cm. This result may suggest the need for higher-order corrections for the regional tectonic deformation, for instance, using interpolated velocities from the entire GPS network.

Finally, we also performed a comparison between the mean LOS velocity field obtained using the SBAS algorithm with the results of optimized stacking procedure in which each SAR acquisition is assigned a weight according to the estimated atmospheric noise contribution (Fialko, 2004). Raw SAR data were processed using JPL/Caltech ROI_PAC software with phase unwrapping using the SNAPHU algorithm (Chen and Zebker, 2002). Figure 11b shows a comparison of the mean LOS velocities from the ERS track 356 across the SAF zone. The profile, highlighted in Fig. 11a, corresponds to that used by Fialko (2006). There is an overall agreement between the LOS velocities obtained using the different DInSAR processing algorithms. The accuracy of both methods is further validated by the good agreement with the mean GPS velocities projected onto the satellite LOS, thus confirming the expected accuracy of about 1–2 mm/year.

We finally remark that, although the presented results are focused on interseismic deformation, the analysis of the computed time series in principle allows one to accurately measure also coseismic displacements. Compared to the analysis of conventional interferograms spanning over the earthquake date, advantages offered by the SBAS technique include removal of the phase artifacts (e.g., due to orbital errors and atmospheric noise) and no need for phase unwrapping. We illustrate this approach by measuring deformation due to the 1992 Landers earthquake using the time series from track 356. The inferred coseismic range changes are shown in Fig. 12a. The earthquake signal is clearly visible in the northwest corner of the image. For comparison, we also present three conventional interferogram spanning the time intervals 26 May 1992–17 November 1992, 26 May 1992–22 December 1992 and 21 April 1992–6 April 1993, respectively (Fig. 12e–g). Figure 12b–d shows the time series for three selected positions within the radar image (see Fig. 12a) with the interseismic trend removed. Coseismic displacements shown in Fig. 12a represents a jump in the inferred LOS displacements between the SAR acquisitions bracketing the earthquake date. Although the Landers event is not optimal for recovery of coseismic deformation as there are few pre-earthquake acquisitions (so that the preseismic rates are not resolved), an overall agreement between the displacement fields shown in Fig. 12a and those in Fig. 12e–g is encouraging. In particular, as expected, the coseismic

Table 3

Results of the comparison between GPS measurements and DInSAR deformation time series relevant to track 127; for the latter we considered the unfiltered results (left column) and those corrected for the orbital artifacts only (right column)

GPS stations	Standard deviation of the difference between LOS-projected GPS and unfiltered DInSAR measurements (cm)	Standard deviation of the difference between LOS-projected GPS and DInSAR measurements corrected for the orbital artifacts only (cm)
AZRY	3.0	1.4
BBRY	4.5	3.0
BILL	3.4	1.3
BMRY	4.5	1.6
CACT	0.9	0.8
COTD	1.2	2.4
CTMS	2.3	1.0
DSSC	3.1	1.6
ESRW	3.0	1.1
MONP	2.5	2.1
MVFD	2.9	1.5
OAES	1.7	1.0
OGHS	5.1	1.0
PSAP	1.9	1.9
ROCH	2.7	1.5
SGPS	2.8	0.8
SIO3	3.8	2.6
SLMS	2.9	3.1
TMAP	1.9	2.1
USGC	3.8	3.0
WIDC	1.7	1.6
WWMT	2.3	0.3
Average value	2.8	1.7

Table 4

Results of the comparison between GPS measurements and DInSAR deformation time series relevant to track 356; for the latter we considered the unfiltered results (left column) and those corrected for the orbital artifacts only (right column)

GPS stations	Standard deviation of the difference between LOS-projected GPS and unfiltered DInSAR measurements (cm)	Standard deviation of the difference between LOS-projected GPS and DInSAR measurements corrected for the orbital artifacts only (cm)
BMHL	1.7	1.1
CACT	0.8	0.6
CRRS	2.3	1.9
DHLG	2.4	2.2
GLRS	2.0	1.6
HNPS	1.2	1.5
IVCO	2.9	1.8
MONP	2.4	2.7
OAES	1.7	1.6
SLMS	2.1	2.2
TMAP	1.2	1.3
USGC	2.3	1.6
Average value	1.9	1.7

displacement map obtained from the time series analysis (Fig. 12a) is less affected by artifacts such as atmospheric noise (cf. Fig. 12e–g) due to the mentioned filtering steps implemented through the SBAS processing. We note that the inferred time series (Fig. 12b–d) appear to have an overall pattern similar to the post-seismic GPS data (e.g., FIALKO, 2004; Fig. 8), although relatively sparse SAR acquisitions following the Landers earthquake do not allow one to determine whether the range changes between 1992 and 1993 evident in Fig. 12b–d represent primarily coseismic deformation or also include a rapid post-seismic response. The nature of a rather broad anomaly in the LOS displacements at the eastern side of the radar image is not clear and might be due to residual orbital or propagation errors. Further investigations are required to validate coseismic applications of the SBAS time series analysis.

5. Conclusions

We have investigated the capability of the SBAS-DInSAR technique to measure secular deformation in active seismogenic areas. In particular, the focus of this work is on a 200 × 200 km area around the Coachella Valley section of the SAF in Southern California (USA). The analysis is carried out by exploiting the data acquired by the ERS-1/2 sensors from two adjacent tracks (127 and 356) during the 1992–2006 time interval. Results presented in this paper provide a clear picture of the ongoing interseismic deformation due to the Southern SAF system, extending the data coverage compared to previous analyses both in space (by adding track 127) and in time (by including data from the 2000–2006 epoch).

To achieve the presented results, we show the relevance of exploiting external measurements from

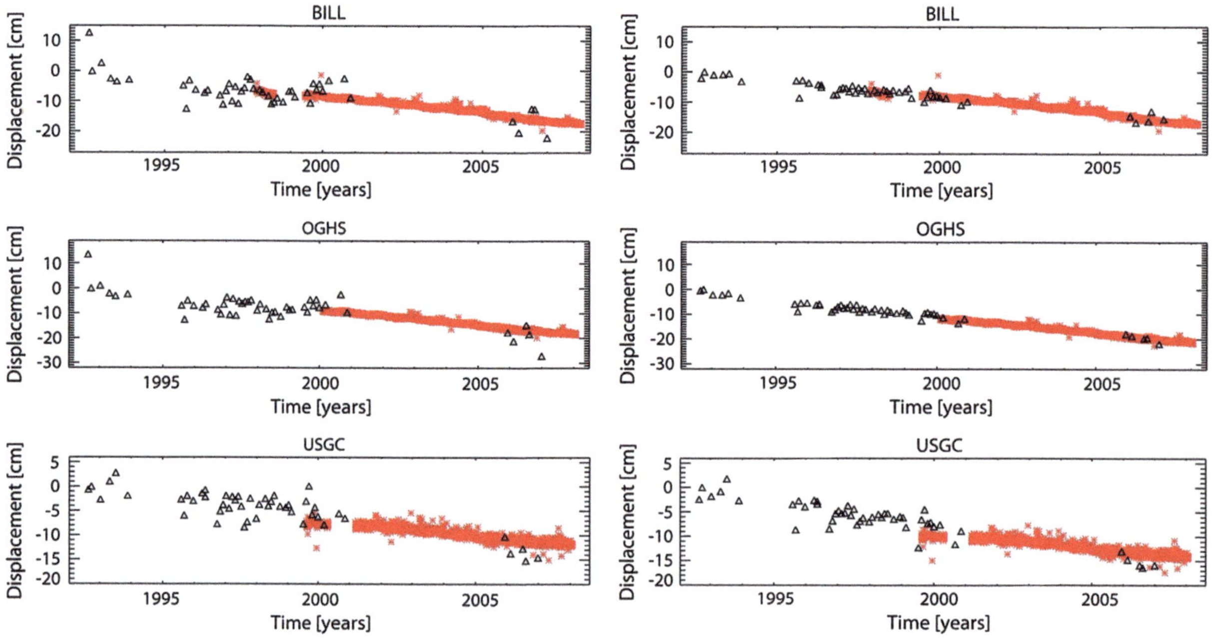

Figure 10
Comparison between the retrieved DInSAR deformation time series for data belonging to track 127 (*black triangles*) and the LOS-projected GPS measurements (*red stars*) at the locations of the continuous GPS sites labeled as BILL, OGHS and USGC (Fig. 4). *Left*: the DInSAR time series are computed without orbital artifact correction; *right*: the DInSAR time series are computed with the orbital ramp correction

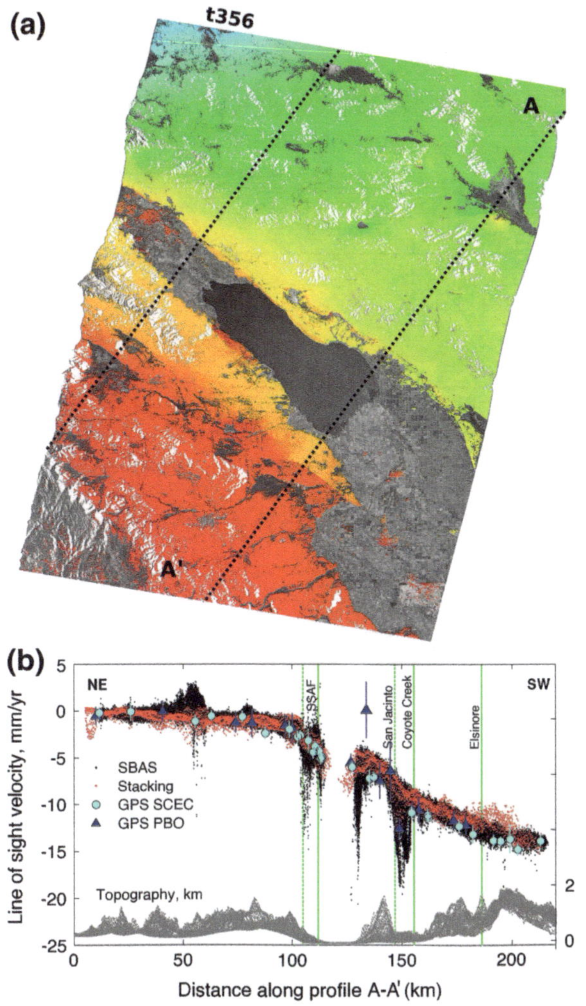

Figure 11
a Mean deformation velocity map for track 356 (same as for Fig. 7b) where the *dashed strip* outlines a profile from which the DInSAR and GPS data are extracted for the comparison shown in Fig. 11b. **b** LOS velocities as inferred from the modified SBAS analysis (*black dots*) and optimized stacking (*red dots*). *Color symbols* represent LOS-projected GPS data. Campaign GPS data from the SCEC velocity model are denoted by *circles* and continuous data from the PBO network by *triangles*. *Vertical green lines* denote the location of active faults (see Fialko (2006) for details)

continuous GPS stations in order to discriminate the interferometric phase signal components due to long-wavelength deformation, caused by interseismic strain accumulation, from that caused by orbital artifacts. In particular, we propose a simple solution, with a minimum impact on the SBAS procedure that is based on pre-filtering of the DInSAR interferograms by using the available measurements of a limited set of continuous GPS stations. In particular, this pre-filtering step allows us to successfully compensate, within the SBAS processing, for orbital artifacts because it drastically mitigates the presence of long-wavelength interseismic displacements that may be wrongly interpreted and filtered out as orbital phase errors.

By comparing these results with those available through the local USGS/PBO continuous GPS network, we show that the orbital phase correction is, at least for one of the two time series, even more relevant than the atmospheric phase filtering. We also show the achieved accuracy of the finally retrieved deformation time series is consistent with that obtained through previous SBAS product validation analyses focused on different deformation scenarios. Moreover, there is also a very good agreement between the estimated mean deformation velocity and that computed using optimized stacking techniques.

We finally remark that the presented work is focused on a specific case study but it can be easily extended to other seismic areas. In this context, the full exploitation of the available huge archives of C-band SAR data, which are largely unused, can be of great importance for carrying out new analysis of occurred seismic events as well as for quantifying subtle interseismic strain accumulation in tectonically active areas. In these scenarios, the spatial distribution and density of the exploited continuous GPS stations can play a significant role. Indeed, a GPS network adequately "covering" the whole study area, with measurements that significantly overlap with the interval of the DInSAR time series, is clearly needed. An extended investigation of the trade-off between the GPS distribution/density and the accuracy of the retrieved DInSAR results is outside the scope of this work, but it is certainly worth for future analysis in order to assess the requirements and

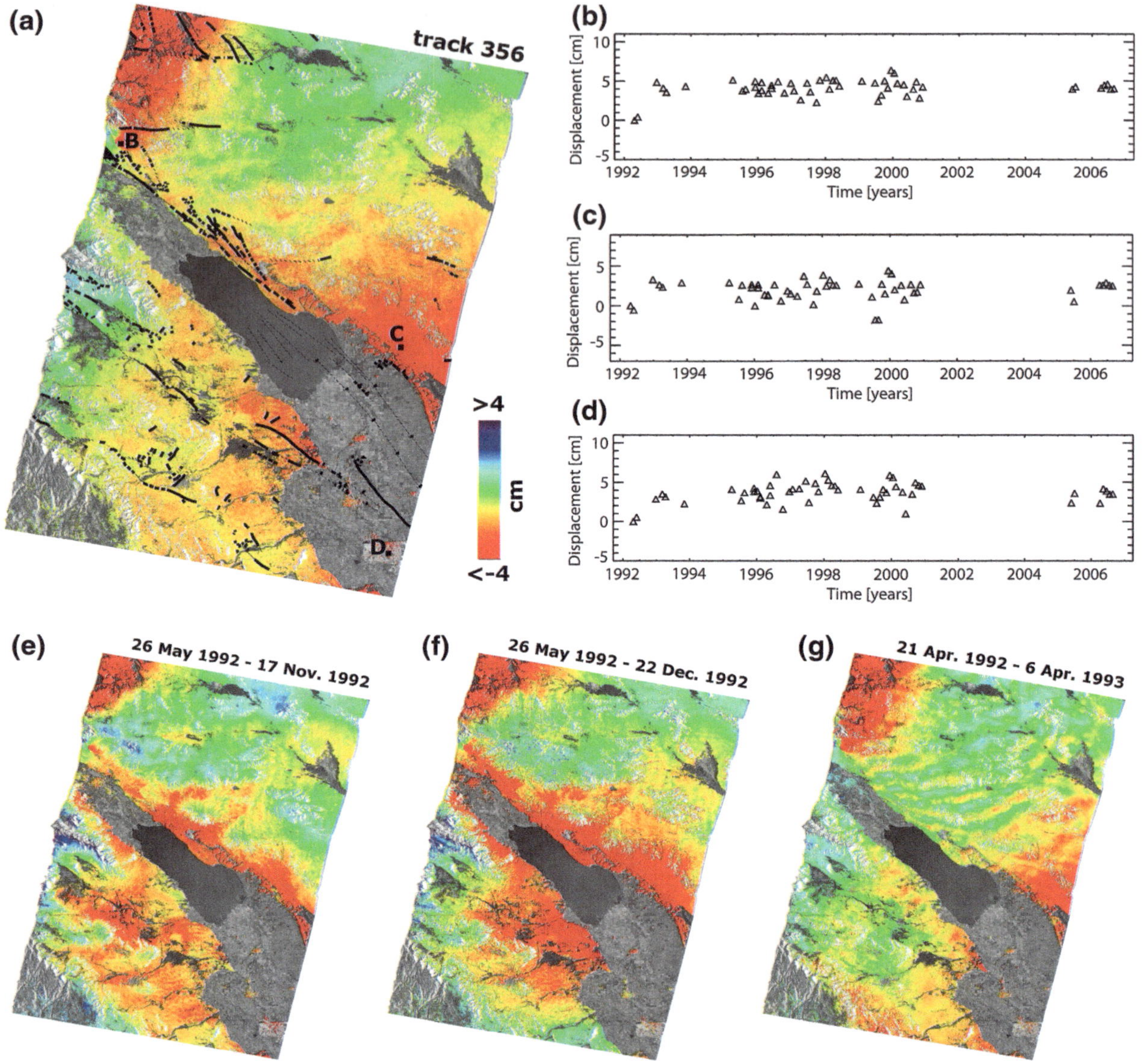

Figure 12
a Deformation map due to the Landers earthquake (28 June 1992) computed from time series of track 356; the trace of the SAF system is identified via black lines. **b–d** Detrended time series computed without atmospheric filtering at the locations referred to as *B*, *C* and *D* in Fig. 12a. **e–g** Unwrapped interferograms relevant to the 26 May 1992–17 November 1992 (**e**), 26 May 1992–22 December 1992 (**f**) and 21 April 1992–6 April 1993 (**g**) time intervals. Note that the exploited ERS image acquired on 17 November 1992 is not included in the generated DInSAR time series because it was available only subsequently

performance of the proposed approach that jointly exploits GPS and DInSAR measurements.

Acknowledgments

This work has partially been supported by ASI, the Italian DPC, NASA (NNX09AD23G) and USGS (G09AP00025). ERS SAR data used in this study are copyright of ESA, acquired via the WInSAR Consortium (http://winsar.unavco.org). We thank the Technical University of Delft, The Netherlands, for precise ERS-1/2 orbits. SRTM digital elevation data were produced by NASA and distributed by USGS. Finally, we thank S. Guarino, F. Parisi and M.C. Rasulo for their technical support.

References

Amelung, F., Galloway, D.L., Bell, J.W., Zebker, H.A., and Laczniak, R.J. (1999), *Sensing the ups and downs of Las Vegas: InSAR reveals structural control of land subsidence and aquifer-system deformation*, Geology, *27*, 483–486.

Bennett, R. A., Rodi, W. and Reilinger, R. E. (1996), *Global positioning system constraints on fault slip rates in southern California and northern Baja, Mexico*, J Geophys Res *101*, 21943–21960.

Berardino, P., Fornaro, G., Lanari, R. and Sansosti, E. (2002), *A new algorithm for surface deformation monitoring based on small baseline differential SAR interferograms*, IEEE Trans. Geosci Remote Sens *40*, 2375–2383.

Biggs, J., Wright, T., Lu, Z., and Parsons, B. (2007), *Multi-interferogram method for measuring interseismic deformation: Denali fault, Alaska*, Geophys J Int *170*, 1165–1179.

Burgmann, R., Hilley, G., Ferretti, A. and Novali F. (2006), *Resolving vertical tectonics in the San Francisco Bay Area from permanent scatterer InSAR and GPS analysis*, Geology *34*, 221–224.

Casu, F., Manzo, M., and Lanari, R. (2006), *A quantitative assessment of the SBAS algorithm performance for surface deformation retrieval from DInSAR data*, Remote Sens Environ *102*(3–4), 195–210, doi:10.1016/j.rse.2006.01.023.

Casu, F., Manzo, M., Pepe, A., and Lanari, R. (2008), *SBAS-DInSAR Analysis of Very Extended Areas: First Results on a 60,000 km2 Test Site*, IEEE Geosci Remote Sens Lett *5* 3, doi: 10.1109/LGRS.2008.916199.

Chen, C. W., and Zebker, H. A. (2002), *Phase unwrapping for large SAR interferograms: Statistical segmentation and generalized network models*, IEEE Transactions on Geoscience and Remote Sensing *40*, pp. 1709–1719.

DeMets, C., Gordon, R. G., Argus, D. F., and Stein, S. (1990), *Current plate motions*, Geophys. J Int *101*, 425–478.

Ferretti, A., Prati, C., and Rocca, F.(2000), *Non-linear subsidence rate estimation using permanent scatterers in differential SAR interferometry*, IEEE Transaction on Geoscience and Remote Sensing *38*, 5.

Fialko, Y., Simons, M., and Agnew, D. (2001), *The complete (3-D) surface displacement field in the epicentral area of the 1999 Mw7.1 Hector Mine earthquake, California, from space geodetic observations*, Geophys Res Lett *28*, 3063–3066.

Fialko, Y. (2004), *Evidence of fluid-filled upper crust from observations of post-seismic deformation due to the 1992 Mw7.3 Landers earthquake*, J Geophys Res, *109*, B08401, doi:10.1029/2003JB002985.

Fialko, Y. (2006), *Interseismic strain accumulation and the earthquake potential on the southern San Andreas fault system*, Nature, *441*, doi:10.1038/nature04797, 968–971.

Gabriel, A. K., Goldstein, R. M., and Zebker, H. A. (1989), *Mapping small elevation changes over large areas: Differential interferometry,* J Geophys Res *94*, 9183–9191.

Goldstein, R. M. (1995), *Atmospheric limitations to repeat-track radar interferometry,* Geophys Res Lett 22, 2517–2520.

Gourmelen, N., Amelung F., and Lanari R. (2010), *InSAR–GPS Integration: inter-seismic strain accumulation across the hunter mountain fault in the eastern California shear zone*, J Geophys Res doi:10.1029/2009JB007064, in press.

Hooper, A., Zebker, H., Segall, P., and Kampes, B. (2004), *A new method for measuring deformation on volcanoes and other natural terrains using InSAR persistent scatterers*, Geophys Res Lett *31*, L23611, doi:10.1029/2004GL021737.

Hooper, A. (2008), *A multi-temporal InSAR method incorporating both persistent scatterer and small baseline approaches*, Geophys Res Lett *35* L16302, doi:10.1029/2008GL034654.

Johnson, H. O., Agnew, D. C., and Wyatt, F. K. (1994), *Present-day crustal deformation in southern California*, J Geophys Res *99*, 23951–23974.

Lanari, R., Mora, O., Manunta, M., Mallorquí, J. J., Berardino, P., and Sansosti, E. (2004a), *A small baseline approach for investigating deformations on full resolution differential SAR interferograms*, IEEE Trans Geosci Remote Sens *42*, 7.

Lanari, R., Lundgren, P., Manzo, M., and Casu F. (2004b), *Satellite radar interferometry time series analysis of surface deformation for Los Angeles, California*, Geophys Res Lett *31*, L23613, doi:10.1029/2004GL021294.

Lanari, R., Casu, F., Manzo, M., and Lundgren, P. (2007a), *Application of the SBAS-DInSAR technique to fault creep: a case study of the Hayward fault, California*, Remote Sens Environ J *109*, 1, 20–28, doi:10.1016/j.rse.2006.12.003.

Lanari, R., Casu, F., Manzo, M., Zeni, G., Berardino, P., Manunta, M., and Pepe, A. (2007b), *An Overview of the Small BAseline Subset Algorithm: a DInSAR Technique for Surface Deformation Analysis*, Pure Appl Geophys (PAGEOPH), *164*, 4, 637–661, doi:10.1007/s00024-007-0192-9.

Lundgren, P. E., Hetland, A., Liu, Z. and Fielding, E. J. (2009), *Southern San Andreas-San Jacinto fault system slip rates estimated from earthquake cycle models constrained by GPS and interferometric synthetic aperture radar observations*, J Geophys Res *114*, B02403, doi:10.1029/2008JB005,996.

Lyons, S. and Sandwell, D. (2003), *Fault creep along the southern San Andreas from interferometric synthetic aperture radar, permanent scatterers, and stacking*, J Geophys Res *108*, doi: 10:1029/2002JB001831.

Manzo, M., Ricciardi, G. P., Casu, F., Ventura, G., Zeni, G., Borgström, S., Berardino, P., Del Gaudio, C., and Lanari, R. (2006), *Surface deformation analysis in the Ischia island (Italy) based on spaceborne radar interferometry*, J Volcanol Geotherm Res *151*, 399–416, doi:10.1016/j.jvolgeores.2005.09.010.

Massonnet, D., Rossi, M., Carmona, C., Adragna, F., Peltzer, G., Feigl, K., and Rabaute, T. (1993), *The displacement field of the Landers earthquake mapped by radar interferometry*, Nature, *364*, 138–142.

Massonnet, D., Briole, P., and Arnaud, A. (1995), *Deflation of Mount Etna monitored by spaceborne radar interferometry*, Nature, *375*, 567–570.

Miranda, N., Rosich, B., Santella, C., and Grion, M. (2003), Review of the impact of ERS-2 piloting modes on the SAR Doppler stability, Proceedings Fringe '03, 1-5 December 2003, Frascati, Italy.

Mora, O., Mallorqui, J. J., and Broquetas, A. (2003), *Linear and nonlinear terrain deformation maps from a reduced set of interferometric SAR images*, IEEE Trans Geosci Remote Sens *41*, 2243–2253.

Neri, M., Casu, F., Acocella, V., Solaro, G., Pepe, S., Berardino, P., Sansosti, E., Caltabiano, T., and Lundgren, P. (2009), *Deformation and eruptions at Mt. Etna (Italy): A lesson from 15 years of observations*, Geophy Res Lett *36*, doi:10.1029/2008GL036151.

PELTZER, G., and ROSEN, P.A. (1995), *Surface displacement of the 17 May 1993 Eureka Valley earhtquake observed by SAR interferometry*, Science *268*, 1333–1336.

PELTZER, G., CRAMPE, F., HENSLEY, S., and ROSEN, P. (2001), *Transient strain accumulation and fault interaction in the Eastern California shear zone*, Geology *29*, 975–978.

PEPE, A., and LANARI, R. (2006), *On the extension of the minimum cost flow algorithm for phase unwrapping of multitemporal differential SAR interferograms*, IEEE Trans Geosci Remote Sens *44*, 9, 2374–2383.

PRATI, C., FERRETTI, A., and PERISSIN, D. (2010), *Recent advances on surface ground deformation measurement by means of repeated space-borne SAR observations*, J Geodynamics *49*, 161–170.

RIGNOT, E. (1998), *Fast recession of a west Antarctic glacier*, Science, *281*, 549–551.

ROSEN, P. A., HENSLEY, S., JOUGHIN, I. R., LI, F. K., MADSEN, S. N., RODRIGUEZ, E., *et al.* (2000). Synthetic aperture radar interferometry, IEEE Proceedings, 88, 333–376.

ROSEN, P. A., HENSLEY, S., GURROLA, E., ROGEZ, F., CHAN, S., and MARTIN, J. (2001), *SRTM C-band topographic data quality assessment and calibration activities*, Proc. of IGARSS'01, 739–741.

THATCHER, W., and LISOWSKI, M. (1987), *Long-term seismic potential of the San-Andreas fault southeast of San-Francisco*, California J Geophys Res *92*, 4771–4784.

TIZZANI, P., BERARDINO, P., CASU, F., EUILLADES, P., MANZO, M., RICCIARDI, G. P., ZENI, G., and LANARI, R. (2007), *Surface deformation of Long Valley caldera and Mono Basin, California, investigated with the SBAS-InSAR approach*, Remote Sens Environ J *108*, 277–289, doi:10.1016/j.rse.2006.11.015.

WELDON, R.J., FUMAL T. E., BIASI G. P., SCHARER K. M. (2005), *Geophysics: past and future earthquakes on the San Andreas fault*, Science, *308*, 5724, 966–967.

WERNER, C., WEGMÜLLER, U., STROZZI, T. and WIESMANN, A. (2003), Interferometric point target analysis for deformation mapping, Proc. IGARSS'03, Toulouse (France), 4362–4364.

WORKING GROUP ON CALIFORNIA EARTHQUAKE PROBABILITIES (1995), *Seismic hazards in southern California: Probable earthquakes, 1994–2024*. Bull Seism Soc Am *85*, 379–439.

WRIGHT, T. J., PARSONS, B., ENGLAND, P. C., and FIELDING, E. J. (2004), *InSAR observations of low slip rates on the major faults of western Tibet*, Science *305*, 236–239.

(Received July 13, 2010, revised February 9, 2011, accepted April 11, 2011, Published online September 30, 2011)

Pure Appl. Geophys. 169 (2012), 1483–1506

DOI 10.1007/s00024-011-0420-1

Pure and Applied Geophysics

Analysis of GPS Measurements in Eastern Canada Using Principal Component Analysis

K. F. TIAMPO,[1] S. MAZZOTTI,[2] and T. S. JAMES[2]

Abstract—Continuous Global Positioning System (CGPS) position time series from eastern North America constrain the pattern and magnitude of regional crustal deformation. Initial analysis delineates consistent uplift patterns, as expected from glacial isostatic adjustment (GIA) predictions, but the associated horizontal deformation is not definitive, in part due to the short time periods for a significant number of the available stations. We employ an eigenpattern decomposition in order to define a unique, finite set of deformation patterns for this continuous GPS data. Similar in nature to the empirical orthogonal functions historically employed in the analysis of atmospheric and oceanographic phenomena, the method derives the eigenvalues and eigenstates from the diagonalization of the correlation matrix using a Karhunen–Loeve expansion (KLE). The KLE technique is used to identify the important modes in both time and space for the CGPS data, modes that potentially include signals such as horizontal and vertical GIA, tectonic strain, and seasonal effects. Here we filter both the vertical and horizontal velocity patterns on different spatiotemporal scales in order to study the potential geophysical sources, after the removal of correlated and random noise. The method is successful in disaggregating the linear vertical signal from more variable and less spatially correlated signals. The vertical and horizontal results are compared to the predictions of the ICE-3G GIA loading model with a number of plausible mantle viscosity profiles. The horizontal velocity analysis allows for qualitative differentiation between several potential GIA models and suggests that, with longer time series, this technique can be employed to remove correlated noise and improve estimates of crustal strain patterns and their sources.

1. Introduction

The large, high-quality Global Positioning System (GPS) networks that exist throughout the world today have provided important information on surface deformation on a variety of spatial and temporal scales, including plate motion (e.g., CALAIS *et al.* 2006), coseismic deformation and slow earthquakes (e.g., HUDNUT *et al.* 1996; DRAGERT *et al.* 2001), tectonic strain (e.g., BAWDEN *et al.* 2001; MAZZOTTI *et al.* 2005), and seasonal deformation associated with groundwater and meteorological effects (e.g., BAWDEN *et al.* 2001; WATSON *et al.* 2002; TIAMPO *et al.* 2004). GPS measurements of crustal motion and strain in eastern Canada and the US comprise a combination of long-wavelength glacial isostatic adjustment (GIA) (JAMES and BENT 1994; PELTIER 1998, 2002), seasonal effects, including those with both annual and semiannual signals, and seismic deformation (MAZZOTTI and ADAMS 2005; MAZZOTTI *et al.* 2005; CALAIS *et al.* 2006; SELLA *et al.* 2007). Discerning and separating the various sources of crustal deformation is necessary in order to make progress in a field as diverse as earthquake and earthquake hazard studies (MAZZOTTI 2007), GIA, and seasonal and interannual surface loading related to climate change (BEVIS *et al.* 2005).

Although stable, eastern North America has much lower seismicity than plate boundary regions to the west, there are areas of substantial seismicity and earthquake hazard. A map of seismicity for the years 1900 through 2001 shows three elongated regions of seismic activity superimposed on a relatively low background rate (Fig. 1). Earthquake statistics in the high-activity regions are associated with seismic strain rates of about $1–20 \times 10^{-9}$/year (MAZZOTTI and ADAMS 2005). However, accurate quantification of this relatively small signal, in conjunction with the others noted above, has proven difficult.

Accurate, distributed CGPS time series that can be used for studies of intraplate deformation have only recently been available (MAZZOTTI *et al.* 2005; CALAIS *et al.* 2006; SELLA *et al.* 2007). CALAIS *et al.* (2006) and SELLA *et al.* (2007) analyzed a combination of more than 300 GPS sites distributed across North America. The vertical motions fit GIA models

[1] Department of Earth Sciences, University of Western Ontario, London, ON, Canada. E-mail: ktiampo@uwo.ca

[2] Geological Survey of Canada, Natural Resources Canada, Sidney, BC, Canada.

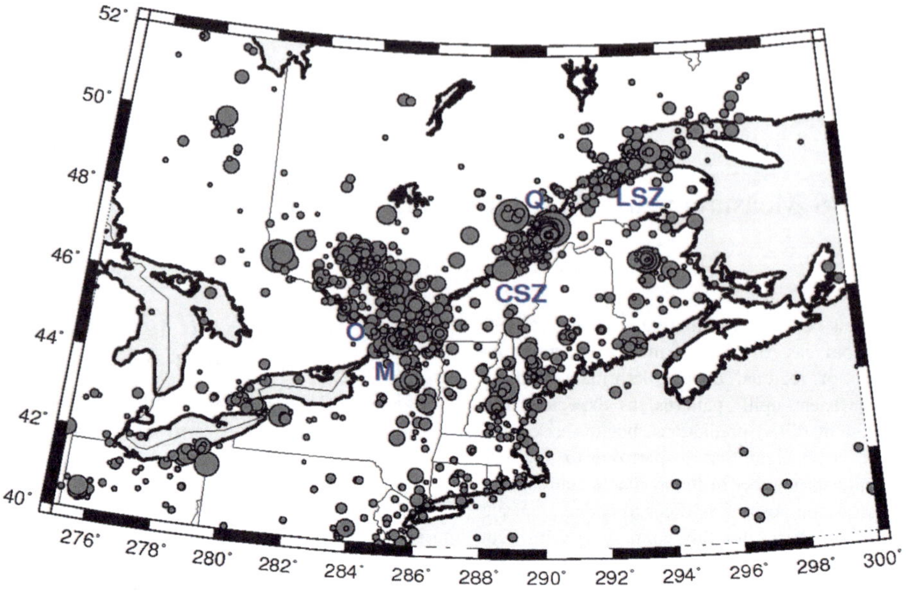

Figure 1
Earthquakes in eastern Canada, $M \geq 3.0$, 1900–2001. *White* and *light grey* colors identify land mass and large water bodies, respectively. *CSZ* Charlevoix Seismic Zone, *LSZ* Lower St. Lawrence Seismic Zone, *Q* Quebec City, *M* Montreal, *O* Ottawa

reasonably well, but the horizontal velocities are an order of magnitude smaller and do not fit the predicted motions well. They conclude that the horizontal scatter is likely a combination of local noise and intraplate tectonic signal.

Filtering techniques have been developed with the goal of reducing or removing the various noise sources in the position time series. For example, Dong *et al.* (2006) utilized common mode error (CME) filtering using a principal component (PC) analysis and the Karhunen–Loeve expansion (KLE) technique in order to improve the identification of both signal and systematic error from GPS regional position time series. They showed that spatiotemporal correlated errors are the dominant error source in daily GPS positions and demonstrated that a KLE technique provides a more general approach to spatial filtering. Tiampo *et al.* (2004) earlier employed a KLE analysis to study spatiotemporally correlated mass loading caused by seasonal deformation in Southern California Integrated GPS Network (SCIGN) position series data. Zerbini *et al.* (2010) applied a similar technique to GPS and gravity data in northeastern Italy and succeeded in identifying hydrology-related correlated variations. Here we will demonstrate that the decomposition of GPS data in eastern Canada and the USA into its spatial eigenvectors and associated temporal signature can be used to separate various modes based upon the important spatial and temporal scales.

In Sect. 2 we give an overview of the geological framework of the region where the CGPS data was obtained in eastern Canada and the United States. Section 3 provides details of the CGPS network and data analysis. Sections 4 and 5 describe the KLE analysis technique and the resulting decomposition of the regional CGPS measurements into their principal components. The last section discusses the implications of this analysis of the CGPS data for constraining different geophysical sources of the modeled deformation.

2. *Geological and Tectonic Setting*

2.1. *Geological Background*

While the area of study is currently in an intraplate tectonic setting, it is the historic location of several major tectonic events over the past ~1 Gyr. The Grenville orogen occurred from ~1,100 to 900 Ma and is associated with the accretion of allochthonous terranes on the southeast margin of the Laurentian craton (KARLSTROM *et al.* 2001). Extension occurred more recently (~700–600 Ma) as a result of the rifting and opening of the Iapetus Ocean in the late Proterozoic, resulting in the formation of a large-scale system of normal faults across much of eastern North America (KUMARAPELI 1985). The Saint Lawrence, Ottawa, and Saguenay grabens of Canada formed during this time (WHEELER 1995).

During the closing of the Iapetus Ocean in the mid- to late-Paleozoic, the Appalachian nappes were thrust over the North American craton as far west as the Saint Lawrence valley (WILLIAMS 1979; FAURE *et al.* 1996). During the following Jurassic rifting and opening of the North Atlantic Ocean, reactivation of Iapetan normal faults saw the last phase of significant tectonic activity in the eastern Canadian region (LEMIEUX *et al.* 2003) and, as a result, the region shown in Fig. 1 is characterized by large eastward dipping normal faults overlain by the westward verging thrust faults and nappes of the Appalachian orogen (KUMARAPELI 1985; TREMBLAY *et al.* 2001). Here the Paleozoic cover is a few kilometers thick throughout most of the region, obscuring the underlying structures. Finally, a meteorite impact in the southern Charlevoix seismic region ~350 Ma created a system of concentric faults and fractures of approximately 60 km in diameter (RONDOT 1968; LEMIEUX *et al.* 2003).

Ice advanced over the Saint Lawrence valley during the late Pleistocene, as far east as the Maritime Provinces and south into New England (DYKE 2004). The weight of this ice depressed the lithosphere beneath the ice. Outside the formerly glaciated region, viscoelastic flow in the mantle generated a peripheral bulge. Between about 7 and 20 ka, the ice sheets retreated and the lithosphere began to rebound upward to regain isostatic equilibrium. The peripheral bulge migrated inward toward the centre of uplift as it gradually dissipated. Present-day GIA uplift rates approach 10 mm/year or more at Hudson Bay and decrease with distance southeastward. Current GIA models forecast that the hinge line between uplift to the northwest and subsidence to the southeast lies somewhere near the Saint Lawrence valley in eastern Canada (TUSHINGHAM and PELTIER 1991; PELTIER 2002). Recent geodetic observations confirm these projections but, to date, the network densification and length of the corresponding time series has not been sufficient for the accurate determination of significant regional or local variations (LAMBERT *et al.* 2001; PARK *et al.* 2002; MAINVILLE and CRAYMER 2005; MAZZOTTI *et al.* 2005; CALAIS *et al.* 2006; SELLA *et al.* 2007).

2.2. *Seismicity*

The seismically active zones in eastern North America generally are associated with large paleotectonic geologic features (e.g., ADAMS and BASHAM 1991). In eastern Canada, these zones correspond to the Mesozoic rifted margin of the Atlantic Ocean, the early Paleozoic rifted margin and associated failed rifts of the Iapetus Ocean along the St. Lawrence and Ottawa valleys, and the Appalachian thrust sheets. The strong correlation between earthquake distribution and geological structures suggests a possible control on the spatial distribution of earthquakes (ADAMS and BASHAM 1991; ADAMS *et al.* 1995; WHEELER and JOHNSON 1992; WHEELER 1995; MAZZOTTI 2007).

The bulk of eastern Canadian seismicity is concentrated in the both the lower Saint Lawrence seismic zone (LSZ) downriver from Québec City and the Charlevoix seismic zone (CSZ) to the west of Québec City (Fig. 1). Since 1938, approximately twelve $4 \leq M \leq 5$ earthquakes have occurred in the LSZ. The CSZ has undergone five $M \geq 6$ earthquakes since 1663 and approximately ten events of $5 \leq M \leq 6$ since the mid-19th century (ADAMS and HALCHUK 2003).

The historic seismicity rates along the Saint Lawrence valley indicates some level of brittle deformation in the crust on spatial and temporal scales significantly different from GIA. MAZZOTTI and ADAMS (2005) estimate that seismic strain rates in most of eastern Canada are about 10^{-13}–10^{-11}/year. Exceptions to these low strain rates can be found in the CSZ and LSZ. Earthquake statistics in these two regions are associated with seismic strain rates of 3–23 $\times$ 10^{-9}/year in the CSZ and 0.1–5 $\times$ 10^{-9}/year in the LSZ. In contrast, seismic strain rates for the entire Saint Lawrence valley are 0.6–12 $\times$ 10^{-11}/year (MAZZOTTI *et al.* 2005).

3. GPS Data

3.1. Data and Processing

We use a subset of 43 CGPS stations in eastern Canada and northeastern USA from western Quebec to southern New England, thus bracketing the St. Lawrence Valley seismicity area (Table 1; Fig. 2). These stations are all GPS dual-frequency receiver-antenna systems, but the antenna monuments vary from geodetic-quality concrete and steel piers (Table 1—station 16) to non-geodetic towers (station 3) and rooftop mounts (station 24). The data span a time period from June 2001 to October 2006, with 16 stations covering the whole period (5.3 year) and 23 stations with shorter time spans ranging from 5.0 to 0.9 year (Table 1). The stations are operated by various national and regional agencies, such as Natural Resources Canada, US National Geodetic Survey, Ministère des resources naturelles du Québec. All the data were archived and processed at the Pacific Geoscience Centre (Natural Resources Canada).

The data was processed with the BERNESE 4.2 software (Hugentobler *et al.* 2001) using an ionospheric-free, double-difference, phase solution in a network least-square adjustment of the daily station positions. Details of the processing are similar to those described in Mazzotti *et al.* (2003), except for the definition of fiducial stations and reference frame alignment. In this study, the daily network position adjustments are aligned to the International Terrestrial Reference Frame (ITRF) by constraining (at 1.0 mm root-mean-square) the position of four fiducial sites to their nominal ITRF2000 values: St. John's (STJO), Schefferville (SCH2), Algonquin Park (ALGO), and Goddard Flight Space Center (GODE). These four stations are long-running, core ITRF sites and provide a regional-scale alignment to the ITRF.

3.2. Position Time Series and Common Mode Filtering

Because they are tightly constrained to their nominal ITRF2000 positions and velocities, the daily position time series of the four fiducial sites (ALGO, GODE, SCH2, and STJO) are forced to follow a linear model (within 1.0 mm RMS) and do not represent the "true" site motions, which include non-linear signals. Thus, we remove these four sites from further analysis of the time series, leaving a 39-station data set.

In the first phase of the analysis, the time series are first corrected for the predicted absolute motion of the North American plate using the ITRF2000 North America rigid rotation vector (Altamimi *et al.* 2002). Linear velocities for the north, east and up components are derived from a least-square model of the residual position time series that include an intercept, linear trend, and 1-year sinusoidal seasonal component. The average daily root-mean-square (RMS) scatters of the positions about this model are 1.7, 1.9, and 4.9 mm in the north, east and up components, respectively (Table 1). These original series are the ones processed using the PC Analysis (Sect. 4).

Common mode filtering (also known as common mode error, or CME filtering, above) consists of stacking the residual time series of all stations, after application of the preliminary model, in order to form a template for the average noise patterns that are common to most stations (e.g., Wdowinski *et al.* 1997; Mazzotti *et al.* 2003). This regional filtering reduces the amplitudes of both the white and colored noise components by a factor of 2-3 (Williams *et al.* 2004) and has been widely employed in the study of various tectonics processes (e.g., Márquez-Azúa and DeMets 2003; Calais *et al.* 2006). However, regional spatial filtering generally is applied in a standard manner, and is not easily adapted to account for correlations in the data or the noise. Márquez-Azúa and DeMets (2003) demonstrated that strong correlations, and hence, nonzero covariances exist between the daily coordinates of GPS stations at distances of several thousand kilometers. Recent work for a variety of GPS networks confirms that, despite advances in monumentation and processing, correlated seasonal signals remain in most refined CGPS time series (Langbein and Johnson 1997; Langbein 2008; Hill *et al.* 2009).

Application of this technique to our 39-position time series leads to a modest reduction in the overall noise level. Average north, east and up daily RMS scatters are reduced by 12, 9, and 20% (i.e., 0.2, 0.2,

Table 1

GPS stations

Site	Monument	Latitude	Longitude	Start date	Time series length (years)	RMS-N (mm)	RMS-E (mm)	RMS-U (mm)	dRMS-N (%)	dRMS-E (%)	dRMS-U (%)
ALGO	Geodetic	45.955	281.928	2001 Jun.	5.3	1.4	2	1.8	–	–	–
ANNE	Roof	49.128	293.505	2005 Sep.	1.1	1.7	1.5	4.1	12	0	17
ATRI	Roof	46.847	288.739	2001 Jun.	5.3	1.7	2	5.2	12	15	21
BAIE	Geodetic	49.186	291.736	2001 Nov.	4.9	2.1	1.9	5.3	10	−5	15
BARH	Roof	44.4	291.78	2001 Jun.	5.3	1.7	1.8	5.2	18	17	23
BARN	Geodetic	44.1	288.84	2001 Jun.	5.3	2.1	2.2	8.2	14	9	17
BRU1	Tower	43.89	290.05	2001 Jun.	5.3	2.2	2.2	6.5	14	14	18
CAGS	Geodetic	45.584	284.192	2001 Jun.	5.3	2	2.3	5.1	15	13	20
CAPL	Roof	48.094	294.347	2002 Oct.	4	1.5	1.6	4.4	13	6	14
CARM	Roof	46.868	291.986	2004 Aug.	2.1	1.5	1.3	4.3	20	8	21
CHIB	Roof	49.913	285.633	2001 Jun.	5.3	2.4	2.1	5.2	4	14	23
CHIC	Roof	48.414	288.745	2001 Oct.	5	1.8	1.9	5.2	11	16	17
ESCU	Geodetic	47.073	295.201	2004 Nov.	1.9	1.4	1.3	4.1	7	0	22
GEOR	Roof	46.13	289.313	2003 Oct.	3	1.7	1.7	4.8	12	0	23
GODE	Geodetic	39.021	283.84	2001 Jun.	5.3	1.8	1.7	1.8	–	–	–
HLFX	Geodetic	44.683	296.389	2001 Nov.	4.9	1.8	2.6	6	11	0	12
HSTP	Roof	50.242	296.394	2001 Nov.	4.9	1.7	2.1	5.4	0	0	13
HULL	Roof	45.426	284.289	2001 Oct.	5	1.8	2.4	4.6	11	8	20
KNGS	Geodetic	44.218	283.482	2002 May	4.4	1.4	1.5	4.2	14	7	21
LAMT	Roof	41.005	286.091	2001 Jul.	5.2	1.8	1.9	5.4	17	16	22
LAUR	Roof	46.546	284.521	2004 Oct.	2	1.4	1.6	4.4	0	0	20
LOUP	Roof	47.827	290.443	2005 Jan.	1.7	1.3	1.4	3.9	8	0	26
LPOC	Geodetic	47.341	289.992	2005 May	1.4	3.3	2	5	9	5	20
MCTN	Roof	46.096	295.166	2005 Oct.	0.9	1.7	1.6	5	12	6	16
MONT	Roof	45.546	286.361	2001 Jun.	5.3	1.6	1.8	4.5	13	17	31
NPRI	Geodetic	41.51	288.67	2001 Jun.	5.3	1.6	1.7	5	19	18	24
OSPA	Roof	43.465	283.488	2002 Apr.	4.5	1.7	1.5	4.5	12	7	20
PARY	Geodetic	45.338	279.964	2002 May	4.4	1.5	1.7	5.1	7	6	12
POR4	Tower	43.07	289.29	2003 Jun.	3.3	2.2	2.2	6.3	14	18	19
PSC1	Roof	44.43	285.75	2001 Jun.	5.3	1.5	1.7	4.8	20	12	23
PWEL	Geodetic	43.236	280.78	2002 May	4.4	1.6	1.7	5.1	13	6	14
RIMO	Roof	48.444	291.479	2004 Sep.	2	1.4	1.3	4	14	8	23
ROUY	Roof	48.241	280.971	2001 Jun.	5.3	1.7	2.3	5	12	13	16
SEPT	Roof	50.205	293.613	2005 Sep.	1	1.5	1.4	4.2	13	7	21
SCH2	Geodetic	54.832	293.167	2001 Jun.	5.3	1.7	2.2	2	–	–	–
STJO	Geodetic	47.595	307.322	2001 Jun.	5.3	1.4	2.3	2.1	–	–	–
SRBK	Roof	45.401	288.102	2001 Oct.	5	1.6	1.7	4.3	19	12	35
TRIV	Roof	46.344	287.461	2001 Oct.	5	1.6	1.9	4.2	13	16	29
UNB1	Geodetic	45.57	293.617	2001 Oct.	5.1	1.7	1.9	5.7	12	11	21
VALD	Geodetic	48.097	282.435	2001 Nov.	4.9	2.3	3	5.4	9	0	9
VCAP	Roof	44.26	287.42	2001 Jun.	5.3	1.7	2.1	6	12	14	22
WES2	Roof	42.61	288.51	2001 Jun.	5.3	1.8	2	5.3	17	20	21
WIL1	Tower	41.305	283.985	2001 Jun.	5.3	1.9	2.2	5.3	16	18	23

No dRMS entries for ALGO, GODE, SHC2, and STJO, which are fiducial

RMS-N, RMS-E, RMS-U RMS in the north, east and up directions

dRMS-N, dRMS-E, dRMS-U reduction in daily RMS due to common-mode

and 1.0 mm), respectively. An example of the filtered time series for stations CAGS is shown in Fig. 3. Similarly, the regional filtering has a negligible effect on the estimated velocities, with average north, east and up changes of ~0.2 mm/year (−1.2 mm/year minimum to +0.7 mm/year maximum).

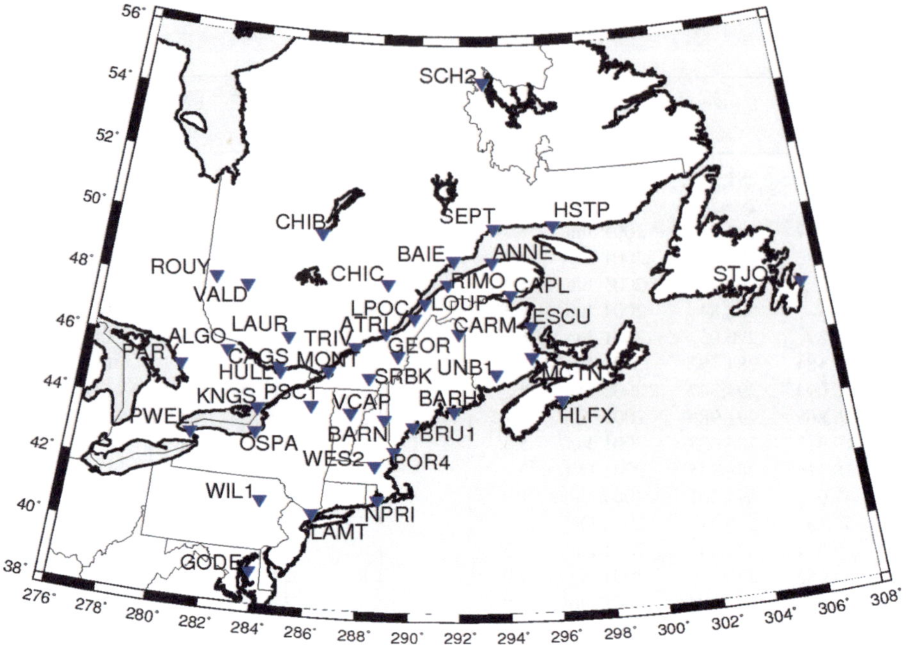

Figure 2
Continuous GPS (CGPS) stations in eastern Canada. Station details given in Table 1

3.3. Patterns of Crustal Deformation

Filtered time series in the north, east and up directions are shown for station CAGS (Fig. 3). The first 6 months of the recorded data, July 1 through December 31, 2001, were removed from the subsequent analysis because only one-third of the stations were consistently recording on July 1, 2001. The large number of initial zeros in the data set, at a large percentage of the stations, resulted in a highly correlated but spurious signal in the original PC analysis that distorted the remaining modes. More than one-half of the stations were active on January 1, 2002, and this was sufficient to provide a robust correlation analysis, as will be shown in Sect. 5.

Figure 4 and Table 2 show the vertical and horizontal velocities for each station for the filtered data sets for the time period January 2002 through June 2006. The associated standard error ellipses are estimated from the RMS errors on the linear fit to the estimated velocities and, as such, are likely underestimated relative to an estimate that would include the correlated noise normally associated with GPS data. However, as demonstrated by Dong *et al.* (2006), spatiotemporal correlated errors are the dominant error source in daily GPS positions. The KLE analysis described in Sects. 4 and 5 separates correlated from uncorrelated signals.

The vertical GIA hinge line appears to run south of the St. Lawrence River in the east and then curls north through Lake Ontario. The vertical velocities are reasonably consistent with what might be expected starting at the St. Lawrence River and continuing northward, that is, uplift rates increase to the north in a generally linear fashion. In contrast, the southern stations do not show a coherent pattern of increased subsidence, perhaps owing to the smaller magnitude of the vertical velocities.

The horizontal velocities are more difficult to interpret. Like the vertical velocities, the stations to the north of the St. Lawrence show a coherent behavior, a generally southeastward motion, on the order of 1 mm/year. However, south of the St. Lawrence the motion becomes much more variable, both in direction and absolute value. Some of these stations, such as CARM, have very short time series, but this cannot explain outliers such as the motion of ANNE, which is several times larger than the overall average and approximately perpendicular to the

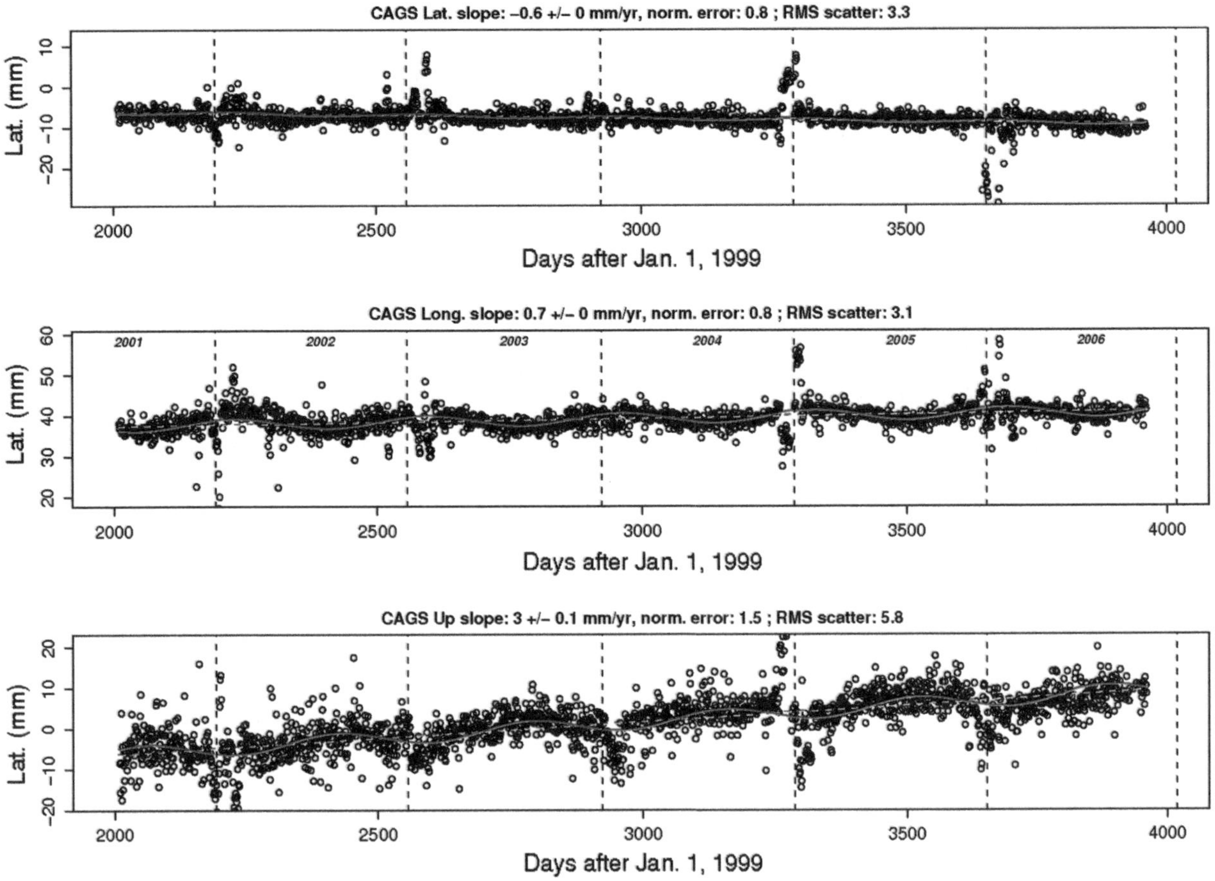

Figure 3
Filtered time series for CGPS station CAGS

general pattern of motion. A correlation principal component analysis was performed on the data set in order to further isolate and identify potential sources of these variations and patterns.

4. *PC Analysis*

The common mode filtering described above is not easily adapted to account for correlations in the time series. Here we show that the decomposition of GPS data in eastern Canada and the USA into its spatial eigenvectors and associated temporal signature can be used to separate out various modes based upon the important spatial and temporal scales. This technique, specifically the KLE analysis, is similar to the empirical orthogonal function (EOF) analysis frequently employed in the atmospheric sciences (PREISENDORFER 1988).

The Karhunen–Loeve method is a linear decomposition technique in which a dynamical system is decomposed into a complete set of orthonormal subspaces. It has been applied to a number of complex nonlinear systems over the last 50 years, including the ocean–atmosphere interface, turbulence, meteorology, biometrics, statistics, and geophysics (HOTELLING 1933; FUKUNAGA 1970; AUBREY and EMERY 1983; PREISENDORFER 1988; SAVAGE 1988; PENLAND 1989; VAUTARD and GHIL 1989; POSADAS *et al.* 1993; PENLAND and SARDESHMUKH 1995; HOLMES *et al.* 1996; MOGHADDAM *et al.* 1998; TIAMPO *et al.* 2002; DONG *et al.* 2006; MAIN *et al.* 2006; SMALL and ISLAM 2007; SMITH *et al.* 2007). In one of the first successful geodetic

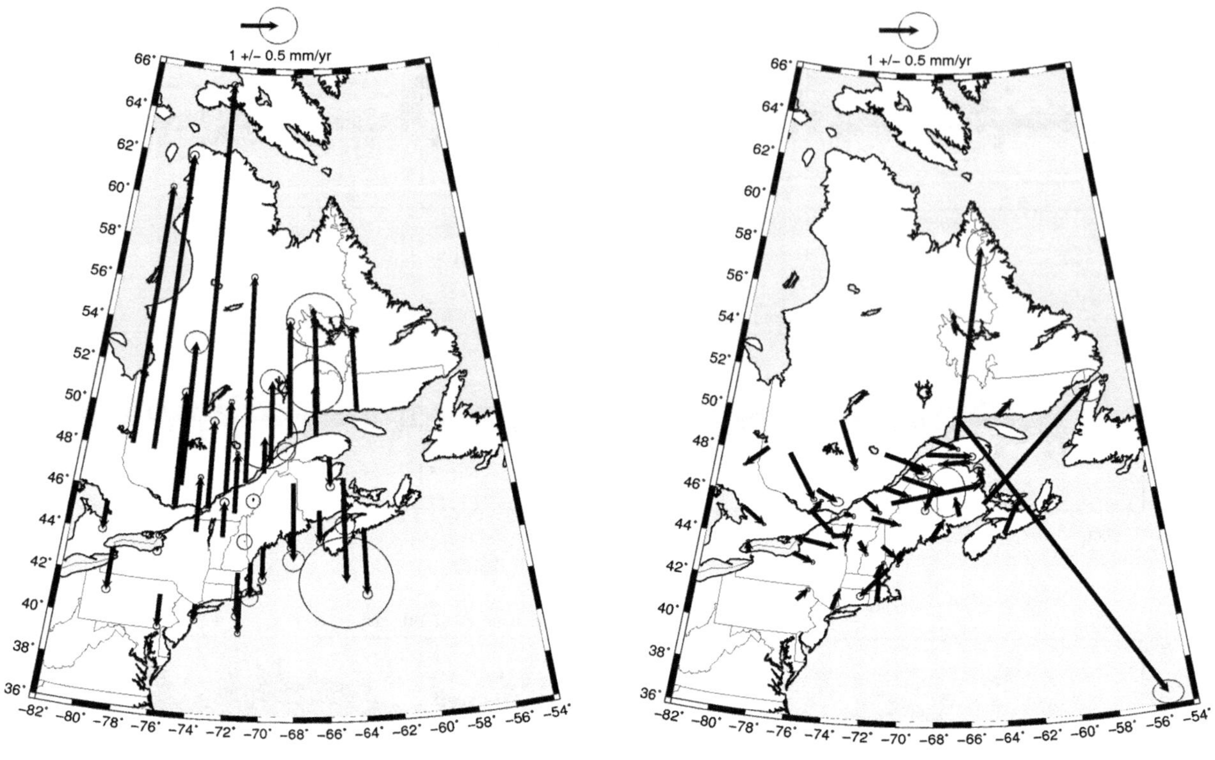

Figure 4
Velocities for filtered CGPS data, in millimetres per year, for January 2002 through June 2006. Vertical velocities are shown on the *left* (north is up) and horizontal velocities on the *right*. Error ellipses show the one sigma error on the estimated velocities

applications, Savage (1988) decomposed the deformation at Long Valley caldera into its predominant modes in order to study only the signal that accounted for the greatest percentage of the variance, the volcanic source below the dome. In addition, he identified the primary error sources in the data using the remaining eigenmodes.

4.1. Method

In an application of the KLE procedure to historic seismic data, Tiampo *et al.* (2002) constructed a correlation operator, $C(x_i,x_j)$, for seismic events over time. Subsequently, $C(x_i,x_j)$ is decomposed into its orthonormal spatial eigenmodes and their associated time series, $a_j(t)$. These spatial–temporal pattern states were used to reconstruct the primary modes of the system, with or without noise, and to characterize the underlying dynamics and the physical parameters such as stress levels and interactions that control the observable patterns of events. The decomposition implicitly assumes that one is dealing with a process that is both Markov and stationary in time. Anghel applied a similar methodology to modeled deformation data in order to capture the coherent structures and their interactions (Anghel 2001; Anghel *et al.* 2004). Tiampo *et al.* (2004) applied the KLE technique to SCIGN data in order to determine the principal modes of deformation for the southern California fault system, while Dong *et al.* (2006) applied a similar decomposition technique to SCIGN data in order to study the CME.

Similar to the EOF technique developed by Preisendorfer (1988) for the atmospheric sciences, the KLE is obtained from the p time series that record the deformation history at particular locations in space. Unlike an EOF analysis, the KLE decomposition is performed on a correlation operator, not a covariance matrix (Fukunaga 1970).

Table 2

Filtered station velocities and associated 1-sigma error (mm/year)

No.	Station	Vertical	Error	Latitude	Error	Longitude	Error
1	ANNE	1.363	0.717	5.260	0.484	0.739	0.367
2	ATRI	2.507	0.085	−0.218	0.042	0.680	0.043
3	BAIE	3.087	0.102	−0.274	0.054	0.719	0.049
4	BARH	−0.324	0.079	0.561	0.027	0.349	0.031
5	BARN	−0.013	0.206	−0.296	0.044	0.370	0.047
6	BRU1	−0.921	0.105	−0.381	0.040	−0.102	0.043
7	CAGS	3.088	0.106	−0.644	0.062	0.680	0.060
8	CAPL	−0.793	0.108	−0.039	0.033	−0.754	0.053
9	CARM	−2.112	0.283	−0.608	0.098	−0.222	0.096
10	CHIB	9.086	0.084	−1.257	0.076	0.475	0.047
11	CHIC	4.729	0.089	−0.390	0.053	1.125	0.040
12	ESCU	−1.277	0.258	0.405	0.137	−0.110	0.124
13	GEOR	−0.076	0.193	0.571	0.070	2.280	0.080
14	HLFX	−1.803	0.129	0.856	0.040	0.392	0.065
15	HSTP	2.217	0.095	0.387	0.039	0.388	0.058
16	HULL	1.301	0.074	0.358	0.033	0.232	0.050
17	KNGS	−0.089	0.074	−0.164	0.029	1.080	0.034
18	LAMT	−0.426	0.085	0.542	0.032	0.162	0.034
19	LAUR	3.916	0.308	−0.285	0.114	0.539	0.162
20	LOUP	2.222	0.337	−0.167	0.150	0.346	0.196
21	LPOC	0.972	0.852	−0.367	0.676	1.128	0.481
22	MCTN	−2.277	1.238	3.125	0.440	2.865	0.430
23	MONT	2.388	0.109	−0.064	0.066	0.246	0.041
24	NPRI	−1.016	0.074	0.634	0.034	0.034	0.029
25	OSPA	−0.079	0.112	−0.194	0.057	0.522	0.047
26	PARY	−0.760	0.101	−0.416	0.036	0.617	0.046
27	POR4	−0.930	0.239	−0.786	0.092	−0.702	0.092
28	PSC1	1.476	0.070	0.073	0.024	0.471	0.032
29	PWEL	−1.161	0.108	0.303	0.034	−0.054	0.044
30	RIMO	−0.040	0.241	−0.039	0.094	1.177	0.097
31	ROUY	7.089	0.079	−0.554	0.033	−0.534	0.048
32	SEPT	2.543	0.740	−7.631	0.316	5.355	0.405
33	SRBK	1.624	0.057	−0.159	0.028	0.716	0.043
34	TRIV	2.483	0.070	−0.398	0.042	0.456	0.050
35	UNB1	−0.852	0.092	0.474	0.030	−0.096	0.038
36	VALD	8.100	0.113	−1.171	0.061	0.771	0.078
37	VCAP	0.936	0.119	−0.331	0.043	0.204	0.045
38	WES2	−1.196	0.104	0.420	0.046	0.192	0.055
39	WIL1	−0.864	0.088	0.304	0.049	0.265	0.058

Each time series, $y(x_s,t_i) = y_i^s$, $s = 1, \ldots p$, consists of n time steps, $i = 1, \ldots n$. The goal is to construct a time series for each of a large number of locations for a given short period of time. If, for example, the time interval was decimated into units of days, the result could be a time series of 365 time steps for every year of data, with values of position for that location at each time step. These time series are incorporated into a matrix, $\boldsymbol{T}$, consisting of time series of the same measurement for p different locations, i.e.,

$$T = [\bar{y}_1, \bar{y}_2, \ldots \bar{y}_p] = \begin{bmatrix} y_1^1 & y_1^2 & \cdots & y_1^p \\ y_2^1 & y_2^2 & \cdots & y_2^p \\ \vdots & \vdots & \ddots & \vdots \\ y_n^1 & y_n^2 & \cdots & y_n^p \end{bmatrix}. \qquad (1)$$

For analysis of GPS data, the values in the matrix $\boldsymbol{T}$ consist of horizontal or vertical position measurements. The covariance matrix, $S(x_i,x_j)$, for these events is formed by multiplying $\boldsymbol{T}$ by $\boldsymbol{T}^T$, where S is a $p \times p$ real, symmetric matrix. The covariance matrix, $S(x_i,x_j)$, is converted to a correlation operator, $C(x_i,x_j)$, by dividing each element of $S(x_i,x_j)$, by the variance of each time series, $y(x_i,t)$ and $y(x_j,t)$,

$$\sigma_p = \sqrt{\frac{1}{n}\sum_{k=1}^{n} (y_k^p)^2}, \qquad (2)$$

and

$$C = \begin{bmatrix} \frac{s_{11}}{\sigma_1\sigma_1} & \frac{s_{12}}{\sigma_1\sigma_2} & \cdots & \frac{s_{1p}}{\sigma_1\sigma_p} \\ \frac{s_{21}}{\sigma_2\sigma_1} & \frac{s_{22}}{\sigma_2\sigma_2} & \cdots & \frac{s_{2p}}{\sigma_2\sigma_p} \\ \vdots & \vdots & \ddots & \vdots \\ \frac{s_{p1}}{\sigma_p\sigma_1} & \frac{s_{p2}}{\sigma_p\sigma_2} & \cdots & \frac{s_{pp}}{\sigma_p\sigma_p} \end{bmatrix}. \qquad (3)$$

This equal-time correlation operator, $C(x_i,x_j)$, is decomposed into its eigenvalues and eigenvectors in two parts. The first employs the trireduction technique to reduce the matrix $\boldsymbol{C}$ to a symmetric tridiagonal matrix, using a Householder reduction. The second part employs a QL algorithm to find the eigenvalues, λ_j, and eigenvectors, e_j, of the tridiagonal matrix (Press *et al.* 1992). These eigenvectors, or eigenstates, are orthonormal basis vectors arranged in order of decreasing variance that reflect the spatial relationship of events in time. If one divides the corresponding eigenvalues, λ_j, by the sum of the eigenvalues, the result is that percent of the correlation accounted for by that particular mode. We then reconstruct the time series associated with each location for each eigenstate by projecting the initial data back onto these basis vectors in what is called a PC analysis (Preisendorfer 1988). These time dependent expansion coefficients, $a_j(t)$, which represent temporal eigenvectors, are reconstructed by multiplying the original data matrix by the eigenvectors, i.e.,

$$a_j(t_i) = \vec{e}^T \cdot T = \sum_{s=1}^{p} e_j y_i^s, \qquad (4)$$

where j, $s = 1,\ldots p$ and $i = 1, \ldots n$. This eigenstate decomposition technique produces the orthonormal spatial eigenmodes for this nonlinear threshold system, e_j, and the associated principal component time series, $a_j(t)$. These principal component time series represent the signal associated with each particular eigenmode over time. For purposes of clarity, the spatial eigenvectors are designated KLE modes and the associated time series PC vectors.

The EOF analysis is often used to filter data through the identification of those modes associated with large percentages of unwanted covariance or those lower modes accounting for random noise (PREISENDORFER 1988; PENLAND 1989; DONG *et al.* 2006). Others have applied the KLE technique to investigate spatiotemporally correlated geophysical signals in the position time series (SAVAGE 1988; POSADAS *et al.* 1993; TIAMPO *et al.* 2002; MAIN *et al.* 2006; SMITH *et al.* 2007), such as mass loading caused by seasonal deformation (TIAMPO *et al.* 2004). For example, DONG *et al.* (2002) and KEDAR *et al.* (2003) demonstrated that some seasonal effects are due to systematic errors in the daily CGPS time series. The first few PCs often represent the biggest contributors to the variance of the network residual time series and the higher-order PCs are related to local site effects. Such a PC approach was successfully applied to the decomposition of geodetic data for the study of interseismic deformation (SAVAGE 1988; SAVAGE and LANGBEIN 2008; SCHERNECK *et al.* 2000; PARKER 2001) and regional filtering (JOHANSSON *et al.* 2002).

4.2. GPS Data

Here, the vertical and horizontal time series are treated as two different data sets. After removing the mean from every time series, for the time period January 1, 2002, to June 30, 2006, the time series matrices T_z and T_{ll} were formed for the vertical and horizontal data, respectively. The resulting T_z matrix has dimensions 39 by 1,718, and the resulting correlation matrix, C_z, represents the correlations between the vertical components at all 39 stations. After decomposing C_z into its corresponding eigenvectors and eigenvalues, ordered by descending eigenvalues, each of the eigenvectors, or KLE eigenmodes, represents the spatial correlations in the data. Generally, those modes with the longest spatial wavelengths tend to occur at the lowest KLE modes, or largest eigenvalues, as these correspond to the greatest correlation values. Figure 5a shows the eigenvalue plot for the vertical decomposition. The associated reconstructed PC modes are formed by multiplying the KLE modes by the original time series matrix, T_z (Eq. 4). These are the time series for each of these modes which, when multiplied by the appropriate eigenmode value for a particular station, give the time series for that mode at each individual station. Velocities are calculated, using a standard linear regression technique, for each of these reconstructed time series, and the formal standard error is estimated from that fit.

In order to correlate the horizontal components at every station, the T_{ll} matrix has dimensions of 78 by 3,436. The first 39 columns consist of the latitude time series followed by the longitude time series for each station, and the second 39 columns are the longitude time series followed by the latitude time series at those same stations. The resulting correlation matrix, C_{ll}, is of dimension 78 by 78, and the resulting KLE eigenmodes, after decomposition, again represent the correlations between the horizontal components. The eigenvalues for this decomposition are shown in Fig. 5b. Velocities at each station for each KLE mode were calculated for these reconstructed time series.

4.3. Model Error

The error associated with the KLE decomposition is normally calculated in two different ways. First, the reconstruction error associated with the decomposition itself can be calculated by projecting the eigenvectors onto the PC vectors and calculating the difference between this projection and the actual data. The square root of the sum of the absolute value of this difference is the L2-norm on the reconstruction, and is called the reconstruction error (PREISENDORFER 1988). Effectively, it is the root-mean-square difference between the original data and the reconstructed data created using the eigenvector decomposition and

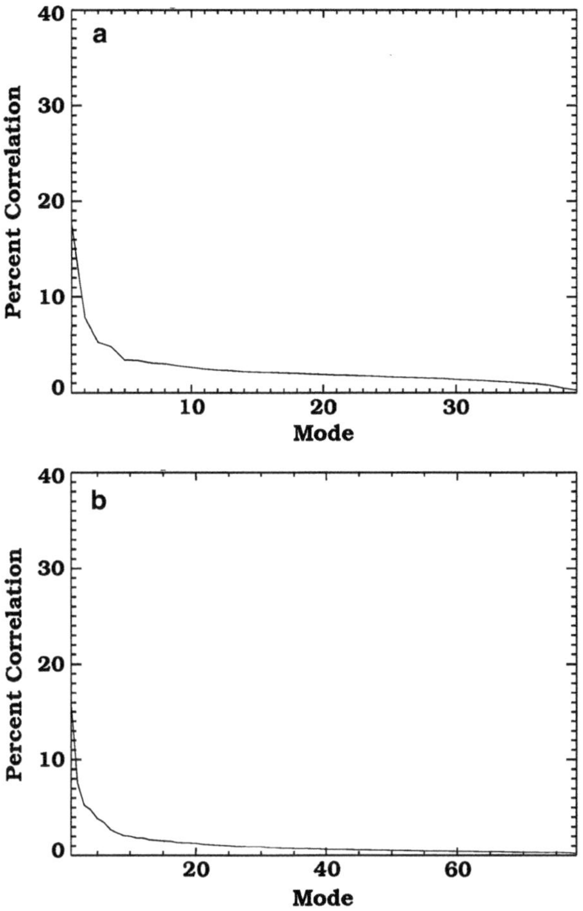

Figure 5
Eigenvalues for **a** vertical and **b** horizontal KLE modes, filtered data set

the associated temporal modes. The average value of this reconstruction error for all position time series is 1.2% in the vertical and 1.7% in the horizontal.

In addition, the percentage of the variance in the data accounted for by each mode can be directly related to the normalized eigenvalues. For a correlation decomposition, the covariance assigned to each mode of the decomposition is equivalent to the squared eigenvalue, λ_j^2. The normalized eigenvalues, $\lambda_j/\Sigma\lambda$, therefore, represent that percentage of the root-mean-square error accounted for by the *j*th mode (Fancourt and Principe 1998; Preisendorfer 1988; Smith *et al.* 2007) and are shown in Fig. 5 and tabulated in Tables 3 and 4. Note that results are shown for only the first 39 horizontal modes (Table 4), as the percent variance accounted for by the last 39 modes was less than 1/1,000 of a percent in every case. In addition, Tables 3 and 4 list the percent variance associated with each mode as well. It can be seen that the first two modes of both the vertical and horizontal decompositions account for more than 65% of the variance in each data set. This information must be taken into account in the analysis used to reconstruct the most significant KLE modes, as well as the geophysical signals that may be responsible for these modes.

5. Results

5.1. Vertical Decomposition

Figure 6 shows the PC time series at station CAGS for first three vertical displacement KLE modes. CAGS was chosen to illustrate the pattern of the reconstituted time series for two reasons. First, it is a long-running IGS reference station operated by the Geodetic Survey Division of Natural Resources Canada (NRCan). Second, it is located north of the Saint Lawrence River and the vertical uplift is expected to be significant. Velocities at each station are reconstructed for each mode by performing a linear regression on the PC time series. The associated standard error ellipses are estimated from the RMS errors on the linear fit to the estimated velocities. Again, these likely underestimate the true error on the CGPS data, but are useful for comparison with the original estimates in Fig. 4. Figure 7 shows the spatial pattern associated with the vertical velocity maps for the first three KLE modes of filtered CGPS data, as shown in Fig. 6.

The first mode is the primary vertical velocity associated with postglacial rebound. The magnitude of the second and third modes is approximately one-tenth of that seen in the first mode. The second mode, examined closely, appears to delineate a residual vertical velocity associated with an east–west tilt. As expected, this mode has a smaller spatial wavelength, approximately half that of the first mode. The PC time series for the third mode (Fig. 6) contains a seasonal signal that may be related to atmospheric pressure, moisture, or some combination as has been observed in other regional networks (Dong *et al.* 2002; Watson *et al.* 2002). The accompanying spatial pattern appears largely random. For the remaining

Table 3

Percent error (vertical decomposition)

Mode no.	Normalized λ	% Variance
1	17.85	56.05
2	7.86	10.87
3	5.26	4.87
4	4.78	4.02
5	3.40	2.03
6	3.35	1.97
7	3.09	1.67
8	3.00	1.59
9	2.78	1.36
10	2.64	1.23
11	2.48	1.08
12	2.37	0.99
13	2.30	0.93
14	2.18	0.83
15	2.13	0.80
16	2.10	0.77
17	2.04	0.73
18	2.01	0.71
19	1.94	0.66
20	1.89	0.63
21	1.83	0.59
22	1.80	0.57
23	1.75	0.54
24	1.70	0.51
25	1.63	0.47
26	1.59	0.44
27	1.55	0.42
28	1.51	0.40
29	1.47	0.38
30	1.38	0.33
31	1.33	0.31
32	1.28	0.29
33	1.19	0.25
34	1.12	0.22
35	1.02	0.18
36	0.94	0.15
37	0.76	0.10
38	0.44	0.03
39	0.26	0.01

Table 4

Percent error (horizontal decomposition)

Mode no.	Normalized λ	% Variance
1	16.17	57.28
2	7.55	12.51
3	5.27	6.08
4	4.76	4.97
5	3.86	3.27
6	3.46	2.63
7	2.70	1.60
8	2.36	1.22
9	2.07	0.94
10	2.02	0.89
11	1.83	0.73
12	1.81	0.72
13	1.64	0.59
14	1.59	0.55
15	1.52	0.51
16	1.51	0.50
17	1.36	0.41
18	1.32	0.38
19	1.31	0.37
20	1.28	0.36
21	1.18	0.31
22	1.14	0.28
23	1.10	0.26
24	1.06	0.25
25	1.03	0.23
26	0.98	0.21
27	0.95	0.20
28	0.94	0.20
29	0.94	0.19
30	0.93	0.19
31	0.86	0.16
32	0.84	0.16
33	0.79	0.14
34	0.78	0.13
35	0.76	0.13
36	0.74	0.12
37	0.74	0.12
38	0.72	0.11
39	0.70	0.11

KLE modes (not shown), the random behavior becomes more pronounced, as expected.

5.2. Horizontal Decomposition

The PC time series at CAGS for the first six KLE modes of the horizontal decomposition are shown in Fig. 8. The associated horizontal velocity maps are shown in Fig. 9. Figure 8 shows only the latitude component of the time series because the longitude components have exactly the same pattern over time, but with a different magnitude and direction. Like the vertical velocity field, the lower modes display significant spatial coherence that decreases with increasing mode number. In addition, the seasonal (meteorological) signal is observed in more than one mode, including those with a significant linear trend. Particular modes can be identified that are predominantly associated with seasonal phenomena, such as mode one, which has a negligible net velocity. None of the signals corresponds to an apparent rigid rotation, so it is unlikely that there is a remnant

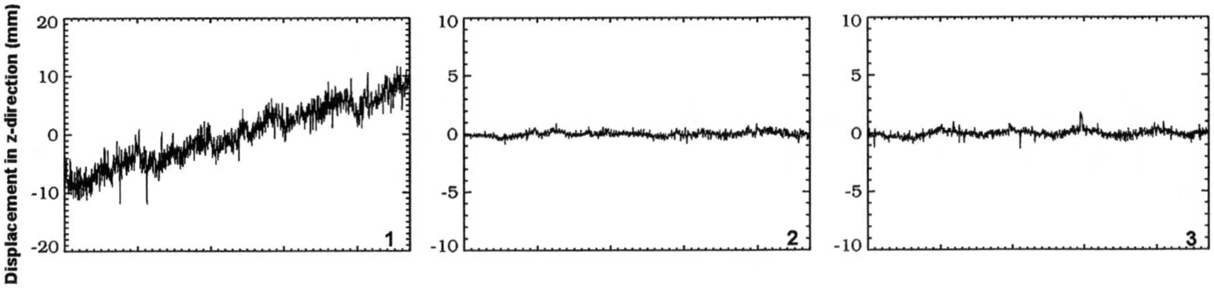

Figure 6
Principal component (PC) time series at CAGS for first three vertical displacement KLE modes (KLE mode 1 *left*; KLE mode 2 *center*; KLE mode 3 *right*)

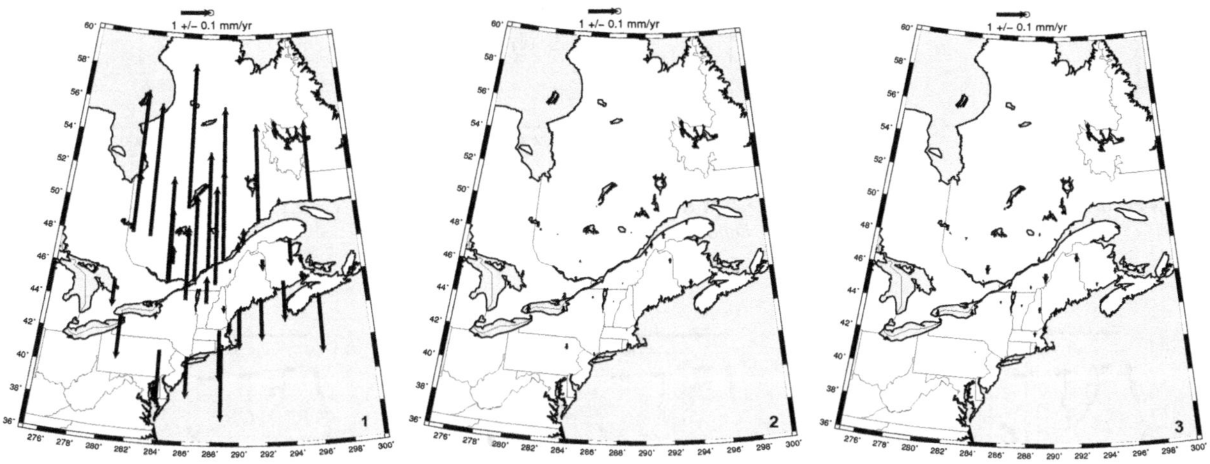

Figure 7
Vertical velocity maps for first three KLE modes of filtered CGPS data, January 2002 through June 2006, as shown in Fig. 6

contribution from the North American plate velocity in the filtered data.

Other interpretations are possible, but determining their significance in an objective manner is difficult. For example, the second mode also seems to be associated with relative east–west motion along the St. Lawrence River, while mode seven might be related to strain perpendicular to the seismic zone underlying the region and responsible for the seismicity shown in Fig. 1.

5.3. *Modal Recombination*

Frequently, the choice of modes to be used for removal of the random error is based upon a determination of the change in slope on the eigenvalues plot, as shown in Fig. 5 (Preisendorfer 1988). Another approach is to calculate an F test for each of the decompositions using the variances listed in Tables 3 and 4 to determine whether or not a particular mode is significant, relative to the remaining modes. For example, if the variances are arranged so that $\sigma_n^2 > \sigma_{n+1}^2$, we can test for significant differences in the variances. In this case, the null hypothesis states that the two variances we are comparing are from the same population (i.e., they are not statistically different). The F value is the ratio of the two variances, and if the calculated value is greater than that of the F-statistic for the degrees of freedom associated with σ_1^2 and σ_2^2, at a particular significance level, then the null hypothesis is not correct and they do not come from the same population.

We test if each mode comes from the same population as the remaining higher numbered, lower

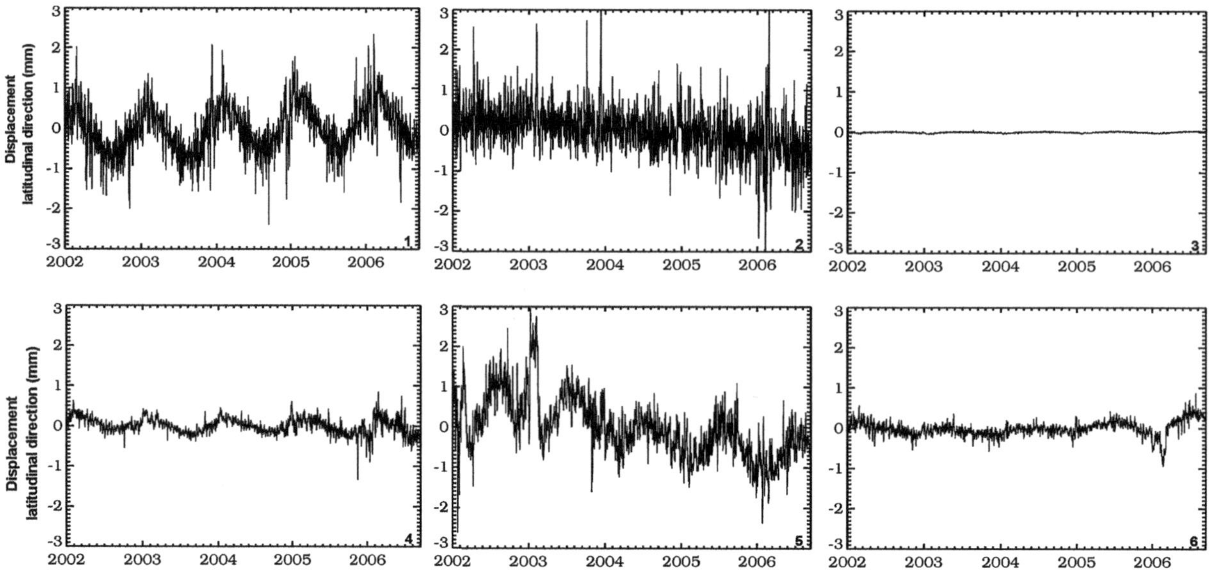

Figure 8
CAGS PC time series for first six horizontal displacement KLE modes in the latitudinal direction. The KLE mode number is shown on the *bottom right corner* of each panel

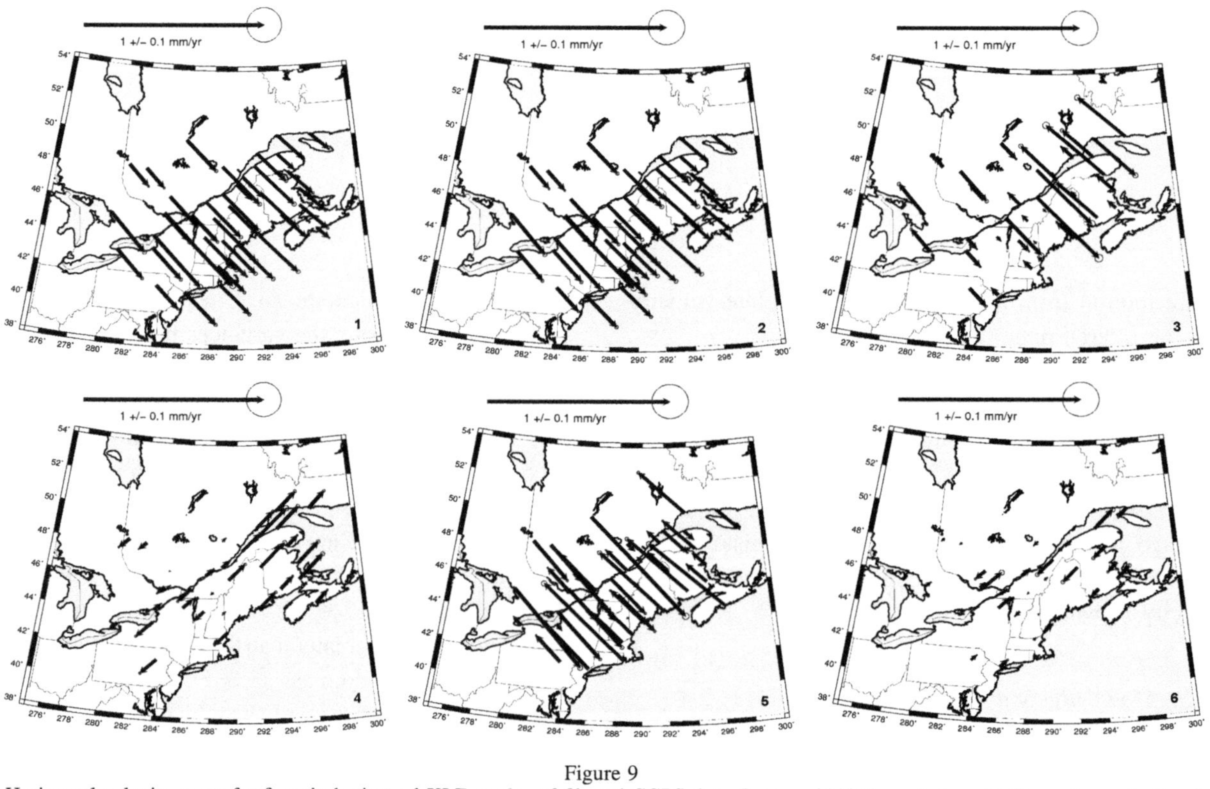

Figure 9
Horizontal velocity maps for first six horizontal KLE modes of filtered CGPS data, January 2002 through June 2006, as shown in Fig. 8

correlation modes. Our analysis is different from a variance decomposition because the modes represent stronger correlations, not necessarily higher variance. Consequently, the remaining modes, when summed, must represent the greatest portion of the random noise in the data. We are, therefore, testing to see which modes are significant above that random noise (Fancourt and Principe 1998; Smith *et al.* 2007). Remembering that the λ_j^2 corresponds to the covariance on each mode, as this is a correlation analysis, the *F* value for each consecutive mode is calculated as follows:

$$F(1, N-j) = \frac{\lambda_j^2}{\sum_{i=j+1}^{N} \lambda_i^2 / N - j}. \quad (5)$$

The resulting analysis shows that, for the vertical decomposition, only the first KLE mode, which accounts for more than 56% of the variance in the vertical time series, is significantly different from the remaining modes at the 95% level (higher mode number, lower correlation). At the 90% level, both modes one and two represent a signal that is significantly different from the remaining modes. For the horizontal decomposition, the first three modes are significant above the remaining modes at the 95% level, accounting for more than 75% of the variance in the horizontal position series. The first six horizontal modes are significant at the 90% probability level.

5.3.1 Vertical Recombination

The first vertical KLE mode is shown in Fig. 7. The velocities, again, are calculated from a linear regression to the PC time series, and the associated standard errors are plotted. *F* test results support the conclusion that this first mode contains the bulk of the correlation associated with the data and samples a different distribution than the remaining nodes, which are largely associated with random, poorly correlated noise. Figure 10 illustrates the reconstructed time series at PARY (Fig. 10a), vertical displacement KLE mode one from Fig. 7 (Fig. 10b), and the time series reconstructed without the first mode included (Fig. 10c). Effectively, Fig. 10c can be thought of as the time series data in Fig. 10a minus the time series data of Fig. 10b. This illustrates several important points. PARY was not fully operational on January 1, 2001. The reconstituted time series, with the mean removed, are an accurate representation of the data and the reconstruction error is small, as noted above. That early feature is clear, and the seasonal signal and random noise component, although small in the original CAGS time series (Fig. 3), is also seen in Fig. 10a. The corresponding vertical KLE mode one contains much less of the random noise component and seasonal signal, providing an effective filter. This is supported by the large random noise component and seasonal effect in the remainder signal, as seen in Fig. 10c. Figure 11 illustrates the variation in the reconstituted time series for this first KLE mode, here at BAIE and CARM. Here the velocity at BAIE is of the same order as that seen at CAGS (Fig. 6), while the velocity at CARM is, like PARY, in the opposite direction but with a smaller amplitude.

The velocity pattern for the first KLE mode, as shown in Fig. 7, is more regular, particularly in the coherence of the uplift pattern in the south, than the filtered data of Fig. 4. While the hinge line still runs approximately parallel to the St. Lawrence River in the east, the uplift in those stations further north is reduced as well. One surprising feature remains in this mode, a feature that existed in the original data as well. Both PARY and PWEL show significant subsidence over this time period, although GIA models of long-term motion generally suggest that PARY and PWEL should rise at rates approaching 2–4 and 0–1 mm/year, respectively (Fig. 12). Vertical GIA rates were computed using ICE-3G for a range of upper mantle viscosity (UMV) and lower mantle viscosity (LMV) of (1) 1×10^{21} Pa s and 2×10^{21} Pa s (commonly known as viscosity model 1—VM1), (2) 1×10^{21} Pa s and 4.5×10^{21} Pa s, (3) 4×10^{20} Pa s and 2×10^{21} Pa s (viscosity model 2—VM2), and (4) 4×10^{20} Pa s and 4.5×10^{21} Pa s.

Mainville and Craymer (2005) used water gauge data in the Great Lakes to constrain long-term GIA uplift rates. The rates shown in Fig. 7 are consistent with the general pattern of their results. They also find subsidence at PARY and PWEL of approximately 3 and 1 mm/year, respectively.

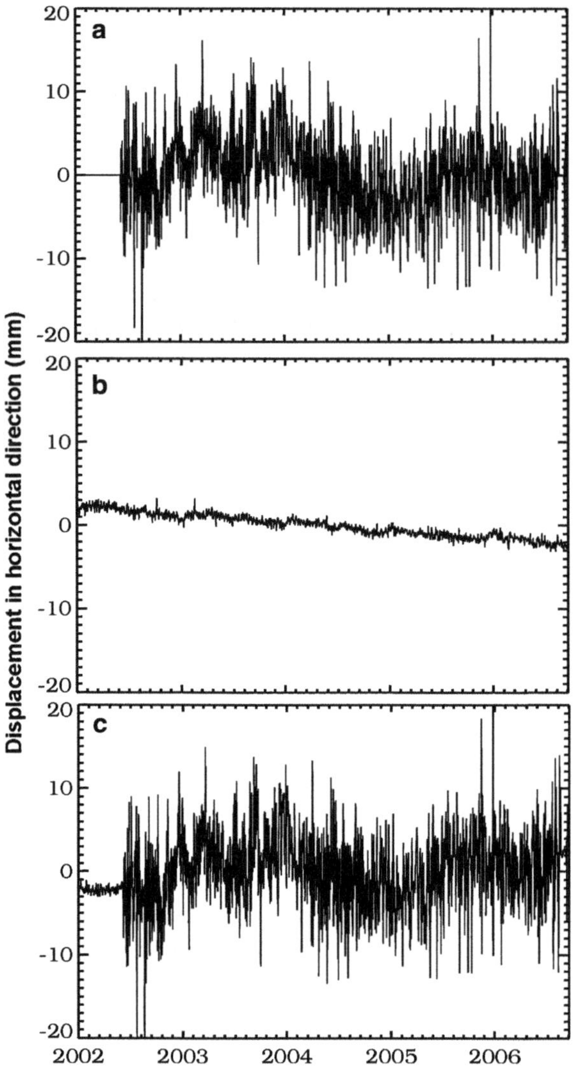

Figure 10
Vertical displacement time series at PARY for **a** the original filtered data, as shown in Fig. 3; **b** KLE mode 1 (see Fig. 6); and **c** all KLE modes except the first mode (KLE1)

One potential explanation for the discrepancy between the GPS/water gauge observations and the GIA model predictions is the recent persistent drought in the Great Lakes basin (cf. NOAA Great Lakes Environmental Research Laboratory, http://www.glerl.noaa.gov/data/now/wlevels/levels.html). For example, water levels in Lakes Michigan and Huron, which have been monitored since 1865, show a 27-year periodicity probably caused by the interaction of two decade long cycles originating in the North Atlantic region (Hanrahan *et al.* 2010). The most recent drought, which began in 1997 and continues with some fluctuations, has significantly lowered the lake levels in Lake Michigan, Lake Huron, and Lake Erie. In 2007, the Lake Michigan–Lake Huron system reached the lowest levels since the Dust Bowl droughts of the mid-1930s. Levels in Lake Huron have dropped as much as one meter since 1997 and fluctuated by approximately 20 cm between 2001 and 2006 (Conroy *et al.* 2009; Hoerling *et al.* 2009; Sellinger *et al.* 2008; Stow 2009). On the other hand, Lake Ontario levels rebounded after the 1997 low and remained relatively stable from 2000 through the present. These lake level changes potentially could affect the local groundwater levels, a phenomenon that has been shown to result in associated surface deformation detectable with CGPS networks, and affect those stations located adjacent to the lakes (Bawden *et al.* 2001; Watson *et al.* 2002). This effect is likely to be less significant for Lake Ontario, as its outflow is regulated by the St. Lawrence River. PARY, which is located on the eastern shore of Georgian Bay, is fastened to both bedrock and a pier at the water's edge. PWEL is slightly further from the shore of Lake Ontario and is anchored to a concrete pier (CSRS database, http://www.geod.nrcan.gc.ca/online_data_e.php).

The hypothesis that vertical motions of these stations are contaminated by ground surface subsidence associated with lowered groundwater levels from the recent drought (superimposed on the long-term GIA rates as recorded by water gauge data) is supported by the fact that both PARY and PWEL appear to have a more significant seasonal component than the other two Great Lake stations. Continued monitoring of these GPS stations is necessary in order to resolve this issue and should improve our knowledge of both the long-term GIA rates and the shorter-term climactic and anthropogenic effects.

Using this PC technique, we can begin to discriminate between the various viscosity models. Figure 12c (VM2 rheology) best fits both the pattern and magnitude of the filtered data for the first KLE mode. The low uplift rates to the north, the location of the hinge line, and the trends to the south are consistent with the data. It should be noted, however, that while Fig. 12a, d show the greatest differences from the result in Figs. 7 and 12b only deviates

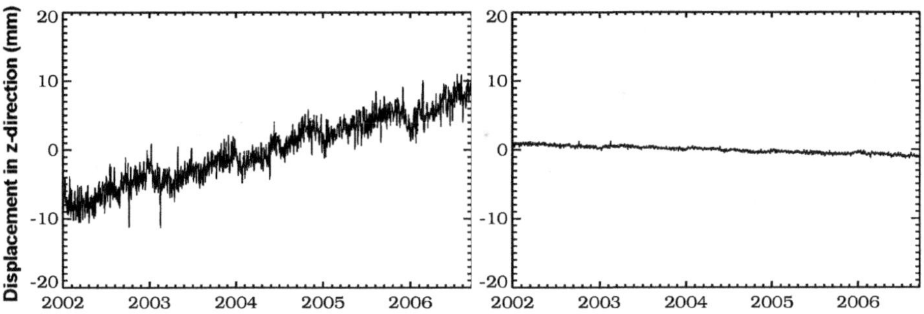

Figure 11
Reconstructed time series for vertical displacement KLE mode 1, as shown for CAGS in Fig. 10b, but for BAIE (*left*) and CARM (*right*)

significantly from the data in the large uplift rates to the north. Given the relatively short length of the data, and the evidence both here and in other CGPS networks that short-term temporal signals can significantly affect the data, it is not possible to definitively conclude which ICE-3G model best fits the data. However, it does provide evidence that CGPS networks in stable continental regions can be used to constrain vertical signals such as GIA over relatively short time periods, and good quality CGPS has the potential to differentiate between particular mantle viscosity models in the near future.

5.3.2 Horizontal Recombination

Figure 13 shows velocity maps for three sets of summed horizontal KLE modes. The three choices were generated by summing: (1) modes one through three, (2) modes two and three only, and (3) modes one through six. The first choice is based on the *F* test and includes the three modes that are significant at the 95% level. The second choice excludes the strongly seasonal mode 1, and thus may isolate other signals. The third choice sums all the modes that are significant at 90%.

For all three combinations, the amplitude of the velocity parallel to the St. Lawrence valley is small, in a narrow band just south of the river. If the first mode is removed (not shown here), that signal becomes even smaller, and in a strictly northeastward direction. In addition, the summation of modes 2 through 6 (also not shown here) produces a spatial velocity pattern almost identical to that seen for the sum of modes 2 and 3 (Fig. 13), and suggests that the trend parallel to the LSZ/CSZ is contained predominantly in mode one and presenting the possibility that it is related to some mechanism other than GIA, such as atmospheric loading effects. Finally, all three combinations remove the outliers and fluctuations in the southern portion of the network that obscured the coherent pattern in the original filtered data (Fig. 4).

Figure 14 shows the reconstructed time series for the sum of the latitudinal KLE modes corresponding to the velocity maps in Fig. 13 at CAGS and PARY. These plots reinforce the conclusion that mode one primarily contains a seasonal signal whose removal has little effect on the magnitude and direction of the horizontal velocities. The sum of modes 2 and 3 is primarily linear at CAGS, providing support for the interpretation that these are modes primarily associated with a horizontal GIA or tectonic signal. The physical influence of the varying modal amplitude at individual stations also is illustrated in this figure. For example, this leads to a much stronger cyclic pattern in the deformation at PARY than at CAGS, most likely due to proximity to the lake. It is apparent that the highly correlated seasonal signal bleeds through into several of these correlation-based modes and the associated reconstructions, although the greatest proportion exists in modes 1 and 5. Unlike the vertical GIA effect, those correlated geophysical sources *not* associated with what is frequently called common mode error can be of the same order or smaller than the correlated noise for the horizontal deformation, and these can occur at the same spatial wavelengths as the CME (Dong *et al.* 2006). This suggests that care should be taken in removing CME

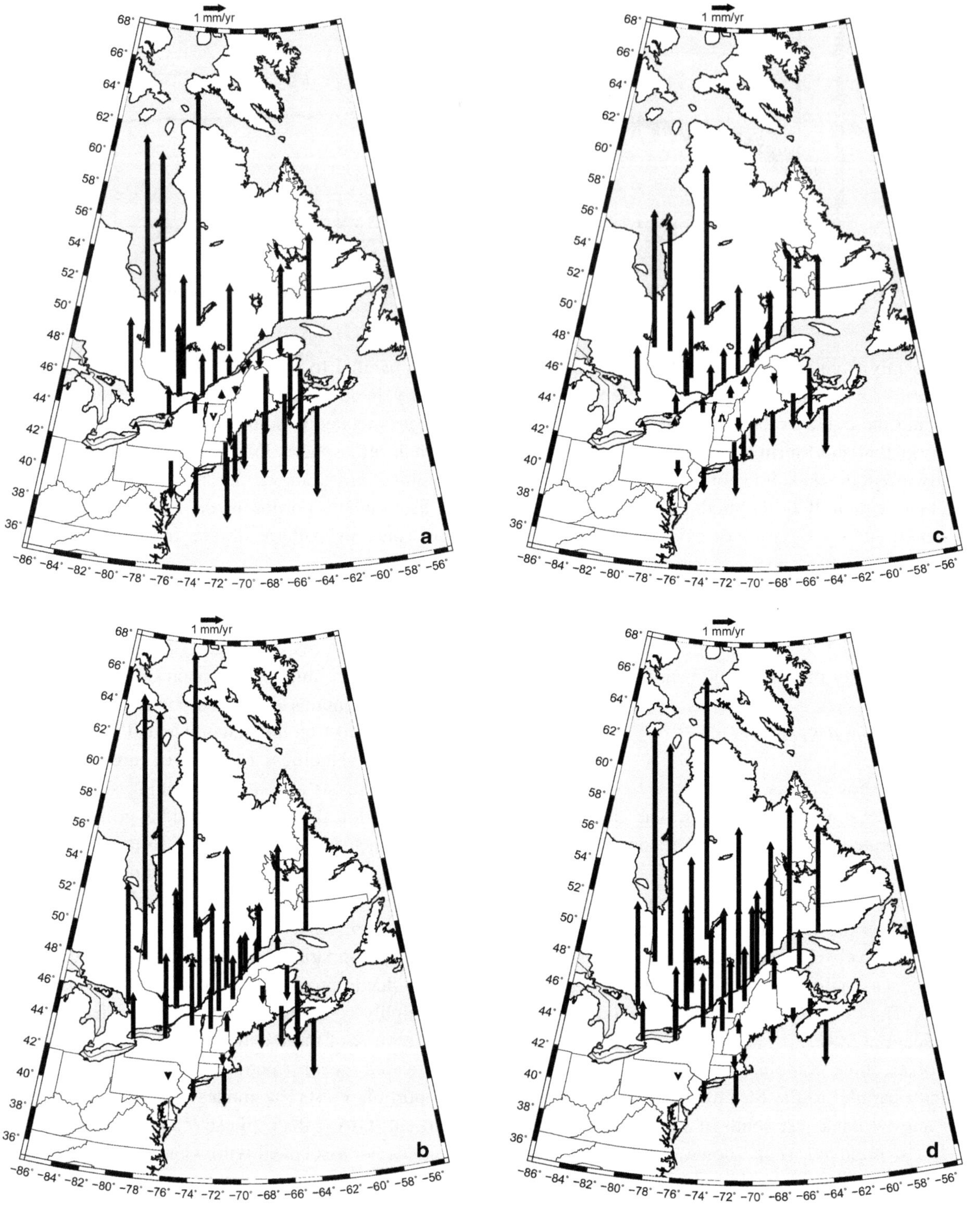

Figure 12
Vertical GIA rates computed using ICE-3G (Tushingham and Peltier 1991) for **a** upper mantle velocity (UMV) of 1×10^{21} Pa s and lower mantle velocity (LMV) of 2×10^{21} Pa s (VM1); **b** UMV of 10^{21} Pa s and LMV of 4.5×10^{21} Pa s; **c** UMV of 4×10^{20} Pa s and LMV of 2×10^{21} Pa s; and **d** UMV of 4×10^{20} Pa s and LMV of 4.5×10^{21} Pa s (VM2)

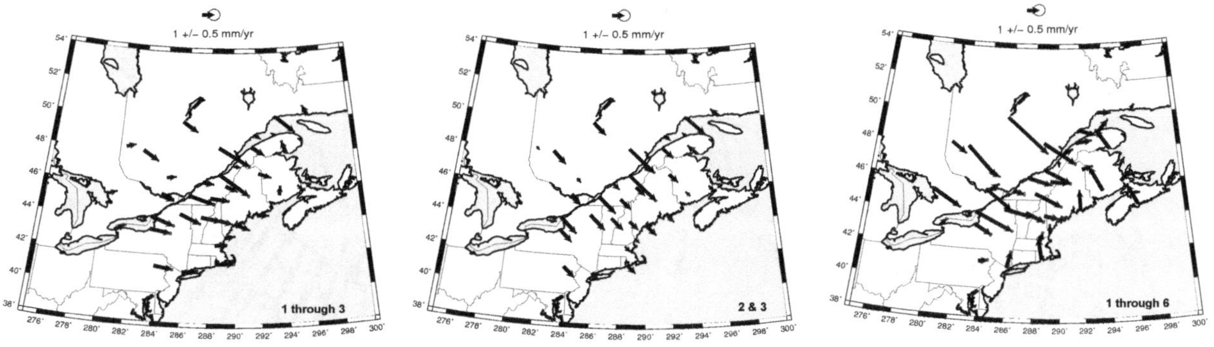

Figure 13
Velocity map for summed horizontal KL modes 1 through 3 (*left*); 2 and 3 (*center*); and 1 through 6 (*right*), as shown in Figs. 8 and 9

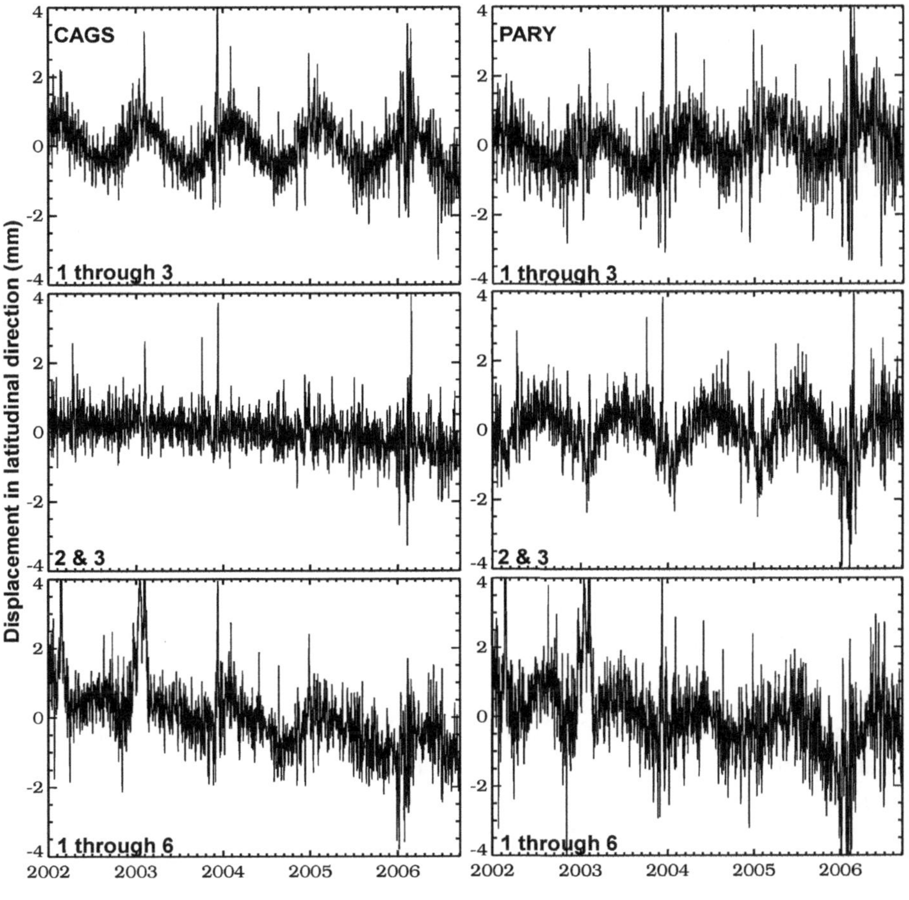

Figure 14
Reconstructed time series for the sum of latitudinal displacement KL modes 1 through 3 (*top*); 2 and 3 (*middle*); and 1 through 6 (*bottom*); as shown in Fig. 12, for CAGS (*left*) and PARY (*right*)

when there is a significant possibility that there exists a strongly correlated geophysical signal with large spatial wavelengths, as in this case.

Figure 15 shows the horizontal GIA rates computed using ICE-3G and the range of viscosity profiles described in Sect. 5.2 above. The model

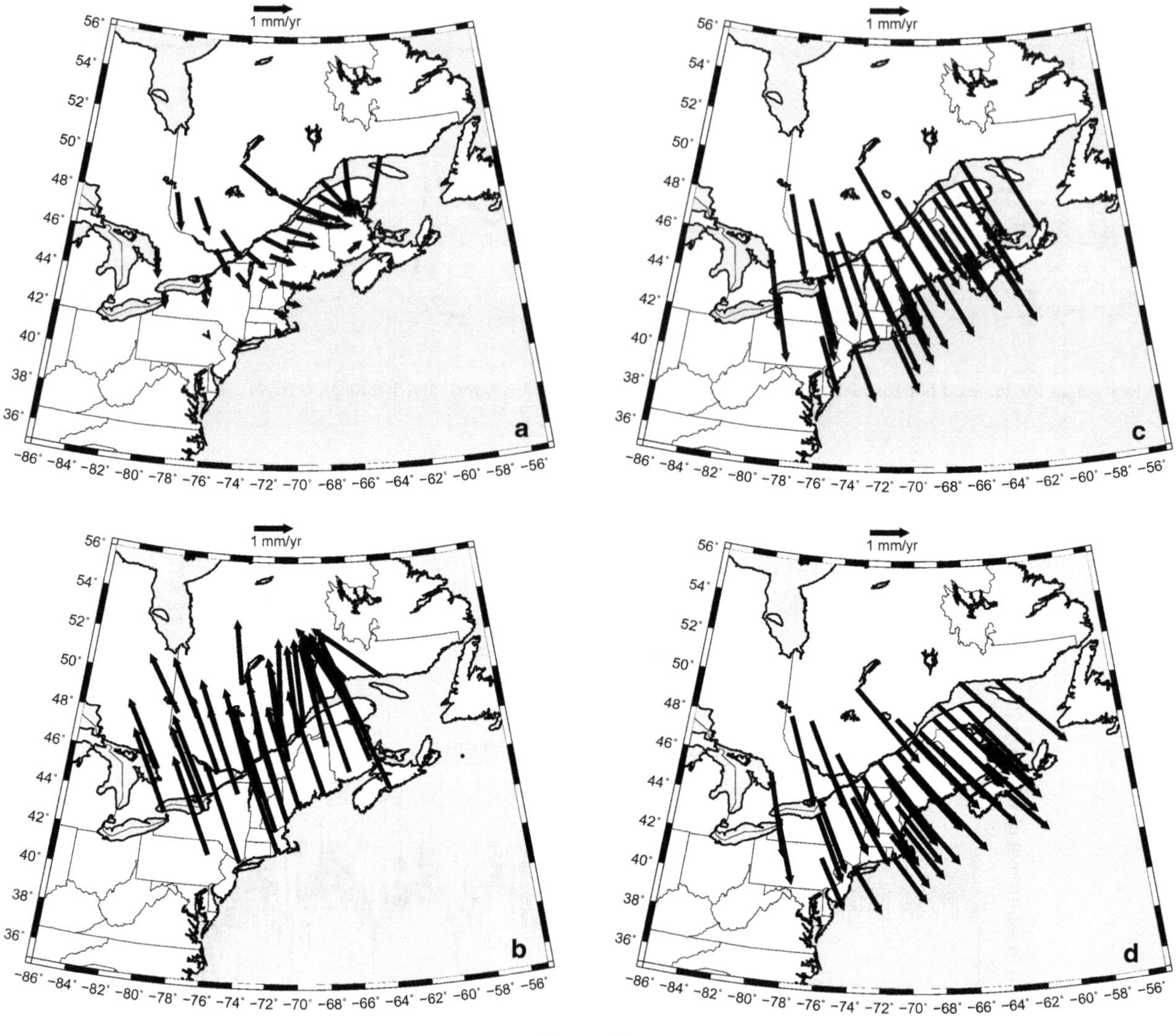

Figure 15

Horizontal GIA rates computed using ICE-3G for **a** UMV of 1 × 10^{21} Pa s and LMV of 2 × 10^{21} Pa s (VM1); **b** UMV of 1 × 10^{21} Pa s and LMV of 4.5 × 10^{21} Pa s; **c** UMV of 4 × 10^{20} Pa s and LMV of 2 × 10^{21} Pa s; and **d** UMV of 4 × 10^{20} Pa s and LMV of 4.5 × 10^{21} Pa s (depth-averaged VM2)

denoted VM1 matches the pattern of the first three summed horizontal KLE modes quite well. However, if the summed KLE modes 2 and 3 alone are considered (Fig. 13), the spatial pattern of Fig. 15c presents a second possibility. Again, as noted above, it is possible that the velocity parallel to the St. Lawrence is related to some other mechanism, potentially tectonic, hydrologic or climactic. For either Fig. 15a or c, the region where the horizontal velocities changes direction in the actual data lies further to the northwest than that of either model. While the magnitude of the reconstruction containing the first three modes is of the order of VM1, the recombination of modes 2 and 3, or any other recombination that lies in that direction (not shown here), is significantly smaller than that estimated by the models.

6. Conclusions

Standard, filtered analysis of CGPS position time series from 39 eastern North American stations show that the vertical velocities range from approximately

2 mm/year subsidence in the south to more than 8 mm/year in the north. After removal of the North American plate velocity, the horizontal velocities show a general southeastward trend that is generally on the order of 1–2 mm/year, but with large variation and significant outliers.

A KLE decomposition analysis performed on the filtered CGPS data identifies those modes with the most significant correlations in the data, at decreasing spatial wavelengths. The associated PC time series can be used to study the potential geophysical sources for those correlated modes, and significance testing allows for the classification of those modes which contribute to the non-random component of the signal. In the vertical decomposition, the first mode is significant at 95% and accounts for approximately 56% of the variance in the data. For the horizontal time series, the first three modes are significant at 95% and account for more than 75% of the variance.

The decomposition of the vertical signal partitions the correlated seasonal signal into a separate mode so that reconstruction of the first vertical KLE mode generates a vertical velocity field in general agreement with previous GPS analyses (Henton *et al.* 2006; Calais *et al.* 2006) and ICE-3G GIA model predictions. The viscosity structure commonly known as VM2 provides the best fitting solution. In the future, with longer CGPS time series, data assimilation using comprehensive inversion modeling will allow for significant, steady improvement in GIA models.

A feature that remains to be explored using these models is the unexpected shorter wavelength tilt pattern revealed by the second vertical KLE mode. For example, performing a similar KLE decomposition on synthetic data produced by GIA models such as ICE-3G may illuminate the particular parameters or shorter wavelength features that could produce a tilt mode with this spatial pattern and amplitude.

The horizontal KLE decomposition distributes the seasonal and atmospheric signals into several modes. Three possible reconstructions of these horizontal KLE modes were examined, using different combinations of the first three modes. Separation of the second and third modes from mode one, which contains the long wavelength correlation associated with regional atmospheric and seasonal signal, reveals a linear signal with a constant slope and a small amount of residual seasonal signal at better quality stations such as CAGS. Longer time series will likely be necessary in order to better differentiate these signals from local or regional signals such as tectonic strain across the LSZ/CSZ.

For all modal combinations the hinge line for the horizontal decomposition is approximately consistent with that of the vertical data. Significantly, the amplitude of the horizontal reconstructions is on the order of 1 mm/year. This is similar to previous results (Calais *et al.* 2006; Henton *et al.* 2006; Sella *et al.* 2007), but the results shown here have greater spatial coherence. In addition, the amplitude is several times smaller than that estimated by GIA models (James and Lambert 1993; James and Bent 1994; Peltier 1998, 2002; Peltier and Drummond 2009). The magnitude of VM1 predictions is similar to the observed horizontal velocities, but the change in the direction of horizontal velocities is too far north. Recent work by Peltier and Drummond (2009) suggests that variations in viscoelastic structure of the upper mantle can have a significant impact on the pattern and magnitude of GIA deformation.

Here, the KLE technique is successful in disaggregating the linear vertical signal from the less correlated and more variable signals. The horizontal decomposition, while not successful in isolating one linear signal, does begin to differentiate between the signals based upon their spatial coverage and provides the means to qualitatively identify and study the potential sources.

Acknowledgments

Research by KFT was funded by an NSERC Discovery Grant. The GPS data used in this study were provided by the National Geodetic Survey (USA) Continuously Operating Reference Stations (CORS), Natural Resources Canada (NRCan) Geodetic Survey Division (GSD), and the Province of Quebec Ministère des Resources Naturelles et Forêts (MRNF). GPS data archiving and processing was supported by the NRCan Canadian Crustal Deformation Service (CCDS). This is ESS contribution number 20110184. Images were plotted with the help of GMT software developed and supported by Paul Wessel and Walter H.F. Smith.

References

Adams, J. and P. W. Basham (1991), The seismicity and seismotectonics of eastern Canada, in *Neotectonics of North America,*

Adams, J., and S. Halchuk (2003), *Fourth generation seismic hazard maps of Canada: Values for over 650 Canadian localities intended for the 2005 Building Code of Canada*, Geol. Surv. Can. Open File, 4459, 155 pp.

Adams, J., P. W. Basham, and S. Halchuk (1995), *Northeastern North America earthquake potential—New challenges for seismic hazard mapping* Geol. Surv. Can. Current Res., 1995-D, 91–99.

Altamimi, Z., P. Sillard, and C. Boucher (2002), *ITRF2000: a new release of the International Terrestrial Reference Frame for earth science applications*, J. Geophys. Res., *107*(B10), 2214, doi:10.1029/2001JB000561.

Anghel, M. and Y. Ben-Zion (2001), *Nonlinear system identification and forecasting of earthquake fault dynamics using artificial neural networks*, EOS Trans., AGU, *82*, F571.

Anghel, M., Y. Ben-Zion, and R.R. Martinez (2004), *Dynamical system analysis and forecasting of deformation produced by an earthquake fault*, Pure and Applied Geophysics, *161*, doi: 10.1007/s00024-004-2547-9.

Aubrey, D. G., and K. O. Emery, *Eigenanalysis of recent United States sea levels*, Cont. Shelf Res., *2*, 21–33, 1983.

Bawden, G.W., W., Thatcher, R.S., Stein, K.W. Hudnut, and G. Peltzer (2001), *Tectonic contraction across Los Angeles after removal of groundwater pumping affects*, Nature, *412*, 812–815.

Bevis M, D. Alsdorf, E. Kendrick, L.P., Fortes, B. Forsberg, R. Smalley, J. Becker (2005), *Seasonal fluctuations in the mass of the Amazon River system and Earth's elastic response*. Geophys. Res. Ltrs., *32*, L16308.

Calais, E., J.Y. Han, C. DeMets, and J.M. Nocquet (2006), *Deformation of the North American plate interior from a decade of continuous GPS measurements* J. Geophys. Res., *111*, B06402, doi:10.1029/2005JB004253.

Conroy, J.L., J.T. Overpeck, J.E. Cole and M. Steinitz-Kannan (2009), *Variable oceanic influences on western North American drought over the last 1200 years*, GRL, *36*, L17703, doi:10.1029/2009GL039558.

Dragert, H., K. Wang, and T. James (2001), *A silent slip event on the deeper Cascadia subduction interface*, Science, *292*, 1525–1528.

Dong, D., P. Fang, Y. Bock, M. K. Cheng, and S. Miyazaki (2002), *Anatomy of apparent seasonal variations from GPS-derived site position time series*, J. Geophys. Res., *107*(B4), 2075, doi:10.1029/2001JB000573.

Dong, D., P. Fang, Y. Bock, F. Webb, L. Prawirodirdjo, S. Kedar, and P. Jamason (2006), *Spatiotemporal filtering using principal component analysis and Karhunen-Loeve expansion approaches for regional GPS network analysis*, J. Geophys. Res. *111*, doi: 10.1029/2005JB003806.

Dyke, A. S. (2004), *An outline of North American deglaciationwith emphasis on central and northern Canada*, in Quaternary Glaciations—Extent and Chronology, Part 2, North America, Dev. Quat. Sci., vol. *2b*, edited by J. Ehlers and P. L. Gibbard, pp. 373–424, Elsevier, New York.

Fancourt C.L. and J. C. Principe (1998), *Competitive Principal Component Analysis for Locally Stationary Time Series*, IEEE Trans. Signal Proc., *46*, 3068–3081.

Faure, S., A. Tremblay, and J. Angelier (1996), *Alleghanian paleostress reconstruction in the northern Appalachians: Intraplate deformation between Laurentia and Gondwana*, Geol. Soc. Am. Bull., *108*, 1467–1480.

Fukunaga, K. (1970), *Introduction to Statistical Pattern Recognition*, Academic, San Diego, Calif.

Hanrahan, J.L., S.V. Kravtsov, P.J. Roebber (2010), *Connecting past and present climate variability to the water levels of Lakes Michigan and Huron*, Geophys. Res. Ltrs., *37*, L01701, doi: 10.1029/2009GL041707.

Henton, J. A., M. R. Craymer, R. Ferland, H. Dragert, S. Mazzotti, and D. L. Forbes (2006), *Crustal motion and deformation monitoring of the Canadian landmass*, Geomatica, *60*, 173–191.

Hill, E.M., J.L. Davis, P. Elo'segui, B.P. Wernicke, E. Malikowski, and N.A. Niemi, (2009) *Characterization of site-specific GPS errors using a short-baseline network of braced monuments at Yucca Mountain, southern Nevada*, J. Geophys. Res., doi: 10.1029/2008JB006027.

Hoerling, M., X.-W. Quan, J. Eischeid, (2009), *Distinct causes for two principal US droughts of the 20th century*, GRL, *36*, L19708, doi:10.1029/2009GL039860.

Holmes, P., J. L. Lumley, and G. Berkooz, *Turbulence, Coherent Structures, Dynamical Systems and Symmetry*, Cambridge Univ. Press, New York, 1996.

Hotelling, H. (1933), *Analysis of a complex of statistical variables into principal components*, J. Educ. Psych., *24*, 417–520.

Hugentobler, U., S. Schaer, and P. Fridez (Eds.) (2001), *Documentation of the Bernese GPS software version 4.2*, 511 pp., Astron. Inst., Univ. of Bern, Bern.

Hudnut, K. W., Z. Shen, M. Murray, S. McClusky, R. King, T. Herring, B. Hager, Y. Feng, P. Fang, A. Donnellan (1996), *Co-seismic displacements of the 1994 Northridge*, California, earthquake, Bull. Seismol. Soc. Am., *86*, s19–s36.

James, T. S., and A. L. Bent (1994), *A comparison of eastern North America seismic strain rates to postglacial rebound strain rates*, Geophys. Res. Lett., *21*, 2127–2130.

James, T. S., and A. Lambert (1993), *A comparison of VLBI data with the ICE-3G glacial rebound model*, Geophys. Res. Lett., *20*, 871–874.

Johansson, J. M., et al. (2002), *Continuous GPS measurements of postglacialadjustment in Fennoscandia: 1. Geodetic results*, J. Geophys. Res., *107*(B8), 2157, doi:10.1029/2001JB000400.

Karlstrom, K. E., K. I. Ahall, S. S. Harlan, M. L. Williams, J. McLelland, and J. W. Geissman (2001), *Long-lived (1.8–1.0 Ga) convergent orogen in southern Laurentia, its extensions to Australia and Baltica, and implications for refining Rodinia*, Precambrian Res., *111*, 5–30.

Kedar, S., G. A. Hajj, B. D. Wilson, and M. B. Heflin (2003), *The effect of the second order GPS ionospheric correction on receiver positions*, Geophys. Res. Lett., *30*(16), 1829, doi: 10.1029/2003GL017639.

Kumarapeli, P. S. (1985), *Vestiges of Iapetan rifting in the craton west of the northern Appalachians*, Geosci. Can., *12*, 54–59.

Lambert, A., N. Courtier, G. S. Sasagawa, F. Klopping, D. Winester, T. S. James, and J. O. Liard (2001), *New constraints on Laurentide postglacial rebound from absolute gravity measurements*, Geophys. Res. Lett., *28*, 109–112.

LANGBEIN, J. (2008), *Noise in GPS displacement measurements from Southern California and Southern Nevada*, J. Geophys. Res., doi:10.1029/2007JB005247.

LANGBEIN, J. and H. JOHNSON (1997), *Correlated errors in geodetic time series: Implications for time-dependent deformation*, J. Geophys. Res., *102*, 591–603.

LEMIEUX, Y., A. TREMBLAY, and D. LAVOIS (2003), *Structural analysis of supracrustal faults in the Charlevoix area, Québec: Relation to impact cratering and the St-Laurent fault system*, Can. J. Earth Sci., *40*, 221–235.

MAIN, I.G., L. LI, K. J. HEFFER, O. PAPASOULIOTIS, and T. LEONARD (2006), *Long-range, critical-point dynamics in oil field flow rate data*, Geophys. Res. Lett., *33*, L18308, doi:10.1029/2006GL027357.

MAINVILLE, A., and M. CRAYMER (2005), *Present-day tilting of the Great Lakes region based on water level gauges*, Geol. Soc. Am. Bull., *117*, 1070–1080.

MÁRQUEZ-AZÚA, B., and C. DEMETS (2003), *Crustal velocity field of Mexico from continuous GPS measurements, 1993 to June 2001: Implications for the neotectonics of Mexico*, J. Geophys. Res., *108*(B9), 2450, doi:10.1029/2002JB002241.

MAZZOTTI, S., and J. ADAMS (2005), *Rates and uncertainties on seismic moment and deformation in eastern Canada*, J. Geophys. Res., *110*, B09301, doi:10.1029/2004JB003510.

MAZZOTTI, S., H. DRAGERT, J. HENTON, M. SCHMIDT, R. HYNDMAN, T. JAMES, Y. LU, and M. CRAYMER (2003), *Current tectonics of northern Cascadia from a decade of GPS measurements*, J. Geophys. Res., *108*(B12),2554, doi:10.1029/2003JB002653.

MAZZOTTI, S., T.S. JAMES, J. HENTON, and J. ADAMS (2005), *GPS crustal strain, postglacial rebound, and seismic hazard in eastern North America: The Saint Lawrence valley example*, J. Geophys. Res., *110*, B11301, doi:10.1029/2004JB003590.

MAZZOTTI, S. (2007), *Geodynamic models for earthquake studies in intraplate North America*, in Stein, S., and Mazzotti, S., eds., Continental Intraplate Earthquakes: Science, Hazard, and Policy Issues, Geol. Soc. Amer. Special Paper 425, p. 17–33, doi: 10.1130/2007.2425(02).

MOGHADDAM, B., W. WAHID, and A. PENTLAND, Beyond eigenfaces: Probabilistic matching for face recognition, paper presented at Third IEEE International Conference on Automatic Face and Gesture Recognition, Nara, Japan, 14–16 April 1998.

PARK, K., R. S. NEREM, J. L. DAVIS, M. S. SCHENEWERK, G. A. MILNE, and J. X. MITROVICA (2002), *Investigation of glacial isostatic adjustment in the northeast US using GPS measurements*, Geophys. Res. Lett., *29*(11), 1509, doi:10.1029/2001GL 013782.

PARKER, J. W. (2001), Analysis and modeling of southern California deformation, in APEC Cooperation for Earthquake Simulation (ACES), 2nd ACES Workshop Proceedings, Univ. of Queensland, Brisbane, Australia.

PELTIER, W. R. (1998), *Postglacial variations in the level of the sea: Implications for climate dynamics and solid-Earth geophysics*, Rev. Geophys., *36*, 603–689.

PELTIER, W. R. (2002), *Global glacial isostatic adjustment: Palaeogeodetic and space-geodetic tests of the ICE-4G (VM2) model*, J. Quat. Sc., *17*, 491–510.

PELTIER, W. R. and R. DRUMMOND (2009), *Rheological stratification of the lithosphere: A direct inference based upon the geodetically observed pattern of the glacial isostatic adjustment of the North American continent*, Geophys. Res. Ltrs., doi:10.1029/2008GL 034586.

PENLAND, C. (1989), *Random forcing and forecasting using principal oscillation pattern analysis*, Mon. Weather Rev., *117*, 2165–2185.

PENLAND, C., and P. D. SARDESHMUKH (1995), *The optimal growth of tropical sea surface temperature anomalies*, J. Clim., *8*, 1999–2024.

POSADAS, A.M., F. VIDAL, F. DEMIGUEL, G. ALGUACIL, J. PENA, J.M. IBANEZ, and J. MORALES (1993), *Spatial-temporal analysis of a seismic series using the principal components method—the Antequera series, Spain, 1989*, J. Geophys. Res., *98*, 1923–1932.

PREISENDORFER, R. W., *Principle Component Analysis in Meteorology and Oceanography*, Elsevier Sci., New York, 1988.

PRESS, W. H., B. P. FLANNERY, S. A. TEUKOLSKY, and W. T. VETTERING, (1992), Numerical Recipes in C, 2nd ed., Cambridge Univ. Press, New York.

RONDOT, J. (1968), *Nouvel impact meteoritique fossile? La structure semicirculaire de Charlevoix*, Can. J. Earth Sci., *5*, 1305–1317.

SAVAGE, J. C. (1988), *Principal component analysis of geodetically measured deformation in Long Valley caldera, eastern California, 1983–1987*, J. Geophys. Res., *93*, 13297–13305.

SAVAGE, J.C., and J. LANGBEIN (2008), *Postearthquake relaxation after the 2004 M6 Parkfield, California, earthquake and rate-and-state friction*, J. Geophys. Res., *113*, B10407, doi:10.1029/2008JB005723.

SCHERNECK, H.-G., J. M. JOHANSSON, and R. HAAS (2000), *BIFROST project: Studies of variations of absolute sea level in conjunction with postglacial rebound in Fennoscandia*, in Towards an Integrated Global Geodetic Observing System (IGGOS), Int. Assoc. Geod. Symp., vol. *120*, edited by R. Rammel et al., pp. 241–244, Springer, New York.

SELLA, G.F., S. STEIN, T.H. DIXON, M. CRAYMER, T.S. JAMES, S. MAZZOTTI, and R.K. DOKKA (2007), *Observation of glacial isostatic adjustment in "stable" North America with GPS*, Geophys. Res. Ltrs., *34*, L02306, doi:10.1029/2006GL027081.

SELLINGER, C.E., C.A. STOW, E.C. LAMON, and S.S. QIAN (2008) *Recent Water Level Declines in the Lake Michigan-Huron System*, Environ. Sci. Technol., *42*, 367–373.

SMALL, D., and S. ISLAM (2007), Decadal variability in the frequency of fall precipitation over the United States, GRL, *34*, L02404, doi:10.1029/2006GL028610.

SMITH, E.G.C., T.D. WILLIAMS, and D.J. DARBY (2007), *Principal component analysis and modeling of the subsidence of the shoreline of Lake Taupo, New Zealand, 1983–1999: Evidence for dewatering of a magmatic intrusion?*, J. Geophys. Res., *112*, B08406, doi:10.1029/2006JB004652.

STOW, C. (2009), Water Levels of the Great Lakes, NOAA Great Lakes Environmental Research Laboratory Brochure.

TIAMPO, K.F., J.B. RUNDLE, S.J. GROSS, S. MCGINNIS, W. KLEIN (2002), *Eigenpatterns in southern California seismicity*, Journal of Geophysical Research, *107*, 2354, doi:10.1029/2001JB000562.

TIAMPO, K.F., J.B. RUNDLE, W. KLEIN, Y. BEN-ZION, S. MCGINNIS, (2004), *Using eigenpattern analysis to constrain seasonal signals in southern California*, Pure and Applied Geophysics, 1991–2003.

TREMBLAY, A., B. LONG, and U. GLASMACHER (2001), *Supracrustal faults of the St Lawrence Rift system, Quebec: Kinematics and geometry as revealed by field mapping and marine seismic reflection data*, Geol. Soc. Am. Abstr. Programs, *33*, A-210.

Tushingham, A. M., and W. R. Peltier (1991), *ICE-3G: A new global model of late Pleistocene deglaciation based upon geophysical predictions of postglacial relative sea level change*, J. Geophys. Res., *96*, 4497–4523.

Vautard, R., and M. Ghil (1989), *Singular spectrum analysis in nonlinear dynamics, with applications to paleodynamic time series*, Physica D, *35*, 395–424.

Watson, K. M., Y. Bock, and D.T. Sandwell (2002), *Satellite interferometric observations of displacements associated with seasonal groundwater in the Los Angeles basin*, J. Geophys. Res., *107*(B4), 2074, doi:10.1029/2001JB000470.

Wdowinski, S., Y. Bock, J. Zhang, P. Fang, and J. Genrich (1997), *Southern California permanent GPS geodetic array: Spatial filtering of daily positions for estimating coseismic and postseismic displacements induced by the 1992 Landers earthquake*, J. Geophys. Res., *102*, 18,057–18,070.

Wheeler, R. L. (1995), *Earthquakes and the cratonward limit of Iapetan faulting in eastern North America*, Geology, *23*, 105–108.

Wheeler, R. L., and A. C. Johnson (1992), *Geological implications of earthquake source parameters in central and eastern North America*, Seismol. Res. Lett., *63*, 491–514.

Williams, H. (1979), *Appalachian orogen in Canada*, Can. J. Earth Sci., *16*, 792–807.

Williams, S. D. P., Y. Bock, P. Fang, P. Jamason, R. M. Nikolaidis, L. Prawirodirdjo, M. Miller, and D. J. Johnson (2004), *Error analysis of continuous GPS position time series*, J. Geophys. Res., *109*, B03412, doi:10.1029/2003JB002741.

Zerbini, S., Raicich, F., Richter, B., Gorini, V., Errico, M. (2010), *Hydrological signals in height and gravity in northeastern Italy inferred from principal components analysis*, J. Geodynamics, doi:10.1016/j.jog.2009.11.001.

(Received March 7, 2011, Published online November 17, 2011)

Pure Appl. Geophys. 169 (2012), 1507–1517

DOI 10.1007/s00024-011-0404-1

Pure and Applied Geophysics

Identification of Glacial Isostatic Adjustment in Eastern Canada Using S Transform Filtering of GPS Observations

Nithin V. George,[1] Kristy F. Tiampo,[2] Sitanshu S. Sahu,[1] Stéphane Mazzotti,[3] Lalu Mansinha,[2] and Ganapati Panda[1]

Abstract—Over the years, a number of different models and techniques have been proposed to both quantify and explain the glacial isostatic adjustment (GIA) process. There are serious challenges, however, to obtaining accurate results from measurements, due to noise in the data and the long periods of time necessary to identify the relatively small-magnitude signal in certain regions. The primary difficulty, in general, is that most of the geophysical signals that occur in addition to GIA are nonstationary in nature. These signals are also corrupted by random as well as correlated noise added during data acquisition. The nonstationary characteristic of the data makes it difficult for traditional frequency-domain denoising approaches to be effective. Time–frequency filters present a more robust and reliable alternative to deal with this problem. This paper proposes an extended S transform filtering approach to separate the various signals and isolate that associated with GIA. Continuous global positioning system (GPS) data from eastern Canada for the period from June 2001 to June 2006 are analyzed here, and the vertical velocities computed after filtering are consistent with the GIA models put forward by other researchers.

Key words: Postglacial rebound, glacial isostatic adjustment, S transform, time–frequency filtering, continuous GPS.

1. Introduction

The last glacial maximum is estimated to have occurred at approximately 20,000 years before present (Richmond and Fullerton, 1986). During this glaciation period, vast areas of North America and Scandinavia were covered under ice sheets of up to 4 km in depth and measuring over thousands of kilometres across. The weight of this ice depressed the lithosphere, and the resulting viscoelastic flow in the mantle caused a peripheral bulge (Mitrovica *et al.*, 2001). More than 10,000 years before present, the ice sheets began their retreat and the lithosphere began to rebound upwards to regain isostatic equilibrium while the peripheral bulge began to migrate inward toward the center of uplift as it gradually dissipated. This phenomenon is called glacial isostatic adjustment (GIA) as well as postglacial rebound, continental rebound, isostatic adjustment, or post-ice-age isostatic recovery. Similar rebounds also affected Northern Europe, Canada, and the USA.

Researchers have used various methods to study GIA, including the study of the changes in global sea levels. Relative sea level curves from geological studies have been employed (Peltier, 1998) for estimating GIA. Mainville and Craymer (2005) employed water level gauges in the Great Lakes to develop a regional model of vertical crustal motion. The vertical velocities have been found to be consistent with the ICE-3G global isostatic model (Tushingham and Peltier, 1991). Absolute measurements are possible only if the movement of these gauges are tracked precisely, which today is done using techniques such as global positioning systems (GPS) or satellite radar altimetry (Lee *et al.*, 2008). Absolute gravity measurements have been used for deriving new constraints on Laurentide postglacial rebound (Lambert *et al.*, 2001).

Precise measurement of crustal motions is possible only when a reference point is taken outside the Earth's surface. GPS is optimal for this purpose, as it uses a reference point in space for measurements. Khan *et al.* (2005) analyzed data from three Danish

[1] School of Electrical Sciences, Indian Institute of Technology Bhubaneswar, Bhubaneswar, Odisha, India. E-mail: nithinvgeorge@gmail.com

[2] Department of Earth Sciences, University of Western Ontario, London, ON, Canada.

[3] Geological Survey of Canada, Pacific Geoscience Centre, Sidney, BC, Canada.

and one Swedish continuously operating GPS stations to observe the crustal deformation resulting from GIA in Denmark. Similar observations have been made for Antarctica (TREGONING *et al.*, 2000), North America (SELLA *et al.*, 2007; PARK *et al.*, 2002), Scandinavia (JOHANSSON, 2002), Greenland (GREGERSEN, 2006), British Colombia (CLAGUE and JAMES, 2002), and the Great Lakes region (BRAUN *et al.*, 2008) using GPS measurements.

Building on the advantages of GPS-based deformation measurements, this paper uses continuous GPS data from eastern Canada to quantify GIA. Continuous GPS time series provide a record of the displacement in the longitudinal, latitudinal, and vertical directions of a GPS station as a function of time. However, these GPS signals are a mixture of a large number of nonlinear geophysical signals. The relatively large seasonal effects, atmospheric signals, and numerous high-frequency components related to site effects make it difficult to observe and quantify the comparatively smaller signals caused by crustal motion (TIAMPO *et al.*, 2004). Most of the geophysical signals are nonstationary in nature, and thus Fourier-domain filtering will fail to remove them precisely. Time–frequency filters are one possible solution to this problem. TIAMPO *et al.* (2008) employed the localized Hartley transform on GPS data in southern California to isolate postseismic deformation after the 1994 Northridge earthquake, with some success that was limited by the number of GPS stations available at that time. This paper employs a time–frequency filter based on the S transform (STOCKWELL *et al.*, 1996), a relatively new time–frequency transform, applied to a larger continuous GPS dataset from eastern Canada.

The paper is organized as follows. Section 2 gives a brief idea of the region of study as well as the nature of the signals observed in that region. The S transform (ST) is described in Sect. 3. An extended S transform filtering is introduced in 4. The need for using an extended filter is also explained. Analysis and discussions are presented in Sect. 5.

2. *Region of Study*

GPS is a constellation of 24 satellites, of which 20 are used for navigation and precise geodetic position measurements. Daily position estimates are determined from satellite signals which are recorded by

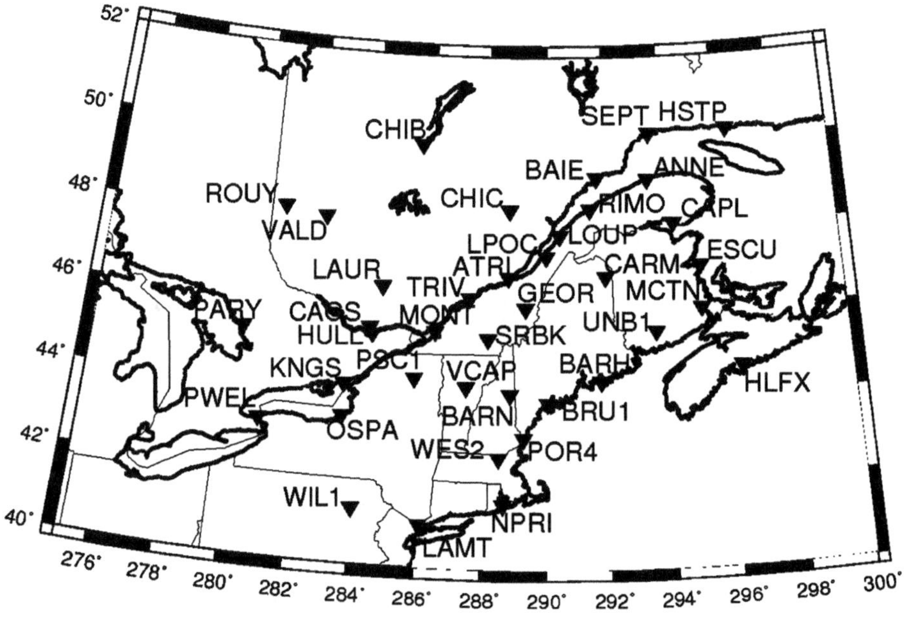

Figure 1
Continuous GPS (CGPS) stations in eastern Canada

GPS receivers on the ground. A subset of 39 GPS stations in eastern Canada from western Quebec to southern New England, including a region of seismicity in the lower Saint Lawrence Valley, was selected for this study (Fig. 1). The time period for this analysis began in June 2001 and finished in June 2006, with 13 stations covering the whole period of 5.3 years, 6 stations with a period in the range 5.0–5.3 years, and the remaining 20 stations with a period ranging from 0.9 to 4.9 years. The reference frame is ITRF2000. The stations are operated by various national and local agencies, and all the data were archived and processed at the Pacific Geoscience Centre (Natural Resources Canada).

The area of study is the historic location of several major tectonic events (Kumarapeli, 1985; Karlstrom *et al.*, 2001; Wheeler, 1995). In addition, ice advanced over the Saint Lawrence Valley during the glaciation of the late Pleistocene (Dyke, 2004). Between about 7 and 20 ka the ice sheets retreated and GIA began. Current uplift rates are 10 mm/year or more at Hudson Bay and decrease with distance southward (Tiampo *et al.*, 2011). Present-day GIA models estimate that the hinge line between uplift to the north and subsidence to the south lies near the Saint Lawrence Valley in eastern Canada (Van der Wal *et al.*, 2009; Koohzare *et al.*, 2006).

Figure 2 shows the GPS daily position time series corresponding to station BAIE. The duration of the time series is 4.9 years from November 2001. The time series appears noisy, containing both outliers and high-frequency noise generally caused by

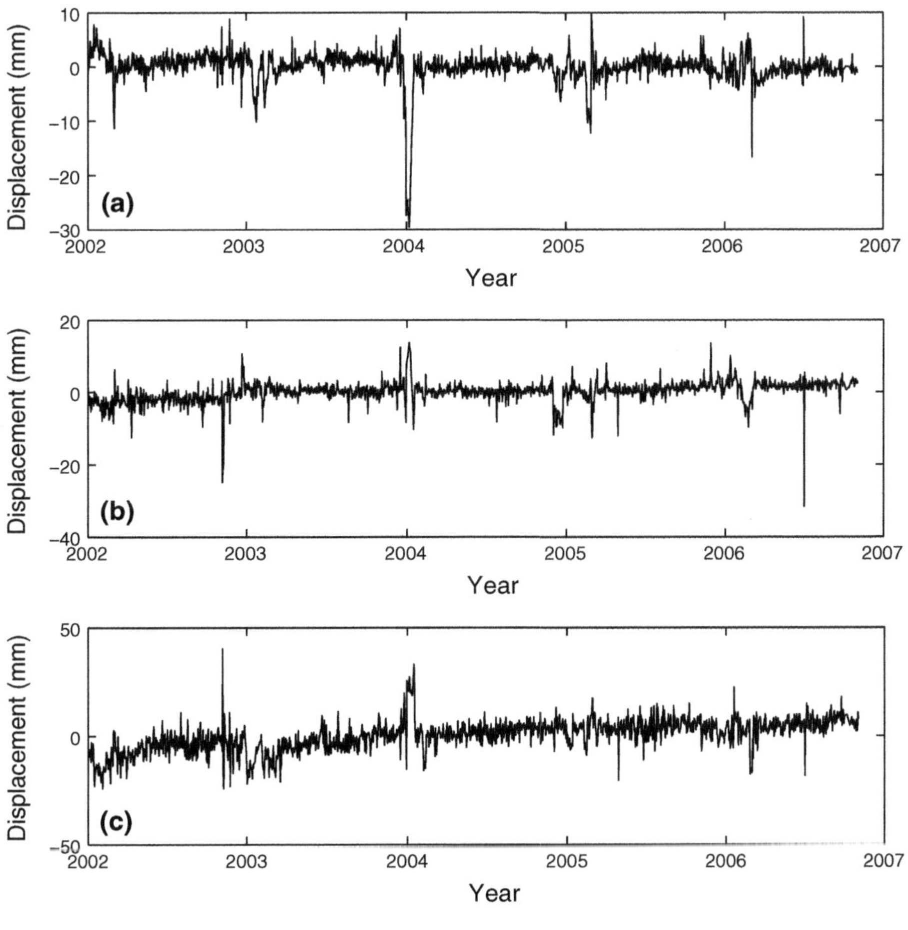

Figure 2
GPS time series for BAIE station: **a** north–south displacement, **b** east–west displacement, and **c** vertical displacement

features such as noise in the data acquisition system and site effects. Improvement in the signal-to-noise ratio of the time series is essential for accurate estimation of the velocity of the GPS station.

3. *The S Transform*

The field of time–frequency analysis got an impetus with the development of the short-time Fourier transform (STFT) (Portnoff, 1980). In STFT, the signal to be transformed is multiplied with a sliding window and the Fourier transform of the resulting signal is computed to generate a two-dimensional time–frequency transform. STFT employs the same window at all frequencies. This has contributed to the poor time–frequency resolution at frequencies shorter or longer than the window. Attempts to improve the resolution of STFT led to the development of the wavelet transform. The wavelet transform uses a scaled and translated wavelet (Daubechies, 1990).

Combining the strengths of STFT and wavelet transform, Stockwell *et al.* (1996) developed the S transform. The S transform uses a Gaussian window whose width scales inversely and whose height scales directly with the frequency, to provide frequency-dependent resolution. The S transform retains the absolute phase of each signal component (Pinnegar and Mansinha, 2003). It can be viewed as a frequency-dependent STFT or a phase-corrected wavelet transform (WT) (Dash *et al.*, 2008). The S transform has found applications in mechanical engineering (McFadden *et al.*, 1999), electrical engineering (Dash *et al.*, 2007), biomedical instrumentation (Huang *et al.*, 2009), and geophysics (Pinnegar and Mansinha, 2003; Tiampo *et al.*, 2008).

The S transform $S(f, t)$ of a signal $x(t)$ is defined as

$$S(f,t) = \int_{-\infty}^{\infty} x(\tau)w(\tau - t)\mathrm{e}^{-j2\pi f\tau}\mathrm{d}\tau, \quad (1)$$

where t represents the time, f the frequency, w the scalable Gaussian window, and the quantity τ is a parameter which controls the position of the Gaussian window on the time axis. The Gaussian window in the time domain is given by

$$w(t,\sigma) = \frac{1}{\sigma\sqrt{2\pi}}\mathrm{e}^{-\frac{t^2}{2\sigma^2}}, \quad (2)$$

where

$$\sigma(f) = \frac{1}{|f|} \quad (3)$$

determines the width and height of the Gaussian window. Combining (1–3) gives

$$S(f,t) = \int_{-\infty}^{\infty} x(\tau)\left\{\frac{|f|}{\sqrt{2\pi}}\mathrm{e}^{-\frac{(\tau-t)^2 f^2}{2}}\mathrm{e}^{-j2\pi f\tau}\right\}\mathrm{d}\tau. \quad (4)$$

The S transform window satisfies the normalization condition

$$\int_{-\infty}^{\infty} w(t,f)\mathrm{d}t = 1, \quad (5)$$

and hence

$$\int_{-\infty}^{\infty} S(f,t)\mathrm{d}t = X(f), \quad (6)$$

where $X(f)$ is the Fourier transform of $x(t)$. Averaging $S(f, t)$ over all values of t yields $X(f)$, providing a direct link between the S transform and Fourier transform. The inverse S transform is given by

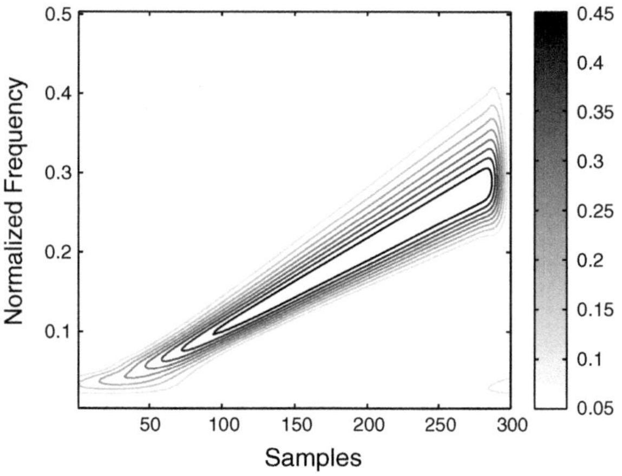

Figure 3
Time–frequency representation of a linear chirp signal using S transform

$$x(t) = \frac{1}{2\pi} \int_{-\infty}^{\infty} \left\{ \int_{-\infty}^{\infty} S(f,t)\mathrm{d}t \right\} e^{j2\pi ft} \mathrm{d}f. \quad (7)$$

Figure 3 depicts the S transform representation of a linear chirp signal (frequency varies linearly with time).

4. *S Transform Filtering*

Filtering of nonstationary signals in the presence of white Gaussian noise has always been a challenging task. A time–frequency (TF) filter is a good candidate for this task (Pinnegar, 2005; Schimmel and Gallart, 2005; George *et al.*, 2009). The basic solution in such filters consists of the data transformation to the TF domain, a data-adaptive weighting of the localized spectra, and a back transformation. The S transform is selected in this paper for the TF transformation because of its competitive advantages over other TF transformations (Pinnegar and Mansinha, 2003). The S transform is sampled at the discrete Fourier frequencies, whereas the sampling is loosely defined in the wavelet transform. In most of the applications, the WT uses octave scaling of frequencies, which leads to an undersampled representation of the high frequencies (the noise in the current study) and oversampled representation of low frequencies (the signal of interest in the current study) (Stockwell, 2007).

Normalization of the WT with the $\frac{1}{\sqrt{f}}$ factor inherently diminishes the amplitude of its high-frequency components. Hence, the high-frequency bursts (i.e., noise) are relatively weak in the WT representation. The unit area localizing function (Gaussian) used for normalization of the S transform

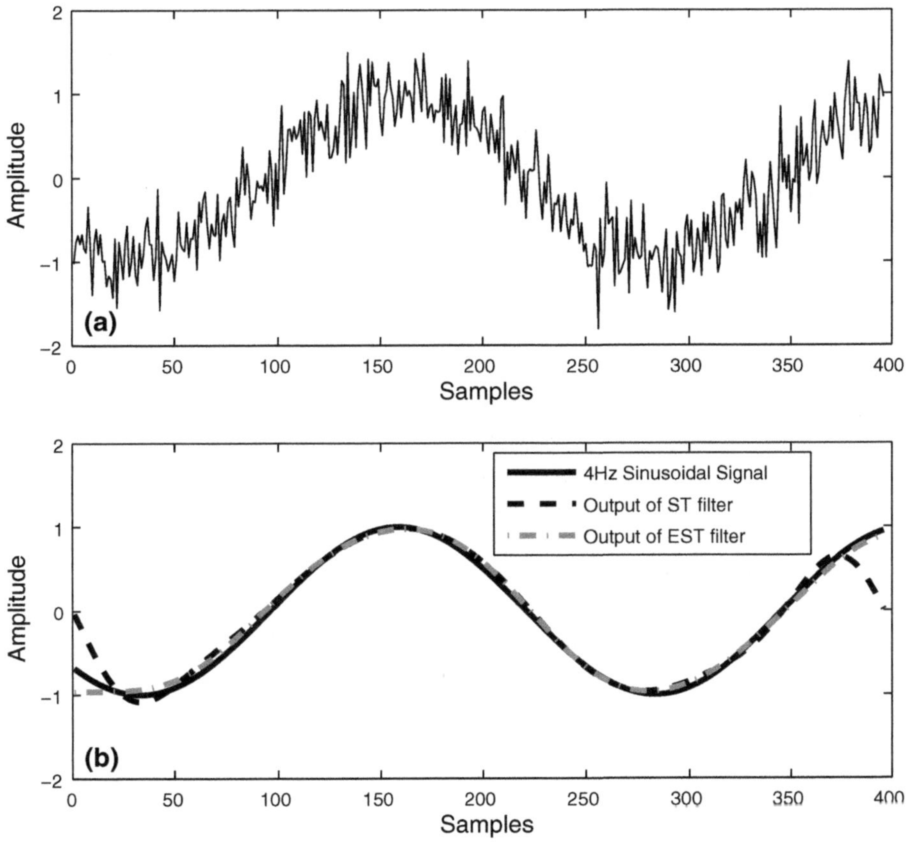

Figure 4
a A 4-Hz sinusoidal signal corrupted with AWGN (SNR = 10 dB). **b** Comparison of results of S transform and extended S transform filtering of the synthetic time series

ensures that the amplitude response of the S transform is invariant of the frequency. Hence the high-frequency bursts are accurate and distinct in the S transform representation, which helps in accurate thresholding of the S transform plane.

The S transform provides absolutely referenced phase information, whereas the WT provides locally referenced phase information. To recognize the absolute referenced phase in the WT, the position of the mother wavelet with respect to the reference point is needed. Because the magnitude and phase are essential for exact recovery of a time series from the TF plane, the S transform is employed in this paper.

Before applying the S transform to a real GPS signal, the validity of the filtering operation is checked using a synthetic signal. A low-frequency sinusoidal signal (4 Hz) corrupted with additive white Gaussian noise (AWGN) (Fig. 4a) is chosen as the input to a TF filter. The output of the filter is shown in Fig. 4b. It is observed that the filtered signal tends to move towards the mean value of the signal at the edges. This happens because of the edge effect in the S transform, similar to that in the WT when used in the treatment of finite-length data sequences (Williams and Amaratunga, 1997). A possible solution to this problem is an extended S transform (EST) filtering approach, which is explained below.

4.1. Extended S Transform Filtering

EST filtering is a crude method of time–frequency filtering. A time series of length N is extended to a time series of length $3N$, using the following approach. The idea here is that, for a time series with a particular nature such as this, the dominant high-frequency noise is stationary over long enough time periods, and the GIA signal is expected to be linear, and as such can be added without significant corruption of the signal of interest. The original time series is preserved in the extended time series from samples $N + 1$ to $2N$. The first sample of the original time series is repeated from samples 1 to N, and the last sample is repeated from samples $2N + 1$ to $3N$. After the time–frequency masking of the extended S transform, an inverse S transform of the TF plane is used to recover a time series of length $3N$. The first N and the last N samples of the recovered time series are discarded to get back the filtered N sample time series.

The effect of extending the length of a time series on the spectrum obtained using the Fourier transform is also examined in this section. For a time series of length N, the frequency axis in the amplitude spectrum ranges from 0 to $(N/2 + 1)$. Suppose that the frequency with highest amplitude is $N/10$, then a clear and distinct peak is seen at frequency $N/10$ in the amplitude spectrum. When the length of the same time series is extended to $3N$, the frequency axis in the amplitude spectrum ranges from 0 to $(3N/2 + 1)$. Hence, the peak which was previously seen at frequency $N/10$ is now seen at frequency $3N/10$ in the new amplitude spectrum. However, the frequency

Table 1

Comparison of the mean square error (MSE) in dB between the desired filter output (4-Hz sinusoidal signal) and the outputs of ST and EST filters for different values of SNR

SNR (dB)	MSE	
	ST filter	EST filter
35	−12.249	−41.337
30	−12.254	−39.475
25	−12.258	−35.972
20	−12.264	−31.594
15	−12.258	−26.667
10	−12.176	−22.816
5	−11.576	−16.758

Only the first 50 and last 50 samples have been compared to study the edge effect. All MSE values are computed as the average of 50 independent iterations. Note that the MSE values are logarithmic

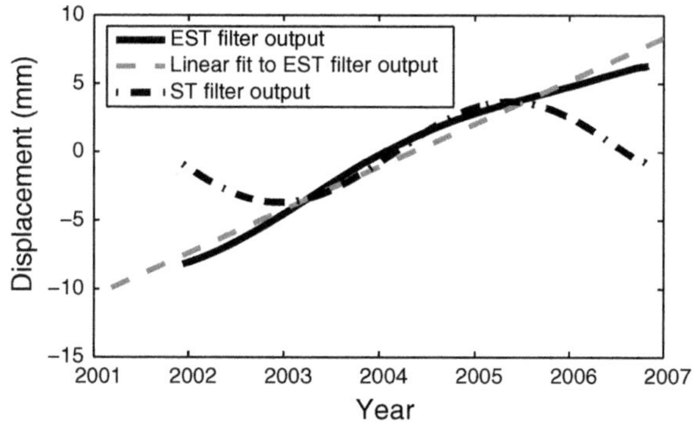

Figure 5
BAIE vertical displacement time series filtered using extended S transform filtering

content in both time series remains the same, i.e., 0 to ($N/2 + 1$), hence the frequency axis in the amplitude spectrum of the extended time series is scaled from 0 to ($3N/2 + 1$) to become 0 to ($N/2 + 1$). With this scaling, the peak will now be seen at frequency $N/10$, as it appears in the amplitude spectrum of the original time series.

The low-pass-filtering output for the synthetic signal using EST filtering is portrayed in Fig. 4b. The EST filter output is a close match with the desired filter output. It also has not added any artifacts in the filtered signal. The mean square error (MSE) between the desired filter output and the ST/EST filter outputs for various AWGN magnitudes are shown in Table 1. For all values of SNR taken, the MSE between the EST filter output and the desired filter output is significantly lower than that between the ST filter output and the desired filter output (note that the values in Table 1 are logarithmic, so that larger negative numbers indicate a better fit to the original synthetic input). Since it has been verified by the above experiment that the EST filtering neither

Table 2

Filtered station velocities and associated 95% confidence intervals (mm/year)

Sl. no.	Station	Latitude	Error	Longitude	Error	Vertical	Error	Samples	Start date
1	ANNE	3.017	0.611	0.007	0.179	0.023	0.172	362	2005 Sep
2	ATRI	−0.047	0.006	0.520	0.005	2.815	0.009	1,892	2001 Jun
3	BAIE	−0.296	0.005	0.704	0.012	3.143	0.024	1,756	2001 Nov
4	BARH	0.680	0.009	0.323	0.002	−0.700	0.017	1,858	2001 Jun
5	BARN	−0.103	0.009	0.249	0.006	−0.016	0.026	1,848	2001 Jun
6	BRU1	−0.425	0.013	−0.047	0.010	−0.145	0.039	1,849	2001 Jun
7	CAGS	−0.524	0.006	0.549	0.007	2.915	0.013	1,920	2001 Jun
8	CAPL	−0.108	0.007	−0.577	0.013	−1.123	0.025	1,423	2002 Oct
9	CARM	−0.481	0.015	0.167	0.010	−2.010	0.123	753	2004 Aug
10	CHIB	−0.971	0.007	0.474	0.005	9.361	0.016	1,885	2001 Jun
11	CHIC	−0.304	0.004	1.098	0.007	4.853	0.027	1,745	2001 Oct
12	ESCU	0.737	0.018	−0.210	0.036	−0.605	0.082	642	2004 Nov
13	GEOR	0.193	0.031	2.317	0.030	−0.459	0.053	1,050	2003 Oct
14	HLFX	0.780	0.009	0.091	0.010	−1.255	0.005	1,734	2001 Nov
15	HSTP	0.396	0.018	0.069	0.013	1.763	0.046	1,691	2001 Nov
16	HULL	0.198	0.021	0.250	0.019	0.922	0.010	1,648	2001 Oct
17	KNGS	−0.087	0.009	0.992	0.012	0.082	0.026	1,559	2002 May
18	LAMT	0.610	0.003	0.322	0.008	−0.076	0.022	1,559	2001 Jul
19	LAUR	0.197	0.016	0.650	0.022	2.031	0.098	691	2004 Oct
20	LOUP	−0.222	0.053	0.857	0.039	0.930	0.075	608	2005 Jan
21	LPOC	−0.483	0.100	0.507	0.093	0.403	0.316	479	2005 May
22	MCTN	2.560	0.072	3.250	0.071	−3.470	0.227	317	2005 Oct
23	MONT	0.311	0.008	0.207	0.003	2.485	0.017	1,844	2001 Jun
24	NPRI	0.637	0.006	0.052	0.002	−0.552	0.019	1,910	2001 Jun
25	OSPA	−0.223	0.024	0.436	0.009	0.076	0.013	1,460	2002 Apr
26	PARY	−0.574	0.005	0.452	0.005	−0.378	0.052	1,551	2002 May
27	POR4	−0.702	0.017	−0.808	0.020	−0.760	0.020	1,175	2003 Jun
28	PSC1	−0.010	0.006	0.443	0.004	1.272	0.019	1,795	2001 Jun
29	PWEL	0.174	0.011	0.011	0.012	−0.954	0.054	1,576	2002 May
30	RIMO	0.267	0.021	1.148	0.040	−1.194	0.077	716	2004 Sep
31	ROUY	−0.448	0.007	−0.559	0.009	7.310	0.018	1,881	2001 Jun
32	SEPT	−7.254	0.136	6.463	0.267	1.830	0.377	350	2005 Sep
33	SRBK	−0.316	0.009	0.709	0.011	1.325	0.030	1,701	2001 Oct
34	TRIV	−0.389	0.004	0.653	0.006	2.191	0.016	1,768	2001 Oct
35	UNB1	0.464	0.004	−0.084	0.020	−0.855	0.033	1,833	2001 Oct
36	VALD	−0.849	0.013	0.461	0.014	8.393	0.032	1,690	2001 Nov
37	VCAP	0.145	0.021	0.294	0.007	0.379	0.039	1,690	2001 Jun
38	WES2	0.569	0.013	0.401	0.005	−1.105	0.011	1,753	2001 Jun
39	WIL1	0.591	0.007	0.087	0.006	−0.901	0.014	1,878	2001 Jun

contributes extra noise nor changes the signal pattern, it can be used to filter physical signals such as the CGPS time series. In particular, it is expected to work well for filtering cyclic signals, for the correct choice of the time period of analysis, without affecting the long-term stationary GIA signal.

5. Analysis and Discussion

The GPS time series for all 39 GPS stations were filtered using the EST filtering method. Most of the time series have outliers. Initial preprocessing was done on the time series using a three-sigma rule to remove outliers from the time series. The time series, after outlier removal, was fed to the EST filter to remove the high-frequency noise components. Figure 5 shows a comparison of the filtered vertical time series at BAIE station using ST and EST filters. The EST filter output has a linear trend, while the ST filter output tends to move towards the mean value at the edges, as expected from the discussion in the previous section.

The filtered GPS position time series were used as a reference for calculating the velocities. A linear curve was approximated by least-squares fitting to the filtered time series, and the slope of that curve represents the velocity of the GPS station movement, which in turn corresponds to the velocity of Earth's surface. The velocities in all three directions along with the 95% uncertainty intervals are listed in Table 2.

The velocity vectors for each station were plotted on the map of eastern Canada for the magnitude (in mm/year) and direction of the velocity at the location of the GPS stations. Figure 6 shows the vertical velocity map for eastern Canada. It can be seen that almost 90% of the velocity vectors above a latitude of 44°N are directed up and most of the vectors below 44°N latitude are directed down. This velocity map is in accordance with other GIA observations for Canada (Mazzotti *et al.*, 2005). There is a clear indication of a hinge line around 44°N that runs approximately parallel to the Saint Lawrence River in the east (Peltier, 2002; Tushingham and Peltier, 1991), where the lithosphere above the hinge line shows uplift and those portions below the hinge line show subsidence.

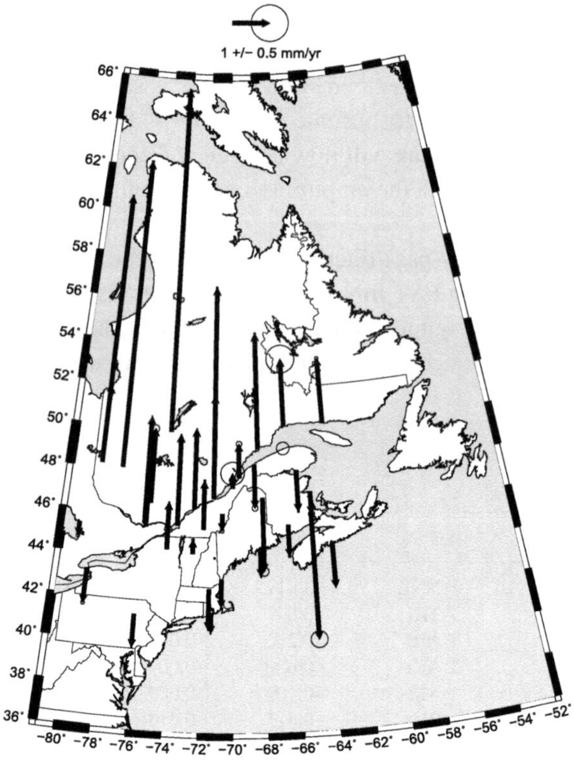

Figure 6
Vertical velocities for filtered CGPS data, in mm/year, for July 2001 through June 2006

Table 3

Comparison of vertical velocities (mm/year) obtained by EST filtering with that obtained by crudely fitting a regression to the despiked time series and that computed using principal component analysis

Station	Tiampo *et al.* (2011)		EST filtering		Crude regression fitting	
	Velocity	Error	Velocity	Error	Velocity	Error
ATRI	2.48	0.035	2.815	0.009	2.753	0.134
BAIE	3.071	0.041	3.143	0.024	2.985	0.147
BARN	−0.022	0.084	−0.016	0.026	0.224	0.145
CAGS	3.078	0.043	2.915	0.013	2.729	0.130
CARM	−2.064	0.116	−2.01	0.123	−1.693	0.409
CHIB	9.067	0.034	9.361	0.016	8.956	0.137
CHIC	4.912	0.096	4.853	0.027	4.589	0.155
MONT	2.417	0.044	2.485	0.017	2.529	0.105
ROUY	6.978	0.058	7.31	0.018	6.959	0.133
SRBK	1.207	0.036	1.325	0.03	1.469	0.093
VALD	8.084	0.046	8.393	0.032	8.202	0.198
WIL1	−0.87	0.036	−0.901	0.014	−0.735	0.119

All errors are computed at 95% percentage confidence interval

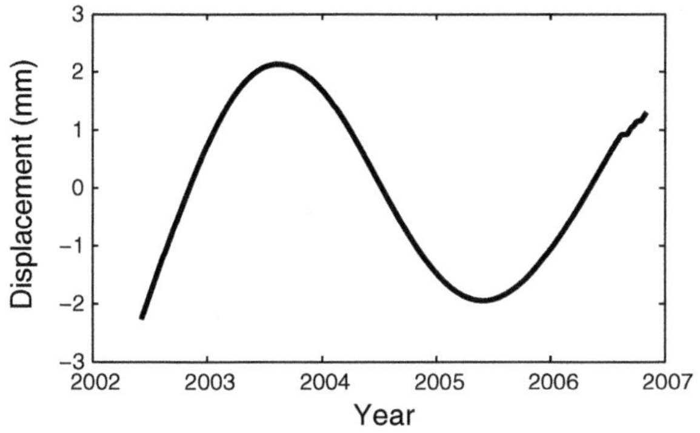

Figure 7
Filtered vertical displacement time series of PARY

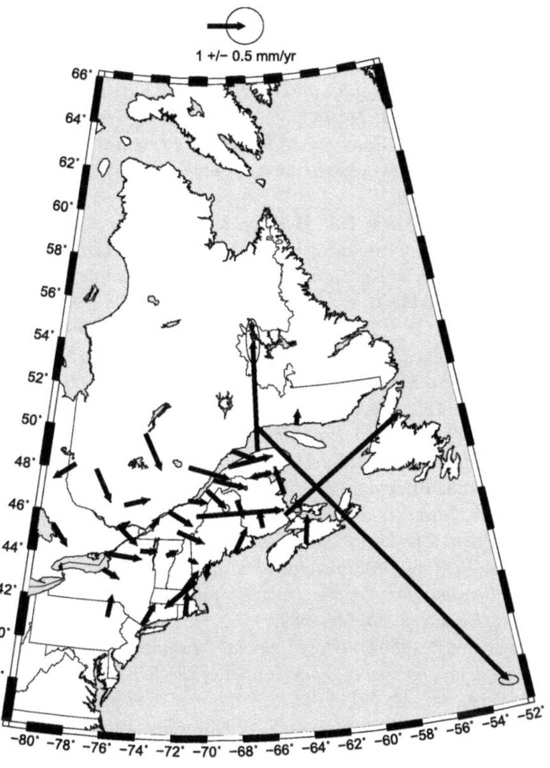

Figure 8
Horizontal velocities for filtered CGPS data, in mm/year, for July 2001 through June 2006

The vertical velocities obtained by EST filtering are compared in Table 3 with the results obtained by Tiampo *et al.* (2011), which employs a principal component analysis technique, and are observed to be in close match. A comparison is also made with the velocities obtained by crude fitting of a line to the despiked vertical position time series. The reduced uncertainties obtained using EST filtering shows the superior performance of the EST filtering method over a linear regression method.

There are some stations near the Great Lakes which behave somewhat differently from the GIA predictions. Figure 7 shows the filtered vertical GPS time series for PARY. One potential explanation for this discrepancy is the recent persistent drought in the Great Lakes Basin which began in 1997, substantially lowering the lake levels in Lake Michigan, Lake Huron, and Lake Erie (Hanrahan *et al.*, 2010). This could potentially lower the local groundwater levels and the adjacent ground surface (Bawden *et al.*, 2001; Watson *et al.*, 2002). However, future work is necessary with longer time series in order to properly evaluate potential mechanisms.

Figure 8 depicts the horizontal velocity map for the region (resultant velocity of the north–south and the east–west velocities). The horizontal velocity pattern is difficult to interpret, in part due to the short time periods for a significant number of the available stations (Sella *et al.*, 2007; Mazzotti *et al.*, 2005; Peltier and Drummond, 2008).

6. Conclusions

This paper has introduced an extended S transform filter as a tool for analysis of GPS time series data in quantifying the GIA process. The S transform filter is suggested as a refinement to other, more prevalent methods to obtain more precise results. The crustal velocities derived from the analysis reveal the presence of a postglacial rebound in eastern Canada. A hinge line is observed, as predicted by the GIA models for eastern Canada. The horizontal velocities exhibit an irregular pattern which is attributed to the fact that many of the GPS stations were installed after 2004, resulting in a relatively short length of the displacement time series in conjunction with the lower amplitude of the horizontal GIA signal.

Acknowledgments

The authors would like to place on record their sincere gratitude to the Department of Foreign Affairs and International Trade (DFAIT), Canada, the University of Western Ontario, Canada, and the National Institute of Technology, Rourkela, India for facilitating this project. The work of K.F.T. was supported by an NSERC Discovery Grant and the NSERC and ICLR/Aon Benfield Industrial Research Chair. GPS data were provided by Natural Resources Canada and Quebec Ministère des Ressources Naturelles. Images were plotted with the help of GMT software developed and supported by Paul Wessel and Walter H.F. Smith.

References

Bawden, G.W., Thatcher, W., Stein, R.S., Hudnut, K.W. and Peltzer, G. (2001), *Tectonic contraction across Los Angeles after removal of groundwater pumping affects*, Nature, *412*, 812–815.

Braun, A., Kuo, C.Y., Shum, C.K., Wu, P., Wal, W.V.D. and Fotopoulos, G. (2008), *Glacial isostatic adjustment at the Laurentide ice sheet margin: models and observations in the Great Lakes region*, J. Geodyn, *46*, 165–173.

Clague, J.J. and James, T.S. (2002), *History of isostatic effects of the last ice sheet in southern British Columbia*, Quat. Sci. Rev., *21*, 71–87.

Dash, P.K., Samantaray, S.R., Panda, G. and Panigrahi, B.K. (2007), *Power transformer protection using S-transform with complex window and pattern recognition approach*, IET Gener. Transm. Distrib., *1*, 278–286.

Dash, P.K., Behera, H.S. and Lee, I.W.C. (2008), *Time sequence data mining using time-frequency analysis and soft computing techniques*, Appl. Soft Comput., *8*, 202–215.

Daubechies, I. (1990), *The Wavelet transform , Time-frequency localization and signal analysis*, IEEE Trans. Informat. Theory, *36*, 961–1005.

Dyke, A.S. (2004), *An outline of North American deglaciation with emphasis on central and northern Canada*, in Quaternary Glaciations Extent and Chronology, Part 2, North America, Dev. Quat. Sci., vol. 2b, edited by Ehlers, J., Gibbard, P. L., 373–424, Elsevier, New York.

George, N.V., Sahu, S.S., Mansinha, L., Tiampo, K.F. and Panda, G. (2009), *Time localised band filtering using modified S transform*, Proc. of ICSPS 2009, Singapore, 42–46.

Gregersen, S. (2006), *Intraplate Earthquakes in Scandinavia and Greenland neotectonics or postglacial uplift*, J. Ind. Geophys. Union, *10*, 25–30.

Hanrahan, J.L., Kravtsov, S.V. and Roebber, P.J. (2010), *Connecting past and present climate variability to the water levels of Lakes Michigan and Huron*, Geophys. Res. Lett., *37*, L01701. doi:10.1029/2009GL041707.

Huang, C.C., Liang, S.F., Young, M.S. and Shaw, F.Z. (2009), *A novel application of the S-transform in removing powerline interference from biomedical signals*, Physiol. Meas., *30*, 13–27.

Johansson, J.M., *et al.* (2002), *Continuous GPS measurements of postglacial adjustment in Fennoscandia*, J. Geophys. Res., *107*(B8), 2157. doi:10.1029/2001JB000400.

Karlstrom, K.E., Ahall, K.I., Harlan, S.S., Williams, M.L., Mc Lelland, J. and Geissman, J.W. (2001), *Long-lived (1.8 1.0 Ga) convergent orogen in southern Laurentia, its extensions to Australia and Baltica, and implications for refining Rodinia*, Precambrian Res., *111*, 5–30.

Khan, S.A., Knudsen, P. and Tscherning, C.C. (2005), *Crustal deformations at permanent GPS sites in Denmark*, in: A window on the future of geodesy, 128, Springer Berlin Heidelberg, Germany, 556–560.

Koohzare, A., Vancek, P. and Santos, M. (2006), *Compilation of a map of recent vertical crustal movements in Eastern Canada using geographic information system*, J. Surv. Engrg., *132*, 160–167.

Kumarapeli, P.S. (1985), *Vestiges of Iapetan rifting in the craton west of the northern Appalachians*, Geosci. Can., *12*, 54–59.

Lambert, A., Courtier, N., Sasagawa, G.S., Klopping, F., Winester, D., James, T.S. and Liard, J.O. (2001), *New constraints on Laurentide postglacial rebound from absolute gravity measurements*, Geophys. Res. Lett., *28*(10), 21092112. doi:10.1029/2000 GL012611.

Lee, H., Shum, C.K., Yi, Y., Braun, A. and Kuo, C.Y. (2008), *Laurentia crustal uplift observed using satellite radar altimetry*, J. Geodyn., *46*, 182–193.

Mainville, A. and Craymer, M.R. (2005), *Present-day tilting of the Great Lakes region based on water level gauges*, Geol. Soc. Am. Bull., *117*, 5–6.

Mazzotti, S., James, T.S., Henton, J. and Adams, J. (2005), *GPS crustal strain, postglacial rebound, and seismic hazard in eastern North America: The Saint Lawrence valley example*, J. Geophys. Res., *110*, B11301. doi:10.1029/2004JB003590.

McFadden, P.D., Cook, J.G., and Forster, L.M. (1999), *Decomposition of gear vibration signals by the generalized S-transform*, Mech. Syst. Signal Pr., *13*, 691–707.

Mitrovica, J.X., Milne, G.A. and Davis, J.L. (2001), *Glacial isostatic adjustment on a rotating Earth*, Geophys. J. Int., 562–578.

Park, K.D., Nerem, R.S., Davis, J.L., Schenewerk, M.S., Milne, G.A. and Mitrovica, J.X. (2002), *Investigation of glacial isostatic adjustment in the northeast U.S. using GPS measurements*, Geophys. Res. Lett., *29*(11), 1509. doi:10.1029/2001GL013782.

Peltier, W.R. (1998), *Postglacial variations in the level of the sea: implications for climate dynamics and solid–Earth geophysics*, Rev. Geophys., *36*, 603–689.

Peltier, W.R. (2002), *Global glacial isostatic adjustment: Palaeogeodetic and space–geodetic tests of the ICE-4G (VM2) model*, J. Quat. Sc., *17*, 491–510.

Peltier, W.R. and Drummond, R. (2008), *Rheological stratification of the lithosphere: A direct inference based upon the geodetically observed pattern of the glacial isostatic adjustment of the North American continent*, Geophys. Res. Lett., *35*, L16314. doi: 10.1029/2008GL034586.

Pinnegar, C.R. and Mansinha, L. (2003), *The S-transform with windows of arbitrary and varying shape*, Geophysics, *68*, 381–385.

Pinnegar, C.R. (2005), *Time-frequency and time-time filtering with the S-transform and TT-transform*, Digit. Signal Process., *15*, 604–620.

Portnoff, M.R. (1980), *Time–frequency representation of digital signals and systems based on short-time Fourier analysis*, IEEE Trans. Acoust. Speech, *28*, 55–69.

Richmond, G.M. and Fullerton, D.S.(1986), *Summation of quaternary glaciations in the United States of America*, Quaternary Sci. Rev., *5*, 183–196.

SCHIMMEL, M. and GALLART, J. (2005), *The inverse S-transform in filters with time-frequency localization*, IEEE Trans. Signal Proces., *55*, 4417–4422.

SELLA, G.F., STEIN, S., DIXON, T.H., CRAYMER, M., JAMES, T.S., MAZZOTTI, S. and DOKKA, R.K. (2007), *Observation of glacial isostatic adjustment in stable North America with GPS*, Geophys. Res. Lett., *34*, L02306.

STOCKWELL, R.G., MANSINHA, L. and LOWE, R.P. (1996), *Localisation of the complex spectrum: the S transform*, IEEE Trans. Signal Proces., *44*, 998–1001.

STOCKWELL, R.G. (2007), *Why to use the S-transform ?* in Fields Institute Communications Pseudo-differential Operators: Partial Differential Equations and Time Frequency Analysis, Edited by L. Rodino, B.-W. Schulze, M.W. Wong, *52*, 279–309.

TIAMPO, K.F., RUNDLE, J.B., KLEIN, W., BEN-ZION, Y. and MCGINNIS, S. (2004), *Using eigenpattern analysis to constrain seasonal signals in Southern California*, Pure Appl. Geophys., *161*, 1991–2003.

TIAMPO, K.F., ASSEFA, D., FERNNDEZ, J., MANSINHA, L. and RASMUSSEN, H. (2008), *Postseismic deformation following the 1994 Northridge earthquake identified using the localized Hartley transform filter*, Pure Appl. Geophys., *165*. doi:10.1007/s00024-008-0390-0, 15771602.

TIAMPO, K.F., MAZZOTTI, S. and JAMES, T.S. (2011), *Analysis of GPS measurements in eastern Canada using principal component analysis*, this issue.

TREGONING, P., WELSH, A., MCQUEEN, H. and LAMBECK, K. (2000), *The search for postglacial rebound near the Lambert Glacier, Antarctica*, Earth Planets and Space, *52*, 1037–1041.

TUSHINGHAM, A.M. and PELTIER, W.R. (1991), *ICE-3G: A new global model of late Pleistocene deglaciation based upon geophysical predictions of postglacial relative sea level change*, J. Geophys. Res., *96*, 4497–4523.

VAN DER WAL, W., BRAUN, A., WU, P. and SIDERIS, M.G. (2009), *Prediction of decadal slope changes in Canada by glacial isostatic adjustment modelling*, Can. J. Earth Sci., *46*(8), 587–595.

WATSON, K.M., BOCK, Y. and SANDWELL, D.T. (2002), *Satellite interferometric observations of displacements associated with seasonal groundwater in the Los Angeles basin*, J. Geophys. Res., *107*(B4), 2074. doi:10.1029/2001JB000470.

WHEELER, R.L. (1995), *Earthquakes and the cratonward limit of Iapetan faulting in eastern North America*, Geology, *23*, 105–108.

WILLIAMS, J.R. and AMARATUNGA, K. (1997), *A discrete wavelet transform without edge effects using wavelet extrapolation*, J. Fourier Anal. Appl., *3*, 435–449.

(Received May 2, 2010, revised February 14, 2011, accepted June 20, 2011, Published online August 27, 2011)

Pure Appl. Geophys. 169 (2012), 1519–1537

DOI 10.1007/s00024-011-0405-0

Pure and Applied Geophysics

Seismic Hazard and Ground Motion Characterization at the Itoiz Dam (Northern Spain)

A. Rivas-Medina,[1] M. A. Santoyo,[2] F. Luzón,[3] B. Benito,[1] J. M. Gaspar-Escribano,[1] and A. García-Jerez[3]

Abstract—This paper presents a new hazard-consistent ground motion characterization of the Itoiz dam site, located in Northern Spain. Firstly, we propose a methodology with different approximation levels to the expected ground motion at the dam site. Secondly, we apply this methodology taking into account the particular characteristics of the site and of the dam. Hazard calculations were performed following the Probabilistic Seismic Hazard Assessment method using a logic tree, which accounts for different seismic source zonings and different ground-motion attenuation relationships. The study was done in terms of peak ground acceleration and several spectral accelerations of periods coinciding with the fundamental vibration periods of the dam. In order to estimate these ground motions we consider two different dam conditions: when the dam is empty ($T = 0.1$ s) and when it is filled with water to its maximum capacity ($T = 0.22$ s). Additionally, seismic hazard analysis is done for two return periods: 975 years, related to the project earthquake, and 4,975 years, identified with an extreme event. Soil conditions were also taken into account at the site of the dam. Through the proposed methodology we deal with different forms of characterizing ground motion at the study site. In a first step, we obtain the uniform hazard response spectra for the two return periods. In a second step, a disaggregation analysis is done in order to obtain the controlling earthquakes that can affect the dam. Subsequently, we characterize the ground motion at the dam site in terms of specific response spectra for target motions defined by the expected values SA (T) of $T = 0.1$ and 0.22 s for the return periods of 975 and 4,975 years, respectively. Finally, synthetic acceleration time histories for earthquake events matching the controlling parameters are generated using the discrete wave-number method and subsequently analyzed. Because of the short relative distances between the controlling earthquakes and the dam site we considered finite sources in these computations. We conclude that directivity effects should be taken into account as an important variable in this kind of studies for ground motion characteristics.

Special issue on Deformation and Gravity Change: Indicators of Isostasy, Tectonics, Volcanism and Climate Change PAGEOPH.

[1] Geodesia y Cartografía, Campus Sur UPM, Universidad Politécnica de Madrid, ETSI Topografia, Ctra. de Valencia, km 7.5, 28031 Madrid, Spain. E-mail: alicia.rivas@upm.es

[2] Departamento de Ciencias de la Tierra, Astronomía y Astrofísica I (Geofísica y Meteorología). Facultad de Ciencias Físicas, Universidad Complutense de Madrid, Plaza de Ciencias s/n, 28040 Madrid, Spain. E-mail: msantoyo@pdi.ucm.es

[3] Departamento de Física Aplicada, Universidad de Almería, Cañada de San Urbano s/n, 04120 Almería, Spain. E-mail: fluzon@ual.es

1. Introduction

The Itoiz dam site is located at the Autonomous Community of Navarre, Northern Spain, 2 km north from Aoiz and 25 km east from Pamplona. The construction was finalized in 2003.

Its impoundment began in January 2004, and 8 months later, on 18 September, a clustered seismic series occurred with epicentre located between the city of Pamplona and the Itoiz reservoir. The mainshock (M_w 4.5) and the largest aftershock were widely felt in this region creating a great social expectation. After that, a large volume of local studies from different fields were performed to assess the dam safety (García-Yagüe 2004; Rueda 2005; Herraiz 2005; Instituto Jaume Almera 2005; Ingenieria del Suelo 2005; Colegio de Ingenieros de Caminos, Canales y Puertos 2005; Colegio de Geologos 2005; García-Mayordomo and Insúa-Arévalo 2010). More recently, several studies about the seismicity and internal geophysical processes were published (Ruiz *et al.* 2006a; Luzón *et al.* 2009, 2010; Jiménez *et al.* 2009; Santoyo *et al.* 2010).

Two new regional-scale seismic hazard studies in regions containing the dam site have been carried out in recent years, in the frame of the ISARD and RISNA projects, respectively. The first is a European Interreg Project (see, Secanell *et al.* 2008), aimed at standardizing the seismic hazard across borders, to resolve differences given by the French and Spanish building codes. The second is a local project (Benito *et al.* 2008) addressing the seismic risk in Navarre province for definition of the emergency plan by Civil Defense. Both studies followed the PSHA methodology (Probabilistic Seismic Hazard Assessment) and provided different zonings and seismic catalogues. These studies gave results of the expected ground motion parameters, ranging between the values given by both seismic codes, and generally higher than the values given by the Spanish building code (see NCSE-02 2002).

Other issues to consider are that the dam is located at a site close to the boundary between France and Spain and the respective building codes provide very different values of PGA and response spectra in such a boundary. There are also differences in the seismic catalogues of both countries for the seismicity of Pyrenees.

This work provides a new contribution to the knowledge of this interesting site by presenting a new ground motion characterization, which may represent some realistic seismic scenarios with implications in the safety of the dam.

In a first part, we propose a methodology with different steps for reaching different approximations to the expected ground motion at the dam site, which may be applied to other cases. Secondly, we revise the geological and seismic characteristics at the dam site, as well as the technical aspects of the dam, in order to obtain the specific parameters for the application of the proposed methodology. This is finally applied at the Itoiz dam, allowing characterization of the ground motion with a detail according to the followed approach. The subsequent sections include particularities of the application and obtained results in each phase.

2. *Methodological Proposal for Ground Motion Characterization at the Dam Site*

As a first objective of this paper, we propose a methodology for characterizing the ground motion at dam sites, which will be applied in our case at Itoiz site, but may be followed for any other dam sites. The approach is structured in three different phases for reaching different approximations to the expected ground motion at the site, based on:

1. A seismic hazard analysis for establishing critical seismic scenarios related to the safety of the dam.
2. A disaggregation analysis for particular target motions linked to the previous scenarios and calculation of site-specific hazard-consistent spectra.
3. A detailed modeling of the source rupture and propagation for the computation of synthetic acceleration time-histories at the dam site.

In the first phase, a Probabilistic Seismic Hazard Assessment (PSHA) is proposed, including a logic tree with two nodes for capturing the epistemic uncertainty related to seismic zoning and ground-motion models. The choice of the return periods for the analysis PSHA must be conditioned by the characteristics of the application, but usually we recommend to take into account two earthquakes with their respectively probabilities: the *project earthquake* (PE), which represents the earthquake taken into account in the design of the dam and the *extreme earthquake* (EE), which may occur during its life time with a low probability.

The PSHA analysis will be carried out for the return periods (RP) corresponding to the PE and EE shocks, giving as results the Uniform Hazard Spectra (UHS) for both RP. Moreover, in each case the mean spectral values μ may be derived by combining the different branches of the logic tree, and the percentiles 16–84% may be also estimated by means of $\mu \pm \sigma$ acceleration values.

The UHS represents the contribution of different seismic actions in the hazard, but not to the ground motion associated to a specific earthquake. In order to characterize more specific ground motions, we propose to consider, for both earthquakes—PE and EE—two dam conditions with different natural vibration periods: one corresponding to an empty dam (De) and another corresponding to a fully water-filled dam (Dwf). We can expect that the ground motions, which may represent some critical situations for the dam, have these spectral periods. The combination of the two earthquakes and the two conditions of the dam gives us four possible cases:

1. PE-De
2. PE-Dwf
3. EE-De
4. EE-Dwf.

Each case represents a seismic scenario, characterized by the spectral acceleration SA at structural periods of the De or Dwf conditions (T_{De}, T_{Dwf}) and return period of the PE and EE earthquakes. The ground motion for each scenario will be obtained in a subsequent step.

The second phase is aimed at determining the controlling earthquakes, which present the highest contribution to the previous seismic scenarios. Then we propose a disaggregation analysis in order to obtain the magnitude–distance–epsilon bin (M_w, r, ε) that provides (with highest probabilities) the target motions given by the obtained values of SA (T_{De}) and SA (T_{Dwf}) (here epsilon refers to the number of standard deviations considered in the ground motion prediction). These bins represent the controlling earthquakes for the dam. In a last step, the specific response spectra will be derived applying the ground motion models chosen in the PSHA analysis to the resulting (M_w, r, ε) bins.

In a last phase we propose to simulate the ground motions due to the controlling earthquakes in time domain, assuming a finite fault rupture embedded in a layered earth, with a given slip distribution, focal mechanism and rupture velocity. The rupture area S for each earthquake may be estimated by some correlation between Magnitude M and S and using a suitable model of rupture. The synthetic accelerograms will be computed using the discrete wavenumber method described by Bouchon and Aki (1977) and Bouchon (1979) for point dislocations. In the modelization of ground motions, directivity plays a key role in the amplitude and time length of the surface motion. This can dramatically change the estimations of the peak ground accelerations and spectral amplitudes for a given site and earthquake depending on the fault orientation. In order to explore this, we propose to compute different sets of synthetic seismograms for several azimuths of the recording sites with respect to the fault.

3. Characterization of the Dam Site

3.1. Geological and Geotechnical Conditions

The Itoiz reservoir is located at central western Pyrenees. This is one of the most active areas in Spain, after the south–southeastern part of the country (Benito and Gaspar-Escribano 2007), with about 35 events of magnitude $m_b \geq 3.0$ recorded in instrumental times (Souriau and Pauchet 1998; Souriau *et al.* 2001). The Pyrenean range (Fig. 1) has resulted from the collision of the Iberian and Eurasian plates with a low convergence rate. The suture between the two plates, the North Pyrenean Fault (NPF), is characterized by a sharp Moho jump. Different geophysical studies since the 1980s reported a complex deep structure (Gallart *et al.* 1981; Ecors Pyrenean Team 1988; Daignières *et al.* 1994; Souriau and Granet 1995; Pous *et al.* 1995) revealing the existence of a 10–15 km Moho jump beneath the trace of the NPF, which pointed to a major role of this structure in controlling the northward underthrusting of the Iberian plate beneath the thinner European plate.

The region where the Itoiz reservoir is located is a Mesozoic and Tertiary cover area composed by anticlines and synclines with the axes trending to the east, but truncated at some places by E–W to ESE–WNW fault systems (Puigdefabregas *et al.* 1978; García-Sansegundo 1993; Ruiz *et al.* 2006a).

Geotechnical studies at the dam site show that it is constructed over a sequence of materials that run from calcareous marlstone to limestone down to about 30 m, where a limestone basement is found. In order to estimate the subsurface S-wave velocity at the Itoiz dam, we performed a field campaign in May 2009 to record seismic noise. We computed the coherence among the signals recorded in pairs of stations which were separated distances from 4.5 m up to 150 m. Thus, using the two stations spatial autocorrelation method (see, e.g., Morikawa *et al.* 2004; Chávez-García *et al.* 2005, 2007; Chávez-García and Luzón 2005; Raptakis and Makra 2010), we computed a phase velocity dispersion curve for Rayleigh waves, which was inverted using the computer codes by Herrmann (1987) to obtain the S-wave velocity model for the subsurface soil at Itoiz dam. In the inversion process we considered one fixed layer with 30 m depth overlaying the halfspace. The results showed S-wave velocities of 902 m/s for the upper layer, and 1,736 m/s for the halfspace, respectively. Thus, following the Spanish earthquake-resistant code (NCSE-02 2002) the soil surface of the

Reprinted from the journal

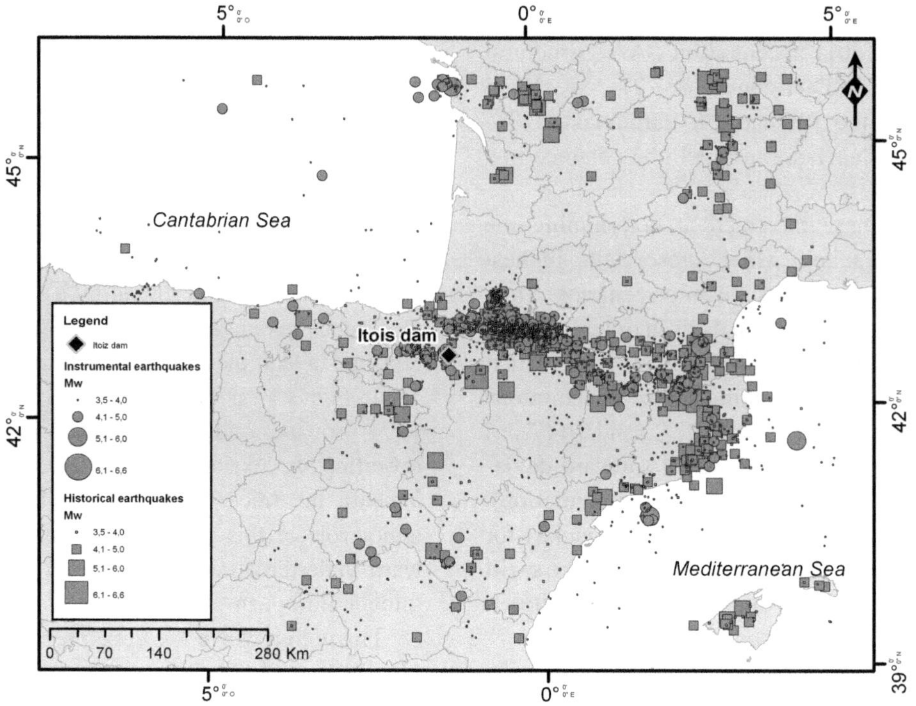

Figure 1
Seismicity catalogue used in the seismic hazard analysis. *Datum European 1950, UTM H30N*

Itoiz dam can be considered as Soil Type I, that is hard rock.

3.2. *Dam Characteristics*

This dam, of concrete gravity type, is located in the Irati river (Navarre) and collects water from both the Urrobi and the Irati rivers. It has a height of 111 m, a length of 525 m at the coronation and a capacity of 418 hm^3.

The fundamental vibration periods of the dam are 0.10 s, for the empty dam (De), and 0.22 s for the water-filled dam (Dwf). These periods are taken from the study of the Colegio de Ingenieros de Caminos and Canales y Puertos (2005) and are obtained by dynamic response analyses of the dam structure, which take into account the seismic action by means of different ground motion time histories. This analysis follows the methodology of Hatanaka (1960), which is based on a cross section model of the dam.

The choice of the return periods for the analysis PSHA must be conditioned by the characteristics of the dam. The Itoiz dam is identified as Category A under the classification established by the Spanish Technical Regulation on Safety of Dams and Reservoirs (Moptma 1996). García-Mayordomo (2007) recommends the identification of two earthquakes: the *project earthquake* (PE), which represents the control earthquake, associated with a return period of 975 years, and another, called the *extreme earthquake* (EE), which identifies the seismic action considering a return period of 4,975 years. These recommendations are in accordance with the values indicated in the *Technical guide for safety of dams* (Guía Técnica de Seguridad de Presas 1999; Guía de Seguridad de Presas 2005) and will be adopted in our case.

4. *Ground Motion Characterization at the Itoiz Site*

According to the proposed methodology (described in Sect. 1) we adopt an approach structured in three phases for reaching different approximations to the expected ground motion at the site. The details and results of each phase are presented in following sections.

4.1. Probabilistic Seismic Hazard Analysis and Definition of Seismic Scenarios

In a first step, a seismic hazard analysis was done following the PSHA methodology (e.g., BUDNITZ *et al.* 1997) based on zonation models. A logic tree with two nodes for capturing the epistemic uncertainty related to seismic zonings and ground-motion models was formulated. As a result, we obtain the PGA and spectral acceleration SA (*T*) values for the two return periods considered in our analysis, 975 and 4,975 years, which characterize the ground motion due to the project and extreme earthquakes, respectively.

As inputs for the application of the PSHA methodology we prepared a seismic catalogue for this study (project catalogue) and selected different zoning and attenuation models after a careful revision of the literature. A logic tree with two nodes for capturing the epistemic uncertainty related to seismic zonings and ground-motion models is formulated.

Next, the weights of the different models that configure the branches of the logic tree are considered, with the criteria explained in the following sections. As a first result of this analysis, we obtain the uniform hazard spectra UHS with the adopted weight scheme.

4.1.1 Seismic Project Catalogue

Navarre is located in a border area between France and Spain. The majority of the seismicity in the western part of Pyrenees, affecting the study site, is concentrated on the French side. Then a combination of data from different catalogues including both countries must be taken in the hazard analysis.

We consider two catalogues developed in the frame of the ISARD project (SECANELL *et al.* 2008) in local magnitude, and RISNA project (BENITO *et al.* 2008) in moment magnitude. A new project catalogue was created taking as initial database the information compiled in both catalogues. Since these catalogues included magnitude data given in different scales, a homogenization process was carried out for obtaining moment magnitude M_w as size parameter for all the events. Applying the correlation (1) obtained with the common earthquakes found in the two catalogues.

Table 1

Reference years estimated with the method by STEPP (1972) for different magnitude ranges derived from the completeness analysis of the seismic catalogue

Magnitude	Reference year
$M_w = 3.5$	1,980
$3.5 < M_w \leq 4.0$	1,962
$4.0 < M_w \leq 4.5$	1,850
$4.5 < M_w \leq 5.0$	1,750
$5.0 < M_w \leq 5.5$	1,743
$5.5 < M_w \leq 6.0$	1,477
$6.0 < M_w$	1,150

$$M_w = 0.8228 \cdot M_L + 1.0105\left(R^2 = 0.7834\right) \quad (1)$$

The completeness of the catalogue was analyzed according to the methodology developed by STEPP (1972). The reference years from which the catalogue may be considered complete for each magnitude interval were estimated. With this purpose, we plotted the cumulative number of events of particular magnitude ranges through time. Abrupt increases of slope are associated with the reference year of completeness for that particular magnitude range (TINTI and MULARGIA 1985). The identified reference years are listed in Table 1. A constant occurrence rate for each magnitude interval, calculated from the reference year to the present is assumed. These rates were used to solve the problem of completeness and to estimate the seismic parameters of the different zones for the whole analyzed period.

In a last step, a depuration process was developed, by removing fore- and aftershocks, because we assume a Poisson model for the seismicity of every zone. The project catalogue was de-clustered using the program SeriesBuster (ÁLVAREZ-GÓMEZ *et al.* 2005) which allows us to identify seismic series considering the magnitude dependent time–space windows proposed by SECANELL *et al.* (2008).

The resulting project catalogue in terms of moment magnitude (M_w) is composed of a total of 3,086 events and is shown in Fig. 1.

Table 2

Recurrence parameters of each zone

Zone	M_{Wmax}	ν_0	β
ISARD			
1	6.3	0.105	1.960
2	5.9	0.317	3.027
3	5.4	0.038	2.520
4	6.5	0.363	2.308
5	6.6	0.141	1.830
6	5.2	0.042	3.035
7	6.5	1.212	2.747
8	6.3	0.302	2.462
9	6.0	0.235	2.714
10	5.0	0.055	3.784
11	4.4	0.018	6.351
12	4.9	0.042	4.157
13	4.5	0.020	5.630
14	6.2	0.095	2.021
15	5.9	0.145	2.826
17	6.1	0.052	1.833
18	6.5	0.047	1.489
NCSE-02			
19	5.5	0.085	2.888
20	5.9	0.248	2.778
21	6.5	1.216	2.792
22	6.8	0.536	2.194
23	6.4	0.322	2.316
24	6.1	0.149	2.264
25	6.7	0.087	1.571
RISNA			
1	5.4	0.154	3.616
2	5.9	0.074	2.223
3	5.4	0.028	2.235
4	5.2	0.063	3.346
5	6.2	0.136	2.151
6	5.3	0.047	2.795
7	6.0	0.149	2.441
8	6.4	0.063	1.660
9	6.7	0.217	1.950
10	5.1	0.155	4.305
11	6.0	0.075	2.067
12	6.3	0.354	2.495
13	6.5	1.235	2.769
14	6.6	0.234	2.078
15	5.4	0.123	3.224
16	5.3	0.071	3.255
17	4.9	0.066	4.398
18	6.0	0.056	1.909
20	4.7	0.032	4.520
21	6.0	0.032	1.620
22	4.4	0.022	6.851
23	6.2	0.321	2.539
24	6.2	0.197	2.338
GAROÑA			
1	5.8	0.017	1.470

Table 2 *continued*

Zone	M_{Wmax}	ν_0	β
2	5.4	0.096	3.166
3	5.5	0.118	3.004
4	6.7	1.342	2.605
5	5.2	0.195	4.149
6	6.0	0.057	1.923
7	6.0	0.094	2.204
8	6.5	0.050	1.523
9	4.7	0.057	5.248
10	5.9	0.064	2.070
PC NAVARRA			
1	6.0	0.131	2.311
2	5.4	0.065	2.846
3	6.2	0.219	2.431
4	6.5	1.235	2.812
5	6.8	0.727	2.325
6	5.3	0.057	2.927
7	6.2	0.140	2.176
8	4.6	0.026	4.942
9	6.2	0.449	2.717

M_{wmax} maximum moment magnitude, ν_0 earthquake occurrence rate above the threshold magnitude ($M_{\mathrm{wmin}} = 4.0$)

4.1.2 Seismic Zoning

The probabilistic zoning method is suitable for areas with low to moderate seismicity such as the environment of the dam, where the lack of knowledge of the fault parameters prevents us from modeling its activity. In consequence, the seismicity is considered to be distributed in seismogenic zones, wherein uniform seismic potentials are assumed.

For this work we consider several suitable models for the area, as those used in: (i) the regulation of Garoña Nuclear Power Plant (Nuclenor 1983) (Fig. 2a), (ii) the ISARD project (Fig. 2b), (iii) the hazard map of the Spanish Building Code (NCSE-02 2002) (Fig. 2c), (iv) the hazard analysis developed by Prospección e Ingeniería (1992) (Fig. 2d) and the RISNA project (Fig. 2e).

The seismicity of each zone and for each zoning is modeled by a doubly truncated Gutenberg–Richter model. The threshold magnitude was set to 4.0 (M_w) and a least-squares approach was used to derive the parameters α and β according to the expression:

$$\mathrm{Ln}\,N = \alpha - \beta \cdot M_w \quad (2)$$

The number N of earthquakes exceeding a certain magnitude value was estimated by extrapolating the

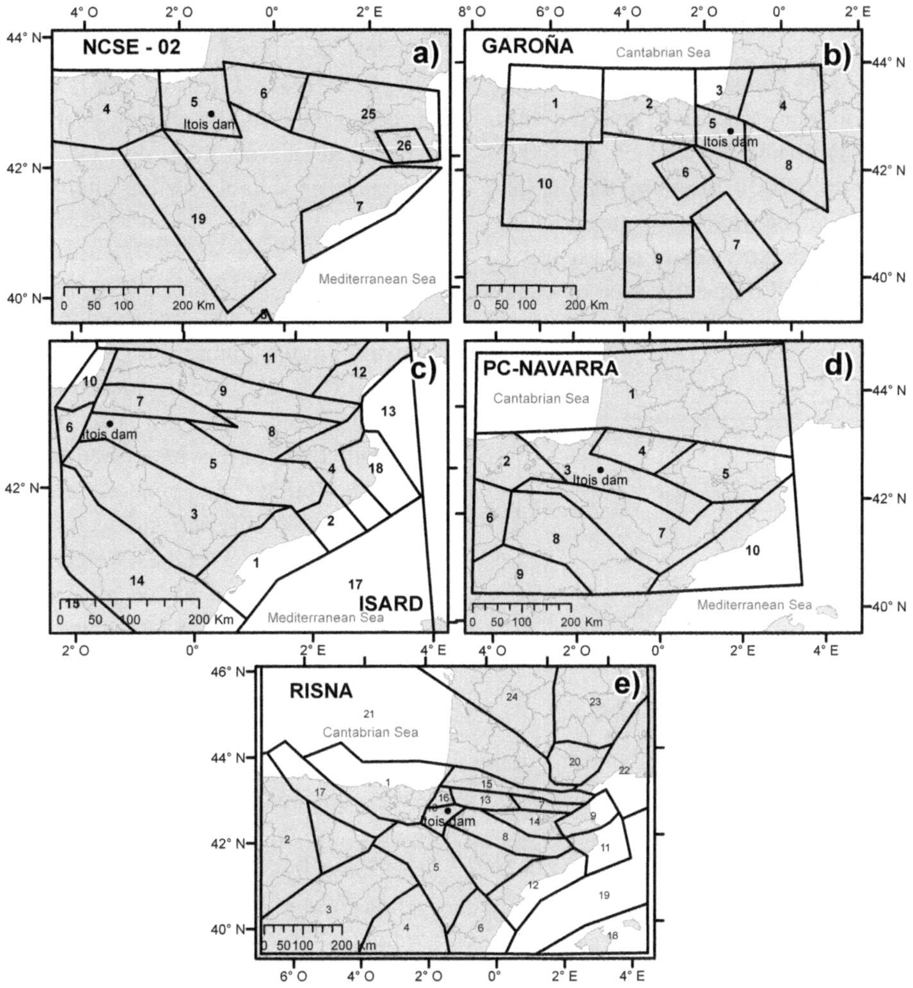

Figure 2

Zoning models included in the logic tree for the hazard analysis. **a** The GAROÑA zoning used for the hazard study at site of Garoña NPP (NUCLENOR SA, 1983). **b** The ISARD zoning. The unified zoning defined for the ISARD Project (Secanell *et al.* 2008). **c** The NCSE-02 zoning. Defined for the hazard map of the Spanish building code (NCSE-02 2002). **d** The PC-NAVARRA zoning. The model used in the hazard analysis developed by Prospección e Ingeniería (1992). **e** The RISNA project zoning (Benito *et al.* 2008)

constant recurrence rates obtained within the completeness period to the entire period of study. The annual rate for the threshold magnitude is:

$$v_0 = N(M_w \geq 4.0) = e^{\alpha - \beta \cdot 4.0} \quad (3)$$

Table 2 summarizes the values of v_0 and β obtained for each zone considering the different zonings.

4.1.3 Selection of Ground Motion Models

The limited amount of strong motion data recorded at the dam and surrounding areas prevents us from developing strong motion models constrained with local data, so it is necessary to resort to models developed for other areas, provided that their characteristics are equivalent to the seismotectonic setting of the study area.

Recent models of strong motion with European data were developed by Ambraseys *et al.* (2005) and Akkar and Bommer (2007a, b), both requiring focal mechanisms. The recent New Generation Attenuation models (NGA—Abrahamson and Silva 2007; Boore and Atkinson 2007; Campbell and Bozorgnia 2007; Chiou and Youngs 2006; Idriss 2007) require also a very specific control of the rupture geometry and other details that are not usually available.

After a revision of the available models, analyzing their suitability for our area, we selected four attenuation laws: (1) Berge-Thierry *et al.* (2003) and (2) Lussou *et al.* (2001) used in the RISNA project, (3) Ambraseys *et al.* (1996) (used in the ISARD project), and (4) Sabetta and Pugliese (1996) based in European data and frequently used in studies of the Spanish territory. The other ground motion models developed with Spanish data, as Cabañas *et al.* (1999) and Tapia (2006), are discarded either because they do not provide coefficients for spectral accelerations, or they do not cover a range of magnitudes suitable for our study (Fig. 3).

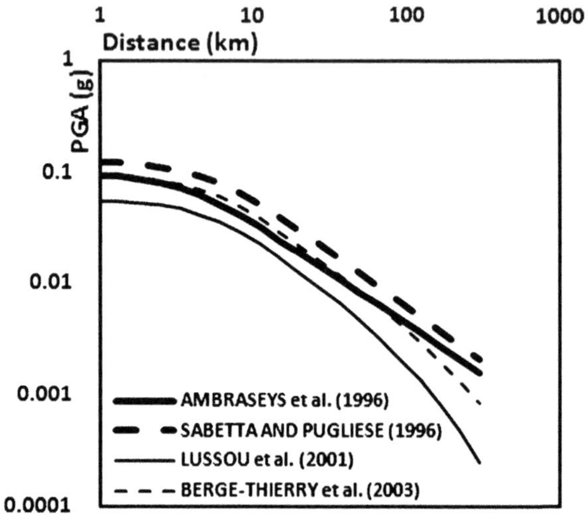

Figure 3
Graphical representation of the loss models for a moment magnitude earthquake of $M_w = 6$ used in the study

4.1.4 Logic Tree

We set up a logic tree composed of two nodes: seismic source zoning and ground-motion attenuation model (Fig. 4). The former node splits into five branches that stand for the five zoning models presented in a previous section. The scheme of weights follows the general criteria: models with more reliable seismic and tectonic information are considered with a higher weight. Thus, the higher weights correspond to the most recently developed models (RISNA and ISARD), and to the model of the Spanish building code.

The attenuation models of Berge-Thierry *et al.* (2003) and Lussou *et al.* (2001) have been recently recommended by Drouet *et al.* (2007) for the Pyrenean area, and therefore, we assigned them a greater weight. On the other hand, the ground motion relations by Sabetta and Pugliese (1996) and Ambraseys *et al.* (1996) were considered with lower weights.

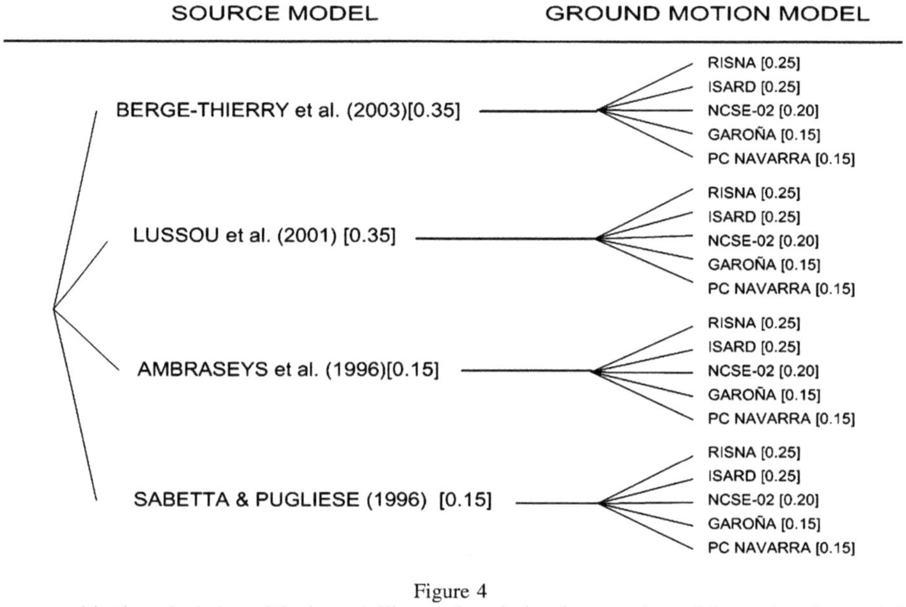

Figure 4
Logic tree used in the calculation of the hazard. The numbers in brackets are the weights assigned to each branch

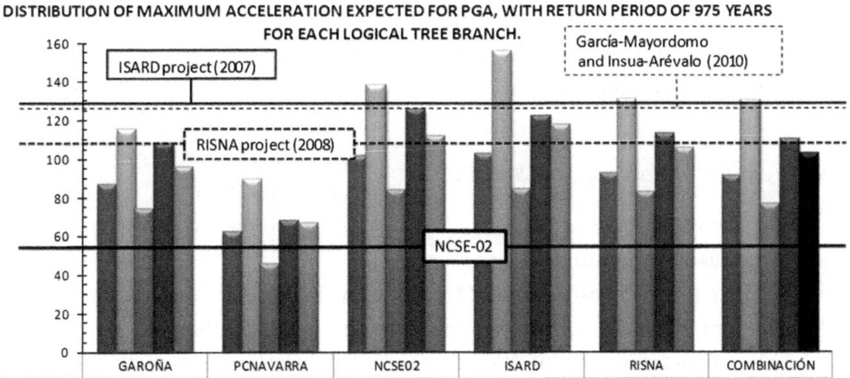

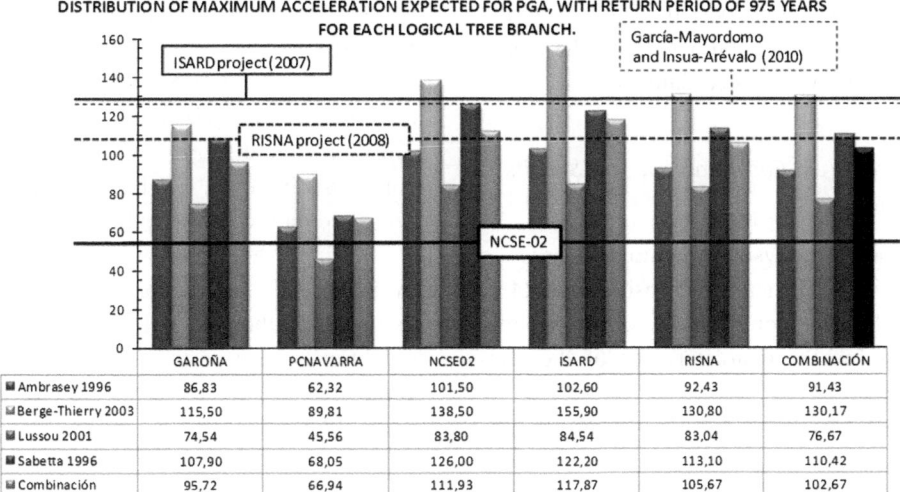

	GAROÑA	PCNAVARRA	NCSE02	ISARD	RISNA	COMBINACIÓN
Ambrasey 1996	86,83	62,32	101,50	102,60	92,43	91,43
Berge-Thierry 2003	115,50	89,81	138,50	155,90	130,80	130,17
Lussou 2001	74,54	45,56	83,80	84,54	83,04	76,67
Sabetta 1996	107,90	68,05	126,00	122,20	113,10	110,42
Combinación	95,72	66,94	111,93	117,87	105,67	102,67

Figure 5
Distribution of maximum acceleration expected for PGA, with return period of 975 years for each logic tree branch (combined two by two). The results were compared with previous studies, NCSE-02 (2002), RISNA PROJECT (Benito *et al.* 2008), ISARD Project (Secanell *et al.* 2008) and García-Mayordomo and Insúa-Arévalo (2010)

Seismic hazard calculations for the different branches of the logic tree were carried out using the CRISIS2007 code (Ordaz *et al.* 2007).

4.1.5 Results: Uniform Hazard Spectra

Figure 5 shows the combination of the final result of the logic tree with the partial results of each branch for PGA and a return period of 975 years. PGA values derived by other studies in the same area are also plotted. It is appreciated that the PGA values calculated in this study are clearly larger than the NCSE-02 (2002) values. This trend has been observed in previous recent studies (Benito *et al.* 2008; Secanell *et al.* 2008; and García-Mayordomo and Insúa-Arévalo 2010).

The uniform hazard spectra (elastic, 5% damping) at the dam site on hard rock sites considering different return periods (975 and 4,975 years) is shown in Fig. 6. They present maximum spectral

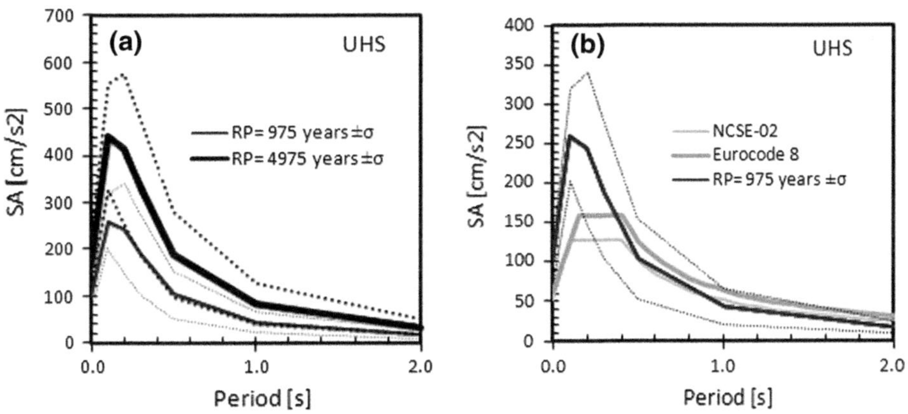

Figure 6
a Uniform hazard spectra at the Itoiz dam site on rock conditions considering different return periods (975 and 4,975 years) and your uncertainties. **b** Comparison the obtained response spectra for 975 years and the corresponding ones given in the Spanish and European seismic codes

accelerations in the short-period range (around 0.15 s), next to the fundamental periods of the empty dam (0.1 s), and the water-filled dam (0.22 s).

4.2. Hazard Disaggregation Analysis, and Specific Site Spectra

A disaggregation analysis is required in order to obtain the magnitude, distance and epsilon bins (M_w, r, ε) with the highest contribution to the seismic hazard to determine the characteristics of the controlling earthquake.

As target motion for the disaggregation we take the expected spectral accelerations of the previous hazard analysis for conditions of the PE and EE shocks (RP 975 and 4,975 years), at the fundamental periods of the dam in empty ($T_{De} = 0.10$ s) and water-filled ($T_{Dwf} = 0.22$ s) conditions. Then, we estimate the density functions of different bins to the motion represented by SA (T) in the following scenarios:

1. PE-De; RP = 975 years and $T_{De} = 0.1$ s
2. PE-Dwf; RP = 975 years and $T_{Dwf} = 0.22$ s
3. EE-De; RP = 4,975 years and $T_{De} = 0.1$ s
4. EE-Dwf; RP = 4,975 years and $T_{Dwf} = 0.22$ s

For each scenario we obtain the two controlling earthquakes which represent the largest hazard contribution and second largest hazard contribution to each ground motion. The resulting parameters are given in Table 3. We can appreciate that all the controlling earthquakes are ranging between two extreme cases: a low magnitude and sort distance event ($M = 4.5$, $R = 5$ km) and a moderate magnitude and intermediate distance shock ($M = 6$, $R = 30$ km).

The specific response spectra of the controlling earthquakes are obtained by application of the selected ground motion models to the values of the corresponding (M_w, r, ε) bins and are shown in Fig. 7. For filled dam condition, the spectrum of the stronger controlling event exceeds the spectrum of the weaker event for all periods for both, the project and the extreme earthquakes. However, for empty dam condition, short-period spectral accelerations are controlled by the lower-magnitude and near event while the long-period spectral accelerations are controlled by the higher-magnitude and more distant event.

4.3. Simulation of Ground Motion in Time Domain

4.3.1 Modelling of Source and Propagation

In the last phase of the proposed methodology we simulate the ground motions due to the controlling earthquakes in a time domain, assuming a finite fault rupture embedded in a layered Earth. For this purpose, a given slip distribution, focal mechanism and rupture velocity are required.

The rupture area for each earthquake was estimated by $\log(S) = M_s - 4.1$ (e.g., Wells 1994), where S is the rupture area in km^2, and M_s is the surface-wave magnitude. Rupture area was assumed rectangular with an aspect ratio of $L = W$ for earthquakes with $M_s \leq 5.0$, and $L = 2W$ for earthquakes with $M_s > 5.0$ (e.g., Dowrick and Rhoades 2004); here L is the length along strike and W is the width

Table 3

Hazard-consistent scenarios from PSHA for the dam

	Largest hazard contribution (CE1)			Second largest hazard contribution (CE2)		
	Magnitude (M_w)	Distance (km)	Epsilon	Magnitude (M_w)	Distance (km)	Epsilon
PE-D_{WF}	4.5	10	1.6	5.0	25	2.2
PE-D_E	4.5	5	1.1	5.5	20	1.5
EE-D_E	4.5	5	1.5	6.0	25	1.5
EE-D_{WF}	5.0	15	2.2	6.0	30	2.0

PE-Dwf project earthquake for the water-filled dam, *EE-Dwf* extreme earthquake for the water-filled dam, *PE-De* project earthquake for the empty dam, *PE-De* extreme earthquake for the empty dam

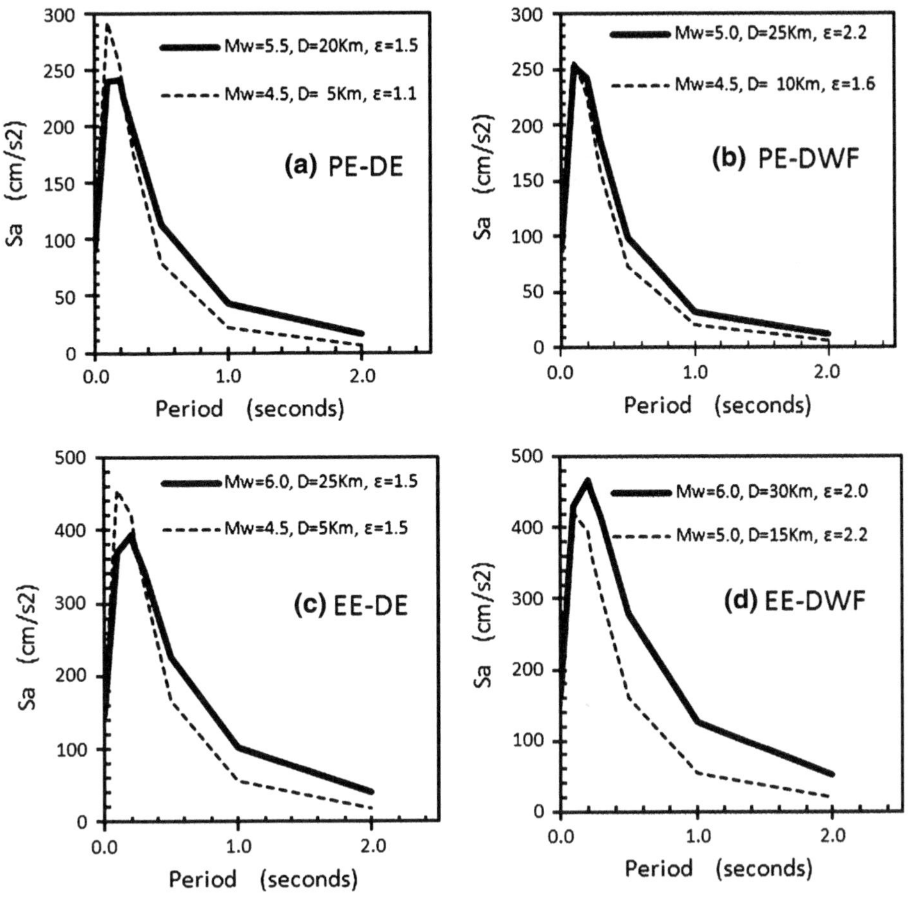

Figure 7
Specific response spectra from control earthquakes studied in this work. **a** *PE-De* project earthquake for the empty dam. **b** *PE-Dwf* project earthquake for the water-filled dam. **c** *EE-De* extreme earthquake for the empty dam. **d** *EE-Dwf* extreme earthquake for the water-filled dam

along dip (Fig. 8a). The rupture area was discretized into (*NL*, *NW*) square subfaults of equal size. For earthquakes $M_s \leq 5.0$, $NL = NW = 5$; for earthquakes $M_s > 5.0$, $NL = 20$ and $NW = 10$. A point dislocation source was located at the center of each subfault that releases the average moment of the entire subfault. The total moment release of each earthquake (M_o) is obtained by $\log(M_o) = 1.5M_w + 16.1$ (e.g., Wells 1994). Here we assumed that $M_w = M_s + 0.2$ given that in the magnitude interval modeled in this study ($4.5 \leq M_s \leq 6.0$) this relation is mostly linear (e.g., Singh and Ordaz 1994). Slip distribution over the fault was assumed semielliptical (Mikumo *et al.* 1998), tapered with a cosine function along the strike and dip directions (Figs. 7c, 8b). In this model, rupture nucleates at the hypocenter and propagates with constant velocity V_r from the nucleation point; here $V_r = 0.8\beta$ and β is the S-wave velocity at the source location (e.g., Mikumo *et al.* 1998). All subfaults share the same focal mechanism, which is also the same used to define the spatial setting of the fault (φ = strike and θ = dip angles). The synthetic accelerograms were computed using the discrete wave-number method described by Bouchon and Aki (1977) and Bouchon (1979) for point dislocations, and the crustal velocity structure used is that proposed for this part of Spain by Ruiz *et al.* (2006b).

In order to explore if directivity plays a relevant role in the amplitude and time length of the surface motion we computed different sets of synthetic seismograms for several azimuth locations of the recording sites with respect to the fault. To assure a smooth rupture, the rise time (τ_r) at each subfault was

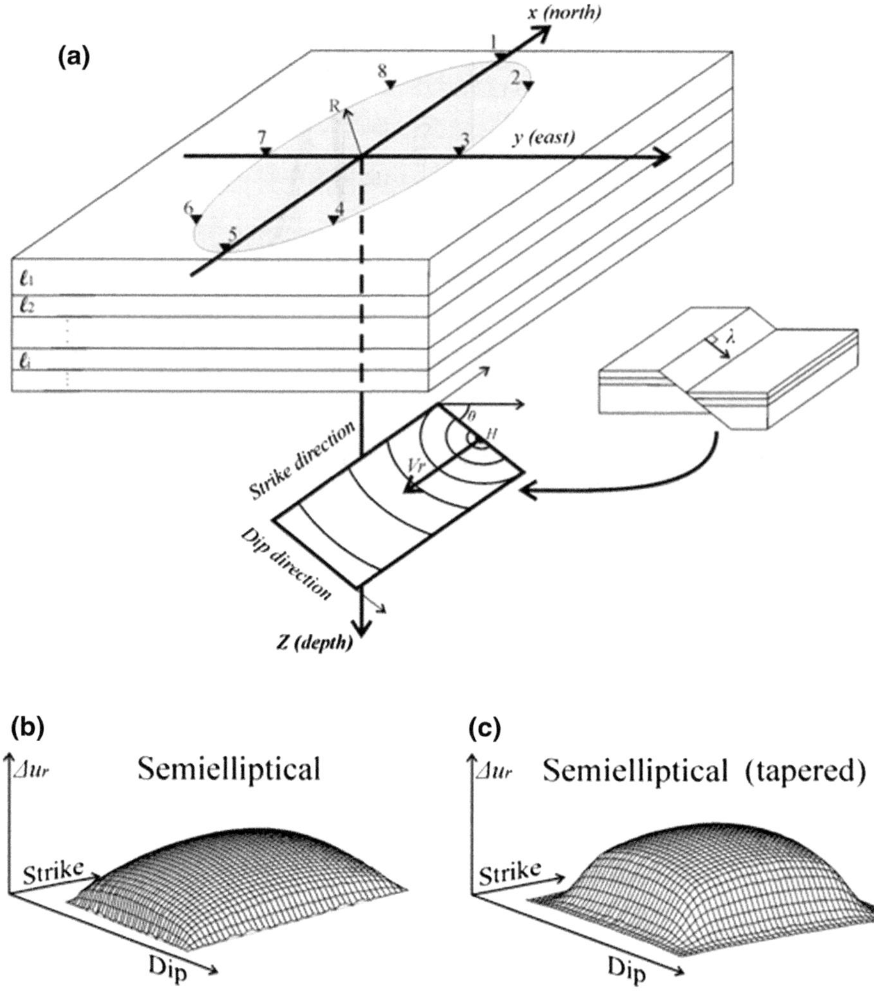

Figure 8
Diagram of the Earth's model, fault geometry and source properties for computations. **a** Diagram of the Earth's layered model and the relative location of the surface station array (numbered from 1 to 8), located on a circle with radio R, centered on the origin. Layers are numbered from top to bottom *l1, l2, l3*, etc. The fault plane (here shown as a rectangle on *z* axis) is located below the array. The hypocenter (H) is located on the northern edge of the fault ($+x$ direction). Rupture nucleates at the hypocenter and propagates with constant velocity V_r from the nucleation point. Rupture fronts propagate as circular patterns from the hypocenter. In this way rupture propagates mainly to the south ($-x$ direction). We assume that all subfaults of the seismic fault share the same focal mechanism along the entire fault, which is also the same used to define the spatial setting of the fault (φ = strike, θ = dip angles). Rake angle of the slip vector $\lambda = -90$ describes a pure dip slip normal fault. **b** Semielliptical slip distribution along the fault. **c** Semielliptical slip distribution tapered on the edges with a cosine function along the fault

assumed to be at least $\tau_r \geq \sqrt{2}(\Delta x/V_r)$, where $\Delta x = \Delta y'$ is the length of each subfault in the strike and dip directions, respectively.

4.3.2 Synthetic Accelerograms

We located eight computation sites, placed at the surface and at the same distance R with respect to the center of the fault (Table 4), every $\pi/4$ rad, numbered in a clockwise sense and beginning with the station placed at the $+x$, $y = 0$ (north) of the fault (Fig. 8a). The distances R were defined based on the extreme earthquakes for an empty and a water-filled dam (Table 4). The hypocentral depth was set to $ZH = 6.0$ km for all the modeled earthquakes, which is the average for the zone (Santoyo *et al.* 2010). Based on focal mechanisms published by Rueda (2005) we set for the model a dip angle of $\theta = 34°$

Table 4

Earthquake source parameters used for the computation of the synthetic accelerograms

	Mo ($\times 10^{23}$ dyn-cm)	Distance (km)	Fault		
			Total area L, W (km)	$\Delta x = \Delta y'$ (km)	τ_r (sec)
EE-D_E CE1	0.71	5.0	1.6, 1.6	0.4	0.2
EE-D_{WF} CE1	3.98	15.0	2.8, 2.8	0.7	0.4
EE-D_E CE2	125.8	25.0	12.6, 6.3	0.7	0.4
EE-D_{WF} CE2	125.8	30.0	12.6, 6.3	0.7	0.4

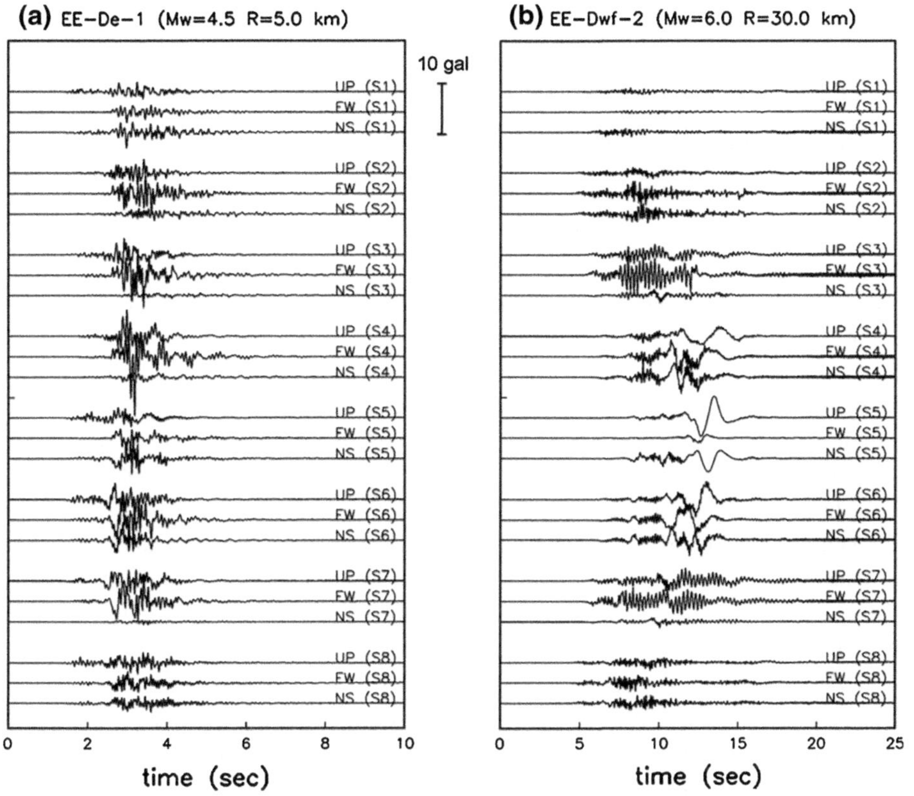

Figure 9

Synthetic accelerograms for the selected earthquakes. **a** Accelerograms for extreme earthquake EE-De-1. Waveforms from stations are numbered from 1 to 8 and components of motion N–S, E–W, Up are on the x, y, and $-z$ directions. **b** Accelerograms for extreme earthquake EE-Dwf-2. Waveforms from stations are numbered from 1 to 8 and components of motion N–S, E–W, Up are on the x, y, and $-z$ directions

and a rake angle of $\lambda = -90°$. Strike angle in this model was set to $\varphi = 0.0°$ (Fig. 8a). The convention used for the component of ground motion (N–S, E–W, Up) is shown in Fig. 8a.

In Fig. 9 we show the synthetic accelerograms obtained for two of the extreme earthquakes considered, showing the station and component of motion (N–S, E–W, Up). We depict here the results for the more nearby and distant controlling earthquakes to the dam. On the left hand side of this figure, we show the accelerations produced by an earthquake with magnitude $M_w = 4.5$ at a distance of 5.0 km, and on the right hand side we show results for an earthquake with magnitude $M_w = 6.0$ at a distance of 30.0 km. As expected, peak ground accelerations at stations located in the rupture direction are larger than in stations located on the opposite side. Amplitudes at stations 4 and 5, are in both cases about three times

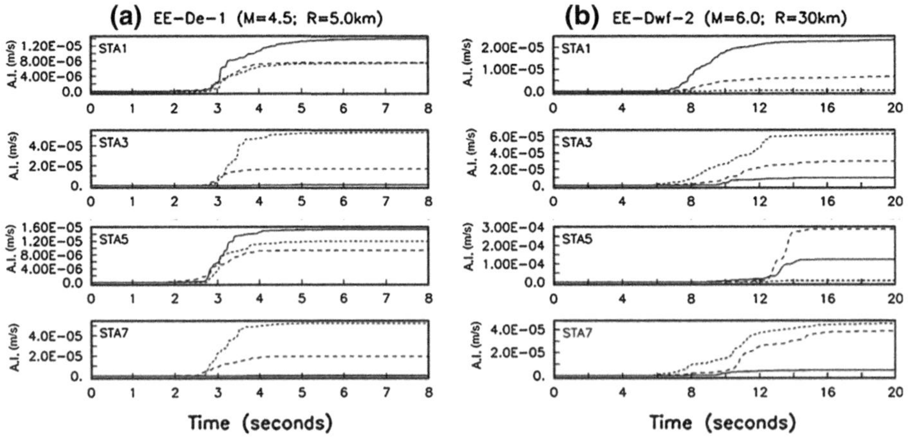

Figure 10
Arias Intensity plot versus time for both earthquakes shown in Fig. 8 and stations 1, 3, 5 and 7. N–S component of motion is shown by solid line, E–W component is shown by a *dashed line* and Up component is shown by *dotted line*. **a** *EE-De-1* extreme earthquake for the empty dam. **b** *EE-Dwf-2* extreme earthquake for the water-filled dam

larger than amplitudes at stations 1 and 8. In the same sense, the pulse width and duration at stations located in the rupture direction are shorter than on stations placed on the opposite direction. From these results it is clearly observed that the pulse width and amplitude of waveforms on accelerograms vary depending on the angle between the rupture direction and the take off angle to the station. In Fig. 10, we show the evolution of the Arias Intensity (AI, Arias 1970) with respect to time in four of the observing stations (1, 3, 5, and 7) computed from the synthetic accelerograms. A rapid change of AI can induce stronger effects at surface than a slower one. For the EE-Dwf-2 earthquake (Fig. 10b), the change of the AI on station 5 is steeper than on the other stations, implying that in this case, both the waveform amplitudes and pulse width have similar importance on the surface motion.

On Fig. 11 we show the response spectra for the selected stations and earthquakes. Here it can be observed that stations located on the rupture direction have larger maximum spectral amplitudes that the ones located at the opposite direction and that spectral amplitudes for periods of interest for the dam (0.1 and 0.22 s) have their largest amplitudes at the perpendicular direction to the rupture plane, which is an unexpected result. Spectral amplitudes can present differences of as much as four times.

5. *Discussion and Conclusions*

We present in this paper a methodological approach for characterizing the ground motion that may represent some realistic seismic scenarios at Itoiz dam. Several forms of hazard-consistent ground motion characterization are presented as result of a full site-specific PSHA analysis, carried out for two return periods, 975 and 4,975 years, following the Spanish Technical Regulation on Safety of Dams and Reservoirs (Moptma 1996). These periods are associated with two earthquakes, which may affect, at different levels, the security of the dam, namely the *project earthquake* (PE) and the *extreme earthquake* (EE). Uniform hazard spectra (UHS) were derived as result of a first part of the analysis, giving values of PGA 0.10 and 0.17 g for these return periods, and maximum spectral acceleration of SA (0.1) = 0.27 g and SA (0.1) = 0.44 g, respectively. These values are higher than the ones given by the Spanish Building Code NCSE-02, and in an approximate agreement with the values obtained from different studies for the region.

A more detailed ground motion characterization was done in a second step, considering specific seismic scenarios that may represent critical actions for the security of the dam. These scenarios were identified taking into account both PE and EE

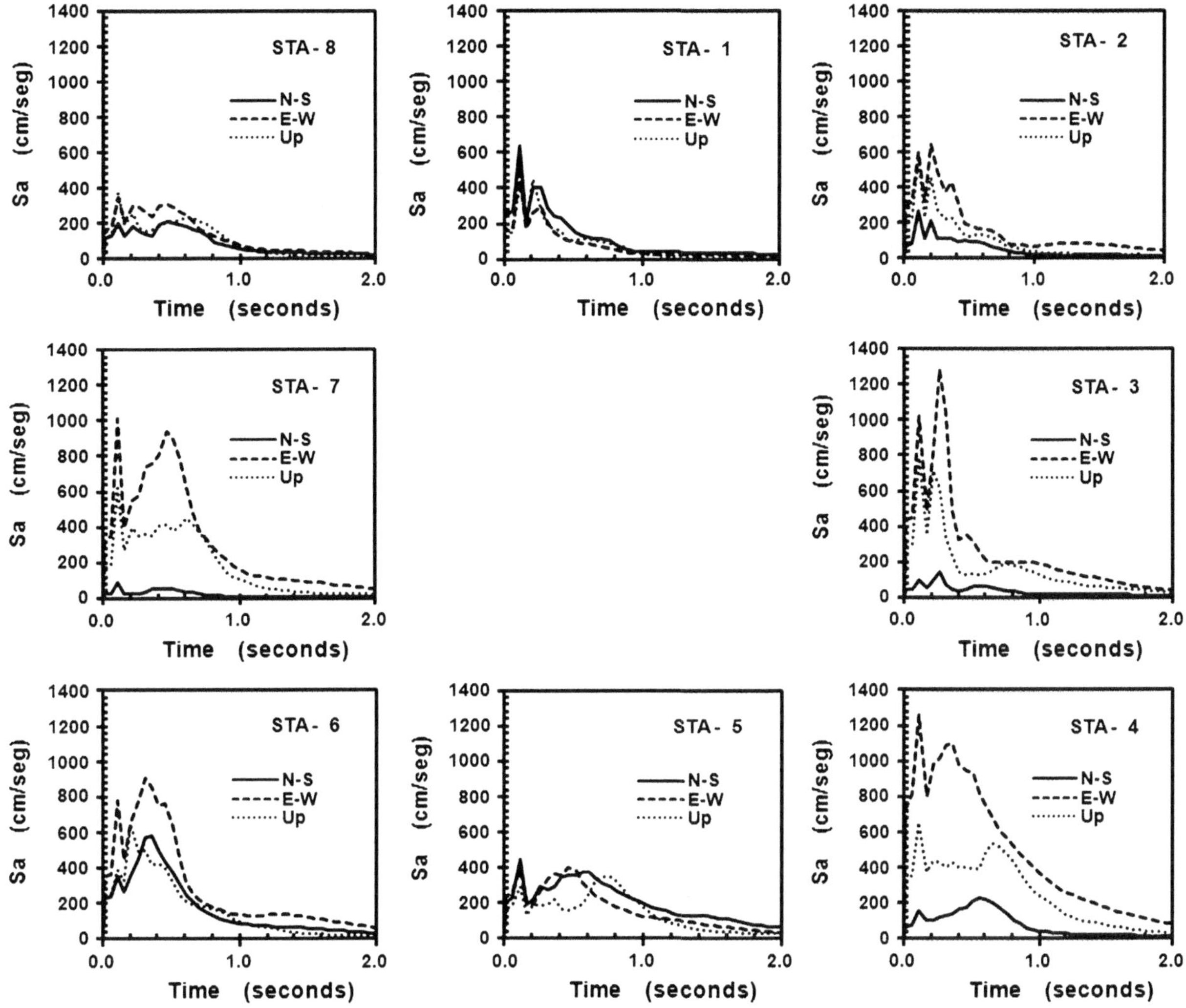

Figure 11
Response spectra for both earthquakes shown in Fig. 8 and all stations. N–S component of motion is shown by solid line, E–W component is shown by a *dashed line* and Up component is shown by *dotted line*. **a** *EE-De-1* extreme earthquake for the empty dam. **b** *EEDwf-2* extreme earthquake for the water-filled dam

earthquakes, and two stages of the dam, empty and water-filled conditions, with fundamental periods of 0.10 and 0.22 s, respectively. A disaggregation analysis was developed for obtaining the control earthquakes linked to the four scenarios, each one identified by a possible combination of return period and structural period. The spectral accelerations SA (T) for the structural periods of $T_{\mathrm{De}} = 0.10$ s and $T_{\mathrm{Dwf}} = 0.22$ s, were taken as target motion for the disaggregation analysis. Earthquakes in a range of magnitude M_{w} 4.5–6 and distance 5–30 km were identified as control earthquakes. We obtained specific response spectra (SRS) for these events, which may be considered a second approximation to the ground motion that may affect the dam.

The SRS are narrower than the UHS previously derived but similar maximum spectral accelerations are found in both types of spectra, around 250 cm/s^2 (PE) and 450 cm/s^2 (EE). These maximum values are found in a period range (0.1–0.2 s) in agreement with the fundamental periods of the dam.

 Reprinted from the journal

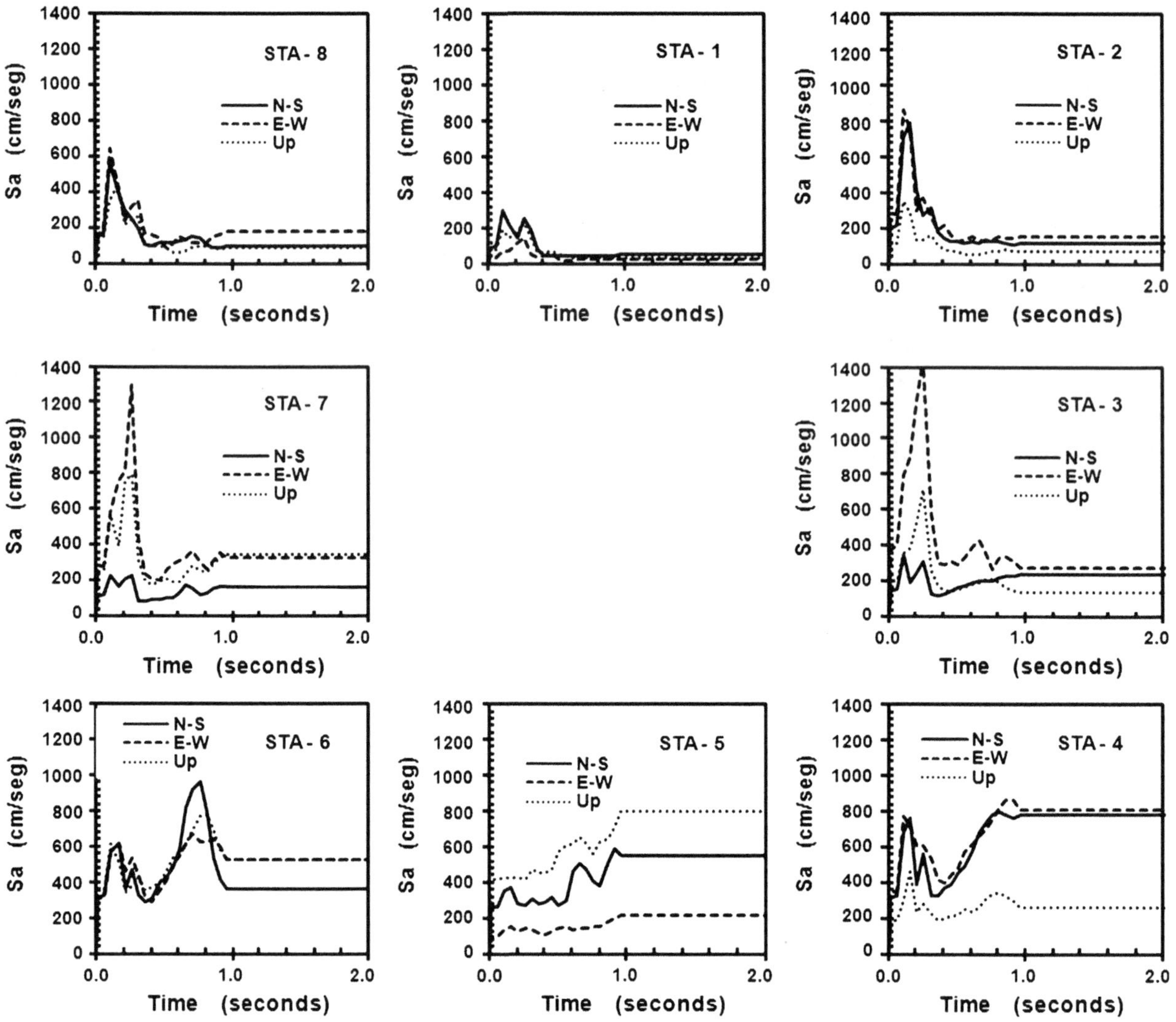

Figure 11
continued

A third order approximation was done assuming that ground motions due to the controlling earthquakes can be generated by a finite fault rupture embedded in a layered Earth, with a given slip distribution, focal mechanism and rupture velocity. A more detailed characterization of the ground motion was carried out by simulation of accelerograms and response spectra considering different locations of the earthquakes relative to the dam site and taking into account possible directivity effects. We then appreciated very important variations in the PGA and SA (*T*) values as a function of the source-site azimuth. PGA is one of the important factors to take into account for seismic hazard assessment; however, these results show the importance of taking into account directivity effects on surface motions and should be considered as an important aspect in the estimations of seismic hazard.

Next steps to be carried out by the regional or national governments, for a more realistic estimation of the ground motion in Itoiz dam due to local earthquakes, should include topographical effects. This problem can be estimated with the knowledge of the geometry of the irregular topography, their elastic properties, and the application of numerical methods for wave propagation modeling in topographical structures such as boundary elements (see, e.g., Sánchez-Sesma *et al.* 1993; Luzón *et al.* 1997) or finite difference methods (see, e.g., Perez-Ruiz *et al.* 2005) among others. Moreover, the influence of strong motion rotations (Trifunac 2009; Godinho *et al.* 2009) produced by possible near earthquakes on the response of the surface topography should be evaluated as well.

Acknowledgments

We thank C. López and S. Limonchi for the help in the field campaigns. We wish to thank Confederación Hidrográfica del Ebro (CHE) for giving access to their facilities and to the seismic data. We also thank Instituto Geográfico Nacional (IGN), Spain, and the Réseau Accélérométrique Permanent (RAP), France, for providing seismic data of the region. This work was partially supported by Secretaría General para el Territorio y la Biodiversidad from Ministerio de Medio Ambiente, Rural y Marino, Spain, under grant 115/SGTB/2007/8.1, by EU with FEDER and by the research team RNM-194 of Junta de Andalucía, Spain. The work done by M.A.S. was under the auspices of a Contract Ramón y Cajal. Z. al Yuncha was in the initial stages of this research. The PSHA study was performed during the research stay of A.R.M. at the University of Almería, Spain.

References

Abrahamson, N.A., and Silva, W.J., (2007), *Abrahamson & Silva NGA ground motion relations for the geometric mean horizontal component of peak and spectral ground motion parameters*, PEER Report Draft v2, Pacific Earthquake Engineering Research Center, Berkeley, CA, pp 380.

Akkar, S., and Bommer, J.J., (2007a), *Prediction of elastic displacement response spectra in Europe and the Middle East*, Earthq Eng Struct Dyn *36*: 1275–1301.

Akkar, S., and Bommer, J.J., (2007b), *Empirical prediction equations for peak ground velocity derived from strong motion records from Europe and the Middle East,* Bull Seism Soc Am *97*(2): 511–530.

Álvarez-Gómez, J.A., García-Mayordomo, J., Martínez-Díaz, J.J., and Capote, R., (2005), *SeriesBuster: a Matlab Program to Extract Spatio-Temporal Series from an Earthquake Database*, Computers and Geosciences, *31*, 521–525.

Ambraseys, N.N., Simpson, K.A., and Bommer, J.J., (1996), *Prediction of horizontal response spectra in Europe*, Earthq Eng Dyn *25*(4): 371–400.

Ambraseys, N.N., Douglas, J., Sarma, S.K., and Smit, P.M. (2005), *Equations for the Estimation of Strong Ground Motions from Shallow Crustal Earthquakes Using Data from Europe and the Middle East: Horizontal Peak Ground Acceleration and Spectral Acceleration*, Bulletin of Earthquake Engineering, *3*, 1–53.

Arias, A. (1970). *A measure of earthquake intensity". In R.J. Hansen, ed. Seismic Design for Nuclear Power Plants, MIT Press, Cambridge, Massachusetts*, pp 438-483.

Benito, B., and Gaspar-Escribano, J. M., (2007), *Ground motion characterization in Spain: context, problems and recent developments in seismic hazard assessment*, Journal of Seismology Vol. *11*: 433–452.

Benito, B., Gaspar-Escribano, J. M., Martínez-Díaz, J.J., Gómez, R., Canora, C. and Álvarez, J.A., (2008), *Evaluación de la peligrosidad sísmica (emplazamientos en roca)* Vol.*1*. *Evaluación del Riesgo Sísmico en Navarra,* Protección Civil de Emergencias de la Comunidad Foral de Navarra. Internal report.

Berge-Thierry, C., Cotton, F., Scotti, O., Griot-Pommera, D.A., and Fukushima, Y., (2003), *New Empirical Response Spectral Attenuation Laws for Moderate European Earthquakes*, J. Earthquake Eng. *7* (2), 193-222.

Budnitz, R.J., Apostolakis, G., and Boore, D.M., (1997), *Recommendations for Probabilistic Seismic Hazard Analysis: Guidance on Uncertainty and Use of Experts*, NUREG/CR-6372, US Nuclear Regulatory Commission.

Boore, D.M., and Atkinson, G.M., (2007), *Boore–Atkinson NGA ground motion relations for the geometric mean horizontal component of peak and spectral ground motion parameters,* PEER Report 2007/01, Pacific Earthquake Engineering Research Center, Berkeley, CA, pp 234.

Bouchon, M., (1979), *Discrete wave number representation of elastic wave field in three space dimensions,* Journal of Geophysical Research, *84*, 3609–3614.

Bouchon, M., and Aki, K., (1977), *Discrete wave number representation of seismic source wave fields,* Bulletin of the Seismological Society of America, *67*, 259–277.

Cabañas, L, Benito, B, Cabañas, C, López, M, Gómez, P, Jiménez, ME, Álvarez, S (1999). *Banco de Datos de Movimiento Fuerte del Suelo MFS*. Aplicaciones, in: Ingeniería sísmica, edited by M. B. Benito, D. Muñoz, Física de la Tierra *11*, 111–137.

Campbell, K.W., and Bozorgnia, Y., (2007), *Campbell–Bozorgnia NGA ground motion relations for the geometric mean horizontal component of peak and spectra ground motion parameters,* PEER Report 2007/02, Pacific Earthquake Engineering Research Center, Berkeley, CA, pp 240.

Chávez-García, F.J., Rodríguez, M., and Stephenson, W.R., (2005), *An alternative approach to the SPAC analysis of microtremors: exploiting the stationarity of noise*, Bull. Seism. Soc. Am., *95*, 277–293.

Chávez-García, F.J., and Luzón, F., (2005), *On the correlation of seismic microtremors,* Journal of Geophysical Research-Solid Earth. Vol. *110*, B11, B11313, doi:10.1029/2005JB003671.

Chávez-García, F.J., Luzón, F., Raptakis, D. and Fernández, J., (2007), *Shear-wave velocity structure around Teide volcano: results using microtremors with the SPAC method and implications for interpretation of geodetic results*, Pure and Applied Geophysics, *164*, 697–720.

Chiou, B., and Youngs, R.R., (2006), *Chiou–Youngs PEER-NGA empirical ground motion model for the average horizontal component of peak acceleration and pseudo-spectral acceleration for spectral periods of 0.01 to 10 seconds*, PEER Report

Draft, Pacific Earthquake Engineering Research Center, Berkeley, CA, 219 pp.

Colegio de Ingenieros de Caminos, Canales y Puertos (2005). *Estudio sobre la Presa de Itoiz,* Internal report. Confederación Hidrográfica del Ebro. http://www.chebro.es.

Colegio de Geologos, (2005). *Informe de supervisión de los estudios Y análisis disponibles sobre la seguridad de la presa de Itoiz,* Internal report. Confederación Hidrográfica del Ebro. http://www.chebro.es.

Daignières, M., Séguret, M., and Ecors Team, (1994). *The Arzacq–western Pyrenees ECORS deep seismic profile*, Publ. Eur. Assoc. Pet. Geol. *4*, 199–208.

Dowrick, D.J., and Rhoades, D. A., (2004), *Relations Between Earthquake Magnitude and Fault Rupture Dimensions: How Regionally Variable Are They?* Bulletin of the Seismological Society of America, Vol. *94*, No. 3, pp. 776–788, June.

Drouet, S., Scherbaum, F., Cotton, F., and Souriau, A. (2007), *Selection and ranking of ground motion models for seismic hazard analysis in the Pyrenees*, J. Seismol. *11*, 87–100.

Ecors Pyrenees Team, (1988). *The ECORS deep reflection seismic survey.* Nature *311*,508–511.

Gallart, J., Banda, E., and Daignières, M., (1981), *Crustal structure of the Paleozoic axial zone of the Pyrenees and transition to the north Pyrenean zone*, Ann. Geophys. *37* (3), 457–480.

García-Yagüe, A., (2004), *Análisis de la sismicidad registrada en el entorno de la presa de Itoiz, Término municipal de Lónguida (Navarra),* Internal report. Confederación Hidrográfica del Ebro. http://www.chebro.es.

García-Mayordomo, J., (2007), *Metodologías modernas en el análisis de peligrosidad sísmica para embalses,* in Jornadas Técnicas sobre Estabilidad de Laderas en Embalses (Ed. Confederación Hidrográfica del Ebro), Memorias, 581–597, Zaragoza.

García-Mayordomo, J., and Insúa-Arévalo, J.M., (2010), *Estudio Sismotectónico y de Actividad Tectónica Reciente en el entorno de la Presa de Itoiz (Navarra): Cálculo de la Peligrosidad Sísmica Mediante Técnicas Modernas,* Internal report of the Instituto Geológico y Minero de España.

García-Sansegundo, J., (1993), *Memoria y mapas geológicos a escala 1: 25.000 de la hoja n° 142 (Aoiz), cuadrantes de Aoiz (I), Irurozqui (II), Monreal (III) y Domeño (IV),* Diputación Foral de Navarra.

Godinho, L., Mendes, P.A., Tadeu, A., Cadena-Isaza, A., Smerzini, C., Sánchez-Sesma, F.J., Madec, R. and Komatitsch, D. (2009). *Numerical simulation of ground rotations along 2D topographical profiles under the incidence of elastic waves,* Bull. Seism. Soc. Am, Vol *99*, No. 2B, pp 1147–1161.

Guía Técnica de Seguridad de Presas. 1999. Volumen 3. Estudios Geológico-Geotécnicos y de prospección de materiales. Comité Nacional Español de Grandes Presas. ISBN: 84-89567-11-5.

Guía de Seguridad de Presas. 2005. *Capítulo 7. Criterios básicos de seguridad.* Colegio de Ingenieros en Caminos, Canales y Puertos (ISBN: 84-380-0298-6) y Comité Nacional Español de Grandes Presas (ISBN: 84-89567-15-8).

Hatanaka, M., (1960). *Study on the earthquake-resistant design of gravity type dams*, 2nd WCEE, Tokyo.

Herraiz-Sarachaga, M., (2005). *Sismicidad Inducida por Embalses.* – Consideraciones generales. Internal report. Confederación Hidrográfica del Ebro. http://www.chebro.es.

Herrmann, R.B., (1987). *Computer programs in Seismology*, Saint Louis University, *7* vols.

Idriss IM (2007). *Empirical model for estimating the average horizontal values of pseudoabsolute spectral accelerations generated by crustal earthquakes*, PEER Report Draft, Pacific Earthquake Engineering Research Center, Berkeley, CA, pp 76.

Ingenieria del Suelo, S.A., (2005). *Informe de auscultación de la presa de Itoiz,* Tomo I: Memoria. (Madrid). Internal report. Confederación Hidrográfica del Ebro. http://www.chebro.es.

Instituto Jaume Almera (2005). *Registro de los sismógrafos instalados en Itoiz,* Internal report. Confederación Hidrográfica del Ebro. http://www.chebro.es.

Jiménez, A., Tiampo, K.F., Posadas, A., Luzón, F., and Donner, R., (2009). *Analysis of complex networks associated to seismic clusters near the Itoiz reservoir dam,* European Physical Journal Special topics, *174*, 181–195.

Lussou, P., Fukushima, Y., Bard, P.Y., and Cotton, F., (2001) *Seismic design regulation codes: contribution of Knet data to site effect evaluation.* J Earthq Eng *5*, 13–33.

Luzón, F., Sánchez-Sesma, F.J., Rodríguez-Zúñiga, J.L., A. M. Posadas, García, J. M., J. Martín, M. D. Romacho and Navarro, M. (1997). *Diffraction of P, S and Rayleigh waves by three-dimensional topographies,* Geophys. J. Int., Vol. *129*, pp. 571–578.

Luzón, F., García-Jerez, A., Santoyo, M.A., and Sánchez-Sesma, F. J., (2009). *A hybrid technique to compute the pore pressure changes due to time varying loads: application to the impounding of the Itoiz reservoir, northern Spain,* In POROMECHANICS-IV, Eds. H. Ling, A. Smyth, and R. Betti, *DEstech* Publications, Inc., Lancaster, Pennsylvania. ISBN:978-1-60595-006-8., pp. 1109–1114.

Luzón, F., García-Jerez, A., Santoyo, M.A., and Sánchez-Sesma, F.J., (2010). *Numerical modelling of pore pressure variations due to time varying loads using a hybrid technique: the case of the Itoiz reservoir (Northern Spain),* Geophysical Journal International, doi:10.1111/j.1365-246X.2009.04408.x.

Mikumo, T., Miyatake, T., and Santoyo, M.A., (1998). *Dynamic rupture of asperities and stress change during a sequence of large interplate earthquakes in the Mexican subduction zone,* Bull. Seism. Soc. Am., *88*, 686–702.

Moptma (1996). *Orden de 12 de marzo de 1996 por la que se aprueba el Reglamento técnico sobre Seguridad de Presas y Embalses,* Boletín Oficial del Estado núm. 78 de sábado 30 de marzo de 1996, Spain.

Morikawa, H., Sawada, S., and Akamatsu, J., (2004), *A method to estimate phase velocities of Rayleigh waves using microseism simultaneously observed at two sites*, Bull. Seism. Soc. Am., *94*, 961–976.

Ncse-02, (2002). Norma de la construcción sismorresistente española (NCSE02) (2002). *Real Decreto 997/2002, de 27 de septiembre, por el que se aprueba la norma de construcción sismorresistente: parte general y edificación (NCSR-02),* Boletín Oficial del Estado (BOE) número 244 de 11/10/2002, páginas 35898 a 35967.

Nuclenor S.A. (1983). *Estudio de Revisión de la Calificación Sísmica del Emplazamiento de la central nuclear de Santa María de Garoña,* Temas SEP II- 4. 3 volúmenes.

Ordaz, M., Aguilar, A., and Arboleda, J. (2007), *Program for computing seismic hazard,* CRISIS2007, Universidad Nacional Autónoma de México.

Pérez-Ruiz, J.A., Luzón F. and García-Jerez, A. (2005). *Simulation of an Irregular Free Surface with a Displacement Finite-Difference Scheme.* Bull. Seism. Soc. Am. Vol. *95*, (6), 2216–2231.

Pous, J., Muñoz, J.A., Ledo, J.J., and Liesa, M., (1995). *Partial melting of subducted continental lower crust in the Pyrenees.* Journal of the Geological Society, London, *152*, 217–220.

Prospeccion e Ingeniería (1992). *Estudio Sísmico de Navarra, Gobierno de Navarra Servicio de Protección Civil,* Memoria 103 pp., 7 planos + 5 Apéndices.

Puigdefabregas, C., Rojas, B., Sánchez-Carpintero, I., and Valle de Lerchundi, J., (1978). *Memoria y Mapa Geológico de España, E. 1:50.000,* 2ª ser., Hoja n° 142 (Aoiz). Inst. Geol. Min. Esp.

Raptakis, D., and Makra, K., (2010). *Shear wave velocity structure in western Thessaloniki (Greece) using mainly alternative SPAC method,* Soil Dynamics and Earthquake Engineering *30* (2010) 202–214.

Rueda, J., (2005). *Informe sobre terremotos ocurridos en Itoiz (Navarra) en septiembre de 2004,* Internal report, Instituto Geográfico Nacional, Madrid.

Rueda, J., and Mezcua, J., 2005. *Near-real-time seismic moment-tensor determination in Spain,* Seismological Research Letters. *76*, 455–465.

Ruiz, M., Gaspà, O., Gallart, J., Díaz, J., Pulgar, J.A., García-Sansegundo, J., Ópezfernández, C., and González-Cortina, J.M., (2006a). *Aftershocks series monitoring of the September 18, 2004, M = 4.6 earthquake at the western Pyrenees: A case of reservoirtriggered seismicity?* Tectonophysics, *424*, 223–243.

Ruiz, M., Gallart, J., Díaz, J., Olivera, C., Pedreira, D., López, C., Gonzálezcortina, J.M., and Pulgar, J.A., (2006b). *Seismic activity at the western Pyrenean edge,* Tectonophysics, *424*, 217–235.

Sabetta, F., and Pugliese, A., (1996) *Estimation of response spectra and simulation of nonstationary earthquake ground motions,* Bull. Seism. Soc. Am., *86*(2): 337–352.

Sánchez-Sesma, F.J., Ramos-Martínez, J. and Campillo, M. (1993). *An indirect boundary element method applied to simulate the seismic response of alluvial valleys for incident-P, incident-S and Rayleigh-waves,* Earthquake Engineering & Structural Dynamics, *22* (4), 279–295.

Santoyo, M.A., García-Jerez, A., and Luzón, F., (2010), *A Subsurface Stress Analysis and its Possible Relation with Seismicity Near the Itoiz Reservoir, Navarra, Northern Spain,* Tectonophysics, doi:10.1016/j.tecto.2009.06.022.

Secanell, R., Bertil, D., Martin, CH., Goula, X., Susagna, TH., Tapia, M., Dominique, P., Carbon, D., and Fleta, J., (2008). *Probabilistic seismic hazard assessment of the Pyrenean region,* J Seismol. doi:10.1007/s10950-008-9094-2.

Singh, S. K., and Ordaz, M., (1994). *Seismic Energy Release in Mexican Subduction Zone Earthquakes,* Bull. Seism. Soc. Am, Vol. *84*, No. 5, pp. 1533–1550.

Souriau, A., and Granet, M., (1995). *A tomographic study of the lithosphere beneath the Pyrenees from local and teleseismic data,* Journal of Geophysical Research, *100*(B9), 18117–18134.

Souriau, A., and Pauchet, H., (1998). *A new synthesis of Pyrenean seismicity and its tectonic implications,* Tectonophysics *290*, 221–244.

Souriau, A., Sylvander, M., Rigo, A., Fels, J.F., Douchain, J. M. and Ponsolles, C., (2001), *Sismotectonique des Pyrénées: principales contrainters sismologiques,* Bull.Soc.géol.France, *172*(1), 25–39.

Stepp, J.C., (1972) *Analysis of completeness of the earthquake sample in the Puget Sound area and its effect on statistical estimates of earthquake hazard,* In: Proceedings the 2nd International Conference on Microzonation, pp 897–910.

Tapia, M. (2006) *Desarrollo y aplicación de métodos avanzados para la caracterización de la respuesta sísmica del suelo a escala regional y local,* PhD Tesis, Universitat Politecnica de Catalunya, Barcelona.

Tinti, S. and Mulargia, F. (1985) *Completeness analysis of a seismic catalog,* Annales Geophysicae *3*(3), 407–414.

Trifunac, M.D. (2009). *The role of strong motion rotations in the response of structures near earthquake fault,* Soil Dynamics and Earthquake Engineering, *29*, 382–393.

Wells D. L. and K. J. Coppersmith (1994). *New Empirical Relationships among Magnitude, Rupture Length, Rupture Width, Rupture Area, and Surface Displacement,* Bull. Seism. Soc. Am, Vol. *84*, No. 4, pp. 974–1002.

(Received June 15, 2010, revised April 15, 2011, accepted June 30, 2011, Published online September 16, 2011)

GPSR Compliance

The European Union's (EU) General Product Safety Regulation (GPSR) is a set of rules that requires consumer products to be safe and our obligations to ensure this.

If you have any concerns about our products, you can contact us on ProductSafety@springernature.com

In case Publisher is established outside the EU, the EU authorized representative is:

Springer Nature Customer Service Center GmbH
Europaplatz 3
69115 Heidelberg, Germany

Batch number: 09479379

Printed by Printforce, the Netherlands